Regression Models for Time Series Analysis

A complete list of the titles in this series appears at the end of this volume.

Regression Models for Time Series Analysis

BENJAMIN KEDEM
University of Maryland

KONSTANTINOS FOKIANOS
University of Cyprus

WILEY-INTERSCIENCE

A JOHN WILEY & SONS, INC., PUBLICATION

This text is printed on acid-free paper. ♾

Published simultaneously in Canada.

For ordering and customer service, call 1-800-CALL WILEY.

Library of Congress Cataloging-in-Publication Data is available.

ISBN 0-471-36355-3

Printed in the United States of America

10 9 8 7 6 5 4 3 2 1

To Carmella
B. K.

To my mother and to the memory of my beloved grandmother
K. F.

Contents

Preface

Regression methods have been an integral part of time series analysis for a long time, dating back at least one hundred years to the work of Schuster (1898) [379]. Schuster's work on sinusoidal regression was applied in the estimation of "hidden periodicities" and led to the invention of the periodogram. Structural regression models for time series have been around for many years and have figured prominently in the econometrics and business literature. Treated rigorously by Anderson (1971) [20], and Fuller (1996) [161] among others, these structural models have been used for years in forecasting and decomposition of time series into "trend", "seasonal", and "irregular" components. Another distinctive example is the class of autoregressive integrated moving average models that came to be associated with Box and Jenkins (1976) [61] but has its roots in the pioneering work of E.E. Slutski and G.U. Yule in the 1920s, and of H.O. Wold in the 1930s. Most of the aforementioned work deals with linear models for time series assuming continuous values. However, there are many instances in practice where the data are not continuous and a linear model is not appropriate. This points to the necessity for alternative modeling.

This book introduces the reader to relatively newer developments and somewhat more diverse regression models and methods for time series analysis. It has been written against the backdrop of a vast modern literature on regression methods for time series and related topics as is apparent from the long list of references.

A relatively recent statistical development is the important class of models known as *generalized linear models* (GLM) that was introduced by Nelder and Wedderburn (1972) [336], and which provides under some conditions a unified regression theory suitable for continuous, categorical, and count data. The theory of GLM was originally intended for independent data, but it can be extended to dependent data under various assumptions. In the first four chapters of this book the GLM methodology is extended systematically to time series where the primary and covariate data are both random and stochastically dependent. There are three notions which enable this [152], [395]. The notion of an increasing sequence of histories relative to an observer, the notion of *partial likelihood* introduced by Cox (1975) [105] and further elaborated on by Wong (1986) [439], and the notion of martingale with respect to a sequence of histories. The latter, under suitable conditions, is applied in asymptotic inference including goodness of fit.

After a general introduction to time series that follow generalized linear models in Chapter 1, Chapters 2, 3, and 4 specialize to regression models for binary, categorical, and count time series, respectively. Chapter 5 is an introduction to various regression models developed during the last thirty years or so, particularly regression models for integer valued time series including hidden Markov models. Chapter 6 summarizes classical and more recent results concerning state space models. The last chapter, Chapter 7, presents a Bayesian approach to prediction and interpolation in spatial data adapted to time series that may be short and/or observed irregularly. We also describe a specially designed software for the implementation of the Bayesian prediction method. A brief introduction to stationary processes can be found in the Appendix. Throughout the book there are quite a few real data applications and further results presented by means of problems and complements.

Parts of the book were taught at the University of Maryland to a mixed audience of beginning and more advanced graduate students. Based on our experience, the book should be accessible to anyone who is familiar with basic modern concepts of statistical inference, corresponding roughly with the master's degree level. A basic course in applied stochastic processes consistent with the level of Parzen (1962) [343] is helpful.

We are very grateful to V. De Oliveira, L. Fahrmeir, R. Gagnon, N.O. Jeffries, C. Kedem, D.E.K. Martin, M. Nerlove, J. Picka, E. Russek-Cohen, T.J. Santner, and others who read parts of the book and provided very helpful suggestions. Special thanks are due to Amy Hendrickson of TeXnology Inc. for many useful LaTeX tips. We would like to acknowledge with thanks the travel support from the University of Cyprus and the grant support, over many years, from the National Aeronautics and Space Administration (NASA). Finally we thank our families for their unlimited support and patience throughout this project.

BENJAMIN KEDEM, College Park, Maryland
KONSTANTINOS FOKIANOS, Nicosia, Cyprus

1

Time Series Following Generalized Linear Models

In ordinary linear regression, a most useful and much dealt with statistical tool, the problem is to relate the mean response of a variable of interest to a set of explanatory variables by means of a linear equation. In many cases this is done under the assumption that the data are normal and independent. There are situations however, regarding non-normal observations such as binary and count data, when ordinary linear regression leads to certain inconsistencies, some of which are resolved very elegantly and successfully by generalized linear models. Emboldened by this success, we wish to import ideas from generalized linear models in modeling time series data. The question then is how to extend the generalized linear models methodology to time series where the data are dependent and the covariates and perhaps even the auxiliary data are time dependent and also random. As we shall see, by using partial likelihood we can transport quite straightforwardly the main inferential features appropriate for independent data to time series, not necessarily stationary, following generalized linear models. An essential component of this is that partial likelihood allows for temporal or sequential conditional inference with respect to a filtration generated by all that is known to the observer at the time of observation. This enables very flexible conditional inference that can easily accommodate autoregressive components, functions of past covariates, and all sorts of interactions among covariates.

In this chapter we provide the necessary background and an overview of generalized linear models by discussing their theoretical underpinnings, having in mind dependent time series data. Specifically, we define what we mean by time series following generalized linear models, introduce the notion of partial likelihood, and discuss in some detail the statistical properties–including large sample results–of the maximum partial likelihood estimator. Examples of special cases are presented at the end of the chapter and in subsequent chapters.

1.1 PARTIAL LIKELIHOOD

The likelihood, defined as the joint distribution of the data as a function of the unknown parameters, lies at the core of statistical theory and practice and its importance cannot be exaggerated. When the data are independent or when the dependence in the data is limited, the likelihood is readily available under appropriate assumptions on the factors in terms of which the joint distribution is expressed. In practice, however, things tend to be more complicated as the nature of dependence is not always known or even understood and consequently the likelihood is not within an easy reach. This gives the impetus for seeking suitable modifications usually by means of clever conditioning. Partial likelihood is an example of such a modification.

To motivate partial likelihood, consider a time series $\{Y_t\}$, $t = 1, \ldots, N$, with a joint density $f_{\boldsymbol{\theta}}(y_1, \ldots, y_N)$ parametrized by a vector parameter $\boldsymbol{\theta}$. In addition, suppose there is some auxiliary information AI known throughout the period of observation. Then the likelihood is a function of $\boldsymbol{\theta}$ defined by the equation

$$f_{\boldsymbol{\theta}}(y_1, \ldots, y_N|\text{AI}) = f_{\boldsymbol{\theta}}(y_1|\text{AI}) \prod_{t=2}^{N} f_{\boldsymbol{\theta}}(y_t \mid y_1, y_2, \ldots, y_{t-1}, \text{AI}). \tag{1.1}$$

When auxiliary information is not available or is not relevant, it can be dropped from the equation as we shall do forthwith to simplify the notation to

$$f_{\boldsymbol{\theta}}(y_1, \ldots, y_N) = f_{\boldsymbol{\theta}}(y_1) \prod_{t=2}^{N} f_{\boldsymbol{\theta}}(y_t \mid y_1, y_2, \ldots, y_{t-1}). \tag{1.2}$$

The main difficulty with (1.2) is that quite generally, if no additional assumptions are made, as the series size N increases so does the size of $\boldsymbol{\theta}$. Hence, instead of getting more and more information about a fixed set of parameters, we obtain information but about an increasing number of parameters, a fact which raises consistency as well as modeling problems. This is rectified when the conditional dependence in the data is limited and the increased amount of information obtained by a growing time series size concerns a fixed set of parameters.

Appropriate assumptions and modifications of the general likelihood (1.2) are called for to accommodate dependent time series data. Helpful clues in the search for a successful definition of "likelihood" can be obtained from Markovian time series, and the notion of partial likelihood advanced by Cox [104], [105].

Markov dependence of some order typifies what we mean by conditional limited dependence. As an example, suppose we observe a first order stationary Markov process, $\{Y_t\}$, at $t = 1, \ldots, N$, and that $f_{\boldsymbol{\theta}}(y_1, \ldots, y_N)$ is the joint density of the observations where $\boldsymbol{\theta}$ is a fixed vector parameter. Due to the Markov assumption the joint density can be factored as

$$f_{\boldsymbol{\theta}}(y_1, \ldots, y_N) = f_{\boldsymbol{\theta}}(y_1) \prod_{t=2}^{N} f_{\boldsymbol{\theta}}(y_t \mid y_{t-1}). \tag{1.3}$$

Ignoring the first factor $f_{\boldsymbol{\theta}}(y_1)$, as it is independent of N, inference regarding $\boldsymbol{\theta}$ can be based on the product term in (1.3). This is an example of *conditional likelihood* resulting from dependent observations expressed as a product of conditional densities. The factorization (1.3), without $f_{\boldsymbol{\theta}}(y_1)$, has some desirable properties worth keeping in mind, such as the fact that the dimension of the factors, as well as that of $\boldsymbol{\theta}$, is fixed regardless of N, and that the derivative with respect to $\boldsymbol{\theta}$ of the logarithm of (1.3) is a zero mean square integrable martingale (see [191].) The latter is useful when studying the asymptotic properties of the resulting maximum likelihood estimator. Important early references where the martingale property was recognized and applied in statistical inference are [50] and [390]. In [50], a central limit theorem for martingales was proved and applied in asymptotic large sample theory.

Next we turn to an idea due to Cox [105] who suggested using only a part of (1.2) such as a factorization that consists only of the odd numbered conditional densities. This suggests an inference based on *partial likelihood.* More precisely, consider an occasion when a time series is observed jointly with some *random time dependent covariates*. Thus, suppose we observe a pair of jointly distributed time series, (X_t, Y_t), $t = 1, \ldots, N$, where $\{Y_t\}$ is a *response* series and $\{X_t\}$ is a *time dependent random covariate*. Employing the rules of conditional probability, as was done in (1.2) and (1.1), the joint density of all the X, Y observations can be expressed as

$$f_{\boldsymbol{\theta}}(x_1, y_1, \ldots, x_N, y_N) = f_{\boldsymbol{\theta}}(x_1) \left[\prod_{t=2}^{N} f_{\boldsymbol{\theta}}(x_t \mid d_t)\right] \left[\prod_{t=1}^{N} f_{\boldsymbol{\theta}}(y_t \mid c_t)\right], \tag{1.4}$$

where $d_t = (y_1, x_1, \ldots, y_{t-1}, x_{t-1})$ and $c_t = (y_1, x_1, \ldots, y_{t-1}, x_{t-1}, x_t)$. The second product on the right hand side of (1.4) constitutes a partial likelihood according to [105] and can be used for inference. Clearly, there is information about $\boldsymbol{\theta}$ in the first product as well, and a question arises as to what happens when this factor is ignored. It turns out that under some reasonable conditions the loss of information due to the ignored factor is small, and in exchange the remaining factor is a simplified yet useful likelihood function. The adjective "partial" also refers to the fact that the remaining factor does not specify the full joint distribution of the response and the covariate data.

The previous discussion points to the potentially useful idea of forming certain likelihood functions by taking products of conditional densities, where the densities depend on a fixed parameter and where the formed products do not necessarily give complete joint or full likelihood information. This motivates the following definition of partial likelihood with respect to a nested sequence of conditioning histories.

Definition 1.1.1 Let $\mathcal{F}_t$, $t = 0, 1, \ldots$ be an increasing sequence of σ-fields, $\mathcal{F}_0 \subset \mathcal{F}_1 \subset \mathcal{F}_2 \ldots$, and let $Y_1, Y_2, \ldots$ be a sequence of random variables on some common probability space such that Y_t is $\mathcal{F}_t$ measurable. Denote the density of Y_t, given $\mathcal{F}_{t-1}$, by $f_t(y_t;\boldsymbol{\theta})$, where $\boldsymbol{\theta} \in R^p$ is a fixed parameter. The partial likelihood (PL) function relative to $\boldsymbol{\theta}$, $\mathcal{F}_t$, and the data $Y_1, Y_2, \ldots, Y_N$, is given by the product

$$\mathrm{PL}(\boldsymbol{\theta}; y_1, \ldots, y_N) = \prod_{t=1}^{N} f_t(y_t;\boldsymbol{\theta}). \tag{1.5}$$

According to Definition 1.1.1, the notion of partial likelihood generalizes both concepts of likelihood and conditional likelihood. Indeed, partial likelihood simplifies to ordinary likelihood when auxiliary information is absent and the data are independent while it becomes a conditional likelihood if the covariate process is deterministic, that is, known throughout the period of observation. Partial likelihood takes into account only what is known to the observer up to the time of actual observation, that is, it allows for *sequential conditional inference*. Closely associated with this is the martingale property alluded to earlier; it manifests itself in Section 1.4.2 on large sample results for generalized linear models. Evidently, partial likelihood does not require full knowledge of the joint distribution–that is, joint statistical dynamics–of the response and the covariates. This enables conditional inference for a fairly large class of "transition" or "transitional" non-Markovian processes where the response depends on its past values *and* on past values of the covariates. See Remark 1.2.1 and compare with [75], [123, Ch. 10].

The vector $\boldsymbol{\theta}$ that maximizes equation (1.5) is called the maximum partial likelihood estimator (MPLE). Its theoretical properties, including consistency, asymptotic normality, and efficiency, have been studied extensively in [439].

Definition 1.1.1 has been extended to continuous time stochastic processes in connection with survival analysis in [392], [393]. Additional references that treat theoretical properties of partial likelihood processes include [224] and [225]. Ramifications of partial likelihood have been considered by several authors. The notion of *marginal partial likelihood* was introduced in [174], and that of *projected partial likelihood* for modeling longitudinal data with covariates subject to drop-out is studied in [333]. Other types of pseudo-likelihoods have been considered by a fairly large number of authors of which we mention the pseudo-likelihood introduced in [45], [46] for spatial data analysis, and the notion of *empirical likelihood* introduced in [341] for nonparametric inference. See [183] for a general treatment of *pseudo-likelihood* and [342, Ch. 4] for a survey of pseudo-likelihoods including *profile* and empirical likelihoods.

1.2 GENERALIZED LINEAR MODELS AND TIME SERIES

Let $\{Y_t\}$ be a time series of interest, called the *response*, and with an eye toward prediction, let

$$\mathbf{Z}_{t-1} = (Z_{(t-1)1}, \ldots, Z_{(t-1)p})',$$

be the corresponding p–dimensional vector of past explanatory variables or covariates, $t = 1, \ldots, N$. We shall refer to $\mathbf{Z}_t$ as the *covariate process*. We denote by $\mathcal{F}_{t-1}$ the σ–field generated by $Y_{t-1}, Y_{t-2}, \ldots, \mathbf{Z}_{t-1}, \mathbf{Z}_{t-2} \ldots$,

$$\mathcal{F}_{t-1} = \sigma\{Y_{t-1}, Y_{t-2}, \ldots, \mathbf{Z}_{t-1}, \mathbf{Z}_{t-2}, \ldots\}.$$

This notation emphasizes the fact that also past values of $\{Y_t\}$ are used in the generation of $\mathcal{F}_{t-1}$, but at times it is convenient to think of $\mathbf{Z}_{t-1}$ as already including past values of the response, $Y_{t-1}, Y_{t-2}, \ldots$. When certain covariates $X_t, W_t, \ldots$, are known at $t-1$, we still write

$$\mathcal{F}_{t-1} = \sigma\{Y_{t-1}, Y_{t-2}, \ldots, X_t, W_t, \ldots, \mathbf{Z}_{t-1}, \ldots\}.$$

Thus, $\mathcal{F}_{t-1}$ is generated by past values of the response series and past and possibly present values (when known) of the covariates, meaning all that is known to the observer at time $t-1$ with the possible inclusion of $X_t, W_t, \ldots$ when known. Again, it is convenient when feasible to include known $X_t, W_t, \ldots$ in $\mathbf{Z}_{t-1}$. The anomaly of known $X_t, W_t, \ldots$ at $t-1$ occurs, for example, when they are deterministic or shifted processes, or when Y_t is a delayed output. A case in point is the relationship between rainfall X, a covariate, and river runoff Y, the response, when the amount of rainfall at a certain location is known at time t but the amount of runoff, perhaps at a different location, can only be determined at a later period. This is a situations where we may include X_t in the "past".

Let us denote by

$$\mu_t = \mathrm{E}[Y_t \mid \mathcal{F}_{t-1}],$$

the conditional expectation of the response given the past. The problem is to relate μ_t to the covariates.

In the classical theory of linear models it is assumed that the conditional expectation of the response given the past of the process is a linear function of the covariates. However, there are problems with this approach when the data are not normal and a linear relationship between the conditional mean μ_t and the covariates leads to nonsensical results. For instance, for binary time series with transition probability π_t, regressing $\pi_t = \mu_t = \mathrm{E}[Y_t \mid \mathcal{F}_{t-1}]$ linearly on the covariates may give estimates $\hat{\pi}_t$ that are negative or greater than 1. Similarly, for Poisson or Gamma data with mean μ_t, linear regression of μ_t on the covariates may lead to negative mean estimates. Generalized linear models as, introduced by [336] and further treated in [313], resolve these problems when the observations follow a distribution from an exponential family, rendering the classical linear model under normality a special case. Exponential families in the context of stochastic processes have been further studied in [249]. As regards time series, the main ideas of generalized linear models, exponential families and monotone links, can be extended quite readily.

We define *time series following generalized linear models* (GLM) by the following stipulations regarding the so called *random and systematic components*.

1. **Random Component.** The conditional distribution of the response given the past belongs to the exponential family of distributions in *natural* or *canonical* form. That

is, for $t = 1, \ldots, N$,

$$f(y_t; \theta_t, \phi \mid \mathcal{F}_{t-1}) = \exp\left\{ \frac{y_t\theta_t - b(\theta_t)}{\alpha_t(\phi)} + c(y_t; \phi) \right\}. \tag{1.6}$$

The parametric function $\alpha_t(\phi)$ is of the form ϕ/ω_t, where ϕ is a *dispersion* parameter, and ω_t is a known parameter referred to as *weight* or *prior weight*. The parameter θ_t is called the *natural* parameter of the distribution.

2. **Systematic Component.** For $t = 1, \ldots, N$, there is a monotone function $g(\cdot)$ such that

$$g(\mu_t) = \eta_t = \sum_{j=1}^{p} \beta_j Z_{(t-1)j} = \mathbf{Z}'_{t-1}\boldsymbol{\beta}. \tag{1.7}$$

The function $g(\cdot)$ is called the *link function* while η_t is referred to as the linear predictor of the model.

Typical choices for $\mathbf{Z}'_{t-1}\boldsymbol{\beta}$, could be

$$\mathbf{Z}'_{t-1}\boldsymbol{\beta} = \beta_0 + \beta_1 Y_{t-1} + \beta_2 Y_{t-2} + \beta_3 X_t \cos(\omega_0 t)$$

or

$$\mathbf{Z}'_{t-1}\boldsymbol{\beta} = \beta_0 + \beta_1 Y_{t-1} + \beta_2 Y_{t-2} + \beta_3 Y_{t-1} X_t + \beta_4 Y_{t-2} X_{t-1}$$

and so on, given a one-dimensional response process $\{Y_t\}$ and a covariate process $\{X_t\}$.

An interesting choice for $\mathbf{Z}'_{t-1}\boldsymbol{\beta}$ is given by the following expression:

$$\mathbf{Z}'_{t-1}\boldsymbol{\beta} = \mathbf{X}'_t\gamma + \sum_{i=1}^{p} \phi_i H_i(Y_{t-i}) + \sum_{i=1}^{q} \theta_i D_i(\mu_{t-i}), \tag{1.8}$$

where $H_i(\cdot)$ and $D_i(\cdot)$ are known functions for all i. It is easy to see that (1.8) is special case of (1.7) upon defining

$$\mathbf{Z}_{t-1} = (\mathbf{X}_t, H_1(Y_{t-1}), \ldots, H_p(Y_{t-p}), D_1(\mu_{t-1}), \ldots, D_q(\mu_{t-q}))'$$

and $\boldsymbol{\beta} = (\gamma', \phi_1, \ldots, \phi_p, \theta_1, \ldots, \theta_q)'$. Consequently, the covariate vector is $\mathcal{F}_{t-1}$ measurable and it follows that partial likelihood inference is applicable to model (1.8).

A model closely related to (1.8) is

$$g(\mu_t) = \eta_t = \mathbf{X}'_t\gamma + \sum_{i=1}^{p} \phi_i (g(Y_{t-i}) - \mathbf{X}'_{t-i}\gamma) + \sum_{i=1}^{q} \theta_i \epsilon_{t-i}, \tag{1.9}$$

where $\epsilon_{t-i} = g(Y_{t-i}) - \eta_{t-i}$. Model (1.9) has been termed Generalized Linear Autoregressive Moving Average [383], GLARMA(p, q), or Generalized Autoregressive Moving Average [35], [36], GARMA(p, q), of order (p, q), respectively, since it contains both autoregressive and moving average terms. The particular GARMA model depends on the conditional distribution (1.6). Thus, when the conditional distribution (1.6) is Poisson we obtain a Poisson GARMA(p,q) and so on. These models have been studied by several authors including [299], [383], [447], and more recently [35], [36], [114], and [185] (see Problems 6 and 7).

In this connection, the use of the Box–Cox power transformation,

$$g(u) = \begin{cases} \frac{u^\lambda - 1}{\lambda} & \text{for } \lambda \neq 0 \\ \log u & \text{for } \lambda = 0 \end{cases} \tag{1.10}$$

has been suggested as a flexible link function where the special cases $\lambda = 1$ and $\lambda = 0$ give linear and log–linear models, respectively [35]. It should be noted that estimation in GARMA models may require profiling when unknown parameters are multiplied by each other.

Remark 1.2.1 *Processes following (1.6) may or may not be Markovian. For example, a nonhomogeneous Markovian dependence can be induced by conditioning on past values of the response while fixing the rest of the covariate data. In general, a Markov process is defined against the backdrop of the* **joint** *dynamics of $\{Y_t, \mathbf{Z}_t\}$, meaning a clear specification of the joint finite dimensional distributions of the response* and *the covariates. However, we do not make any assumption regarding the joint distribution of the response and its covariates.*

The literature offers quite a few parametric models closely related or at least resembling the canonical parametrization (1.6). For example, it is possible to parametrize the exponential family directly via the mean as done in Problem 5. The latter, it may be argued, is perhaps more natural since we are mainly concerned with modeling the conditional mean. But this is only ostensibly true since the canonical parameter θ is a function of the conditional mean and can be written explicitly as such. A closely related model is the exponential dispersion model defined by the probability density

$$f(y; \theta, \lambda) = a(\lambda, y) \exp\{\lambda(y\theta - \kappa(\theta))\},$$

where λ and θ are parameters, and $1/\lambda$ signifies dispersion. When λ is known, $f(y; \theta, \lambda)$ is a member of the exponential family. Exponential dispersion models and their variants are discussed in [234].

Since equation (1.6) introduces a probability density function for every $t = 1, \ldots, N$, we obtain

$$\int f(y; \theta_t; \phi \mid \mathcal{F}_{t-1}) dy = 1$$

for all t. Assuming that differentiation and integration can be interchanged, differentiation of both sides of the previous equation with respect to θ_t yields

$$\int \left(\frac{y - b'(\theta_t)}{\alpha_t(\phi)} \right) f(y; \theta_t, \phi \mid \mathcal{F}_{t-1}) dy = 0, \tag{1.11}$$

which in turn implies the following relationship between the conditional mean and the natural parameter θ_t:

$$\mu_t = \mathrm{E}[Y_t \mid \mathcal{F}_{t-1}] = b'(\theta_t). \tag{1.12}$$

By further differentiation of equation (1.11) we obtain

$$\int \frac{(y - b'(\theta_t))^2}{\alpha_t(\phi)} f(y; \theta_t, \phi \mid \mathcal{F}_{t-1}) dy = b''(\theta_t),$$

which gives

$$\text{Var}[Y_t \mid \mathcal{F}_{t-1}] = \alpha_t(\phi) b''(\theta_t). \tag{1.13}$$

The function $V(\mu_t) = b''(\theta_t)$ is called the *variance function*. It clearly depends on μ_t since, as shown next, θ_t is a function of μ_t.

Since $\text{Var}[Y_t \mid \mathcal{F}_{t-1}] > 0$, it follows that b' is monotone. Therefore, equation (1.12) implies that

$$\theta_t = (b')^{-1}(\mu_t). \tag{1.14}$$

We see that θ_t itself is a monotone function of μ_t and hence it can be used to define a link function. The link function

$$g(\mu_t) = \theta_t(\mu_t) = \eta_t = \mathbf{Z}'_{t-1}\beta \tag{1.15}$$

is called the *canonical link* function. Thus from (1.14) and (1.15), the "natural" choice

$$g = \mu^{-1} \equiv (b')^{-1} \tag{1.16}$$

is the explicit form of the canonical link function. In general, however, for any link

$$\theta_t = (b')^{-1}(g^{-1}(\eta_t)) = \mu^{-1}(g^{-1}(\eta_t)), \tag{1.17}$$

where $\mu^{-1}(\cdot) \equiv (b')^{-1}(\cdot)$, and $\partial\theta/\partial\mu = 1/V(\mu)$.

It is instructive to consider some standard examples in order to clarify the definitions and also elucidate the relevance of generalized linear models for time series analysis. The examples make clear that the notion of generalized linear models is useful for quantitative as well as qualitative dependent data.

Time Series with Conditional Gaussian Marginals

The probability density function (pdf) of the normal distribution with parameters μ_t and σ^2 admits, for $t = 1, \ldots, N$, the form

$$f(y_t; \theta_t, \phi | \mathcal{F}_{t-1}) = \exp\left\{ \frac{y_t\mu_t - \mu_t^2/2}{\sigma^2} - (y_t^2/\sigma^2 + \log 2\pi\sigma^2)/2 \right\}, \tag{1.18}$$

from which we see that $\text{E}[Y_t \mid \mathcal{F}_{t-1}] = \mu_t = \theta_t$, $b(\theta_t) = \theta_t^2/2$, $V(\mu_t) = 1$, $\phi = \sigma^2$, and $\omega_t = 1$, by direct comparison with (1.6). The canonical link is the identity function,

$$g(\mu_t) = \theta_t(\mu_t) = \mu_t = \eta_t = \mathbf{Z}'_{t-1}\beta.$$

Thus in the case of time series with conditional Gaussian marginals of the form (1.18) and canonical link, the conditional mean is a linear function of the covariates and GLM regression reduces for all practical purposes to the classical linear regression model [416].

Time Series of Counts

Observations of dependent counts can in many cases be modeled successfully through the Poisson distribution. The conditional density of the Poisson distribution with mean μ_t can be written as

$$f(y_t; \theta_t, \phi \mid \mathcal{F}_{t-1}) = \exp\left\{(y_t \log \mu_t - \mu_t) - \log y_t!\right\}, \quad t = 1, \ldots, N,$$

so that $\mathrm{E}[Y_t|\mathcal{F}_{t-1}] = \mu_t$, $b(\theta_t) = \mu_t = \exp(\theta_t)$, $V(\mu_t) = \mu_t$, $\phi = 1$, and $\omega_t = 1$. The canonical link is

$$g(\mu_t) = \theta_t(\mu_t) = \log \mu_t = \eta_t = \mathbf{Z}'_{t-1}\boldsymbol{\beta}.$$

As an example , if $\mathbf{Z}_{t-1} = (1, X_t, Y_{t-1})'$, then

$$\log \mu_t = \beta_0 + \beta_1 X_t + \beta_2 Y_{t-1}$$

with $\{X_t\}$ standing for some covariate process, or a possible trend, or a possible seasonal component. At a later chapter we shall treat a more general case of counts following a truncated Poisson distribution.

Binary Time Series

Suppose we observe a sequence of dependent observations $\{Y_t\}$ taking only two values which we call success and failure. Denote by π_t the probability of success given $\mathcal{F}_{t-1}$. Then for $t = 1, \ldots, N$,

$$f(y_t; \theta_t, \phi \mid \mathcal{F}_{t-1}) = \exp\left\{ y_t \log\left(\frac{\pi_t}{1-\pi_t}\right) + \log(1-\pi_t)\right\},$$

with $\mathrm{E}[Y_t|\mathcal{F}_{t-1}] = \pi_t$, $b(\theta_t) = -\log(1-\pi_t) = \log(1+\exp(\theta_t))$, $V(\pi_t) = \pi_t(1-\pi_t)$, $\phi = 1$, $\omega_t = 1$. Thus, what is regressed linearly on some covariate vector $\mathbf{Z}_{t-1}$ is a transformation of the probability of success. The most commonly used model is the one with the canonical link called the *logistic* model,

$$g(\pi_t) = \theta_t(\pi_t) = \log \frac{\pi_t}{1-\pi_t} = \eta_t = \mathbf{Z}'_{t-1}\boldsymbol{\beta}. \tag{1.19}$$

A closer look at (1.19) shows that π_t is related to η_t through the standard logistic cumulative distribution function (cdf) $F_l(x) = \exp(x)/\{1+\exp(x)\}$,

$$\pi = F_l(\eta) \tag{1.20}$$

and hence the term *logistic regression*. Here the logistic cdf serves as the inverse of the canonical link. We can however resort to other inverse links of the form (1.20) for other useful cdf's such as the cdf of the standard normal distribution $\Phi(x)$,

$$\pi = \Phi(\eta). \tag{1.21}$$

Model (1.21) is called *probit regression* (see [147]), in which case the link function is Φ^{-1}. Another possible link for binomial/Bernoulli data is the so called *complementary log-log* link

$$\eta = \log\{-\log(1-\pi)\}. \tag{1.22}$$

We shall examine binary, and more generally categorical time series, more closely in later chapters.

1.3 PARTIAL LIKELIHOOD INFERENCE

We now turn to the inference problem for time series following generalized linear models. Given a time series $\{Y_t\}$, $t = 1, \ldots, N$, conditionally distributed as (1.6), we discuss partial likelihood estimation of the vector of regression coefficients, $\boldsymbol{\beta}$. In doing so, we make repeated use of the chain rule applied to functionally interdependent quantities. It is assumed that $\{\mathbf{Z}_{t-1}\}$ stands for a p–dimensional vector of random time dependent covariates, g is a given link function, and that the dispersion parameter ϕ is known.

By definition 1.1.1, the partial likelihood of the observed series is

$$\mathrm{PL}(\boldsymbol{\beta}) = \prod_{t=1}^{N} f(y_t; \theta_t, \phi \mid \mathcal{F}_{t-1}). \tag{1.23}$$

Then from (1.6), the log–partial likelihood, $l(\boldsymbol{\beta})$, is given by

$$\begin{aligned} l(\boldsymbol{\beta}) &= \sum_{t=1}^{N} \log f(y_t; \theta_t, \phi \mid \mathcal{F}_{t-1}) \\ &= \sum_{t=1}^{N} \left\{ \frac{y_t\theta_t - b(\theta_t)}{\alpha_t(\phi)} + c(y_t, \phi) \right\} \\ &= \sum_{t=1}^{N} \left\{ \frac{y_t u(\mathbf{z}'_{t-1}\boldsymbol{\beta}) - b(u(\mathbf{z}'_{t-1}\boldsymbol{\beta}))}{\alpha_t(\phi)} + c(y_t, \phi) \right\} \equiv \sum_{t=1}^{N} l_t, \end{aligned} \tag{1.24}$$

where, in order to emphasize the dependence of the partial likelihood on $\boldsymbol{\beta}$, we have used the notation

$$u(\cdot) \equiv (g \circ \mu(\cdot))^{-1} = \mu^{-1}(g^{-1}(\cdot)) \tag{1.25}$$

so that from (1.17), $\theta_t = u(\mathbf{z}'_{t-1}\boldsymbol{\beta})$.

It is convenient to introduce the notation

$$\nabla \equiv \left(\frac{\partial}{\partial \beta_1}, \frac{\partial}{\partial \beta_2}, \cdots, \frac{\partial}{\partial \beta_p} \right)'.$$

The *partial score* is defined as $\nabla l(\beta)$, the first derivative of the log–partial likelihood function with respect to the vector of unknown parameters β. Calculation of the latter is carried out upon observing that from the chain rule,

$$\frac{\partial l_t}{\partial \beta_j} = \frac{\partial l_t}{\partial \theta_t}\frac{\partial \theta_t}{\partial \mu_t}\frac{\partial \mu_t}{\partial \eta_t}\frac{\partial \eta_t}{\partial \beta_j} \tag{1.26}$$

for $j = 1, \ldots, p$. From equations (1.12) and (1.13),

$$\frac{\partial l_t}{\partial \theta_t} = \frac{(y_t - b'(\theta_t))}{\alpha_t(\phi)} = \frac{(y_t - \mu_t)}{\alpha_t(\phi)}$$

and

$$\frac{\partial \theta_t}{\partial \mu_t} = \frac{1}{b''(\theta_t)} = \frac{\alpha_t(\phi)}{\mathrm{Var}[Y_t \mid \mathcal{F}_{t-1}]}.$$

It is also clear that since $\eta_t = \sum_{j=1}^{p} z_{(t-1)j}\beta_j$,

$$\frac{\partial \eta_t}{\partial \beta_j} = z_{(t-1)j}.$$

Hence, equation (1.26) becomes

$$\frac{\partial l_t}{\partial \beta_j} = \frac{(y_t - \mu_t)}{\mathrm{Var}[Y_t \mid \mathcal{F}_{t-1}]}\frac{\partial \mu_t}{\partial \eta_t} z_{(t-1)j}$$

for $j = 1, \ldots, p$. Assembling all the above relations, we see that the *partial score* is a p–dimensional vector given by

$$\mathbf{S}_N(\beta) \equiv \nabla l(\beta) = \sum_{t=1}^{N} \mathbf{Z}_{t-1}\frac{\partial \mu_t}{\partial \eta_t}\frac{(Y_t - \mu_t(\beta))}{\sigma_t^2(\beta)} \tag{1.27}$$

with $\sigma_t^2(\beta) = \mathrm{Var}[Y_t \mid \mathcal{F}_{t-1}]$. The *partial score vector process* $\{\mathbf{S}_t(\beta)\}$, $t = 1, \ldots, N$, is defined from the partial sums

$$\mathbf{S}_t(\beta) = \sum_{s=1}^{t} \mathbf{Z}_{s-1}\frac{\partial \mu_s}{\partial \eta_s}\frac{(Y_s - \mu_s(\beta))}{\sigma_s^2(\beta)}. \tag{1.28}$$

By iterated expectation (see Problem 1-b), the fact

$$\mathrm{E}\left[\mathbf{Z}_{t-1}\frac{\partial \mu_t}{\partial \eta_t}\frac{(Y_t - \mu_t(\beta))}{\sigma_t^2(\beta)} \mid \mathcal{F}_{t-1}\right] = \mathbf{0}$$

implies that $\mathrm{E}[\mathbf{S}_N(\beta)] = \mathbf{0}$, and similarly the terms in (1.27) are orthogonal,

$$\mathrm{E}\left[\mathbf{Z}_{s-1}\frac{\partial \mu_s}{\partial \eta_s}\frac{(Y_s - \mu_s(\beta))}{\sigma_s^2(\beta)}\mathbf{Z}'_{t-1}\frac{\partial \mu_t}{\partial \eta_t}\frac{(Y_t - \mu_t(\beta))}{\sigma_t^2(\beta)}\right] = \mathbf{0}, \quad s < t.$$

The solution of the score equation,

$$\mathbf{S}_N(\boldsymbol{\beta}) = \nabla \log \mathrm{PL}(\boldsymbol{\beta}) = \mathbf{0} \tag{1.29}$$

is denoted by $\hat{\boldsymbol{\beta}}$, and is referred to as the maximum partial likelihood estimator (MPLE) of $\boldsymbol{\beta}$. The system of equations (1.29) is nonlinear and is customarily solved by the Fisher scoring method, an iterative algorithm resembling the Newton–Raphson procedure. Before turning to the Fisher scoring algorithm in our context of conditional inference, it is necessary to introduce several important matrices.

An important role in partial likelihood inference is played by the *cumulative conditional information matrix*, $\mathbf{G}_N(\boldsymbol{\beta})$, defined by a sum of conditional covariance matrices,

$$\begin{aligned} \mathbf{G}_N(\boldsymbol{\beta}) &= \sum_{t=1}^{N} \mathrm{Cov}\left[\mathbf{Z}_{t-1} \frac{\partial \mu_t}{\partial \eta_t} \frac{(Y_t - \mu_t(\boldsymbol{\beta}))}{\sigma_t^2(\boldsymbol{\beta})} \mid \mathcal{F}_{t-1}\right] \\ &= \sum_{t=1}^{N} \mathbf{Z}_{t-1} \left(\frac{\partial \mu_t}{\partial \eta_t}\right)^2 \frac{1}{\sigma_t^2(\boldsymbol{\beta})} \mathbf{Z}_{t-1}' \\ &= \mathbf{Z}'\mathbf{W}(\boldsymbol{\beta})\mathbf{Z}, \end{aligned} \tag{1.30}$$

with

$$\mathbf{Z} = \begin{bmatrix} \mathbf{Z}_0' \\ \mathbf{Z}_1' \\ \vdots \\ \mathbf{Z}_{N-1}' \end{bmatrix},$$

an $N \times p$ matrix, and $\mathbf{W}(\boldsymbol{\beta}) = \mathrm{diag}(w_1, ..., w_N)$ where

$$w_t = \left(\frac{\partial \mu_t}{\partial \eta_t}\right)^2 \frac{1}{\sigma_t^2(\boldsymbol{\beta})}, \quad t = 1, ..., N, \tag{1.31}$$

also an $N \times N$ diagonal matrix. The *unconditional information matrix* is simply

$$\mathrm{Cov}(\mathbf{S}_N(\boldsymbol{\beta})) = \mathbf{F}_N(\boldsymbol{\beta}) = \mathrm{E}[\mathbf{G}_N(\boldsymbol{\beta})]. \tag{1.32}$$

It is clear that the above quantity cannot be calculated explicitly, however, we will notice that under some suitable conditions on the covariate process it can be well approximated by the conditional information matrix. Next, let $\mathbf{H}_N(\boldsymbol{\beta})$ denote the matrix of second derivatives of the log–partial likelihood multiplied by -1:

$$\mathbf{H}_N(\boldsymbol{\beta}) \equiv -\nabla\nabla\,' l(\boldsymbol{\beta}).$$

Interestingly, $\mathbf{H}_N(\boldsymbol{\beta})$, the so called observed information matrix, admits a decomposition in terms of a difference between the conditional information matrix and a remainder term,

$$\mathbf{H}_N(\boldsymbol{\beta}) = \mathbf{G}_N(\boldsymbol{\beta}) - \mathbf{R}_N(\boldsymbol{\beta}), \tag{1.33}$$

where

$$\mathbf{R}_N(\beta) = \frac{1}{\alpha_t(\phi)} \sum_{t=1}^{N} \mathbf{Z}_{t-1} d_t(\beta) \mathbf{Z}'_{t-1}(Y_t - \mu_t(\beta)) \tag{1.34}$$

and $d_t(\beta) = [\partial^2 u(\eta_t)/\partial \eta_t^2]$.

Proof of (1.33) and (1.34): Recall the various relationships between the parameters, and use $\mu_t = \mu_t(\beta), \sigma_t^2 = \sigma_t^2(\beta)$. By direct differentiation, we have for $j, k = 1, \ldots, p$

$$\begin{aligned}
\frac{\partial^2 l_t}{\partial \beta_j \partial \beta_k} &= \frac{\partial}{\partial \beta_k}\left(\frac{\partial l_t}{\partial \beta_j}\right) \\
&= \frac{\partial}{\partial \beta_k}\left(\frac{\partial l_t}{\partial \eta_t} z_{(t-1)j}\right) \\
&= z_{(t-1)j} z_{(t-1)k} \frac{\partial^2 l_t}{\partial \eta_t^2},
\end{aligned} \tag{1.35}$$

where

$$\begin{aligned}
\frac{\partial^2 l_t}{\partial \eta_t^2} &= \frac{\partial}{\partial \eta_t}\left(\frac{\partial l_t}{\partial \theta_t}\frac{\partial \theta_t}{\partial \eta_t}\right) \\
&= \frac{\partial}{\partial \eta_t}\left[\frac{(y_t - \mu_t)}{\alpha_t(\phi)}\frac{\partial \theta_t}{\partial \eta_t}\right] \\
&= \frac{1}{\alpha_t(\phi)}\left\{\left[\frac{\partial}{\partial \eta_t}(y_t - \mu_t)\right]\frac{\partial \theta_t}{\partial \eta_t} + (y_t - \mu_t)\left[\frac{\partial}{\partial \eta_t}\left(\frac{\partial \theta_t}{\partial \eta_t}\right)\right]\right\}.
\end{aligned} \tag{1.36}$$

Considering each summand of (1.36) separately, we obtain on account of (1.12) and (1.13),

$$\begin{aligned}
\left(\frac{\partial}{\partial \eta_t}(y_t - \mu_t)\right)\frac{\partial \theta_t}{\partial \eta_t} &= \left(\frac{\partial (y_t - \mu_t)}{\partial \theta_t}\frac{\partial \theta_t}{\partial \mu_t}\frac{\partial \mu_t}{\partial \eta_t}\right)\left(\frac{\partial \theta_t}{\partial \mu_t}\frac{\partial \mu_t}{\partial \eta_t}\right) \\
&= \frac{\partial (y_t - \mu_t)}{\partial \theta_t}\left(\frac{\partial \theta_t}{\partial \mu_t}\right)^2\left(\frac{\partial \mu_t}{\partial \eta_t}\right)^2 \\
&= -\frac{\sigma_t^2}{\alpha_t(\phi)}\frac{\alpha_t^2(\phi)}{\sigma_t^4}\left(\frac{\partial \mu_t}{\partial \eta_t}\right)^2 \\
&= -\frac{\alpha_t(\phi)}{\sigma_t^2}\left(\frac{\partial \mu_t}{\partial \eta_t}\right)^2.
\end{aligned} \tag{1.37}$$

Similarly, the second summand of (1.36) becomes

$$\begin{aligned}
(y_t - \mu_t)\frac{\partial}{\partial \eta_t}\left(\frac{\partial \theta_t}{\partial \eta_t}\right) &= (y_t - \mu_t)\frac{\partial^2 \theta_t}{\partial \eta_t^2} \\
&= (y_t - \mu_t)\frac{\partial^2 u(\eta_t)}{\partial \eta_t^2}.
\end{aligned} \tag{1.38}$$

Substitution of equations (1.37) and (1.38) into (1.36) shows that the (j, k) element of (1.35) with a negative sign is

$$-\frac{\partial^2 l_t(\beta)}{\partial \beta_j \partial \beta_k} = z_{(t-1)j} z_{(t-1)k} \left[\frac{1}{\sigma_t^2} \left(\frac{\partial \mu_t}{\partial \eta_t} \right)^2 - \frac{(y_t - \mu_t)}{\alpha_t(\phi)} \frac{\partial^2 u(\eta_t)}{\partial \eta_t^2} \right],$$

and hence equations (1.33) and (1.34) follow.

Canonical links entail certain simplifications. To see this, recall that when the link is canonical, $\eta_t = \theta_t$. Then,

$$\frac{\partial \mu_t}{\partial \eta_t} = \frac{\partial \mu_t}{\partial \theta_t} = b''(\theta_t).$$

Thus, the partial score (1.27) becomes

$$\mathbf{S}_N(\beta) = \frac{1}{\alpha_t(\phi)} \sum_{t=1}^{N} \mathbf{Z}_{t-1}(Y_t - \mu_t(\beta)), \tag{1.39}$$

and the conditional information matrix (1.30) reduces to

$$\mathbf{G}_N(\beta) = \frac{1}{\alpha_t^2(\phi)} \sum_{t=1}^{n} \mathbf{Z}_{t-1} \sigma_t^2(\beta) \mathbf{Z}'_{t-1}. \tag{1.40}$$

Moreover, for canonical links (1.25) implies that $u(\eta_t) = \eta_t$ in which case $d_t = 0$. Therefore, $\mathbf{R}_N(\beta)$ vanishes and we have the useful result

$$\mathbf{H}_N(\beta) = \mathbf{G}_N(\beta). \tag{1.41}$$

1.3.1 Estimation of the Dispersion Parameter

In our discussion so far we have assumed that the dispersion parameter ϕ is known. However, there are occasions when the dispersion parameter is unknown and needs to be estimated. Then the following method of moments based estimator

$$\hat{\phi} = \frac{1}{N-p} \sum_{t=1}^{N} \frac{\omega_t (Y_t - \hat{\mu}_t)^2}{V(\hat{\mu}_t)} \tag{1.42}$$

can be used in applications. An alternative estimator–based on the so called *deviance*– is given in Section 1.6.1.

1.3.2 Iterative Reweighted Least Squares

As mentioned earlier, due to the nonlinear nature of the partial likelihood equations (1.29), their solution is obtained iteratively by the method of Fisher scoring, a modification of the well-known Newton–Raphson method. Accordingly, the observed information matrix $\mathbf{H}_N(\beta)$ is replaced by its conditional expectation giving the iteration

$$\hat{\beta}^{(k+1)} = \hat{\beta}^{(k)} + \mathbf{G}_N^{-1}(\hat{\beta}^{(k)})\mathbf{S}_N(\hat{\beta}^{(k)}). \tag{1.43}$$

Equation (1.41) implies that the Newton–Raphson method is identical to the Fisher scoring algorithm in the case of canonical link function.

Fisher scoring may be viewed as a weighted least squares estimation method. Indeed, assuming that the inverse of the conditional information matrix exists, (1.43) can be rewritten as

$$\mathbf{G}_N(\hat{\beta}^{(k)})\hat{\beta}^{(k+1)} = \mathbf{G}_N(\hat{\beta}^{(k)})\hat{\beta}^{(k)} + \mathbf{S}_N(\hat{\beta}^{(k)}). \tag{1.44}$$

But the right-hand side of (1.44) is a p–dimensional vector whose l'th element is

$$\sum_{j=1}^{p}\left[\sum_{t=1}^{N}\frac{Z_{(t-1)j}Z_{(t-1)l}}{\sigma_t^2}\left(\frac{\partial\mu_t}{\partial\eta_t}\right)^2\right]\hat{\beta}_j^{(k)} + \sum_{t=1}^{N}\frac{(Y_t-\mu_t)Z_{(t-1)l}}{\sigma_t^2}\left(\frac{\partial\mu_t}{\partial\eta_t}\right)$$
$$= \sum_{t=1}^{N} Z_{(t-1)l}w_t\left\{\eta_t + (Y_t-\mu_t)\frac{\partial\eta_t}{\partial\mu_t}\right\}$$

for $l = 1, \ldots, p$, and μ_t, η_t, $(\partial\mu_t/\partial\eta_t)$ evaluated at $\hat{\beta}^{(k)}$ and w_t defined in (1.31). Thus, if we define for $t = 1, \ldots, N$,

$$\begin{aligned} q_t^{(k)} &= \sum_{j=1}^{p} Z_{(t-1)j}\hat{\beta}_j^{(k)} + (Y_t-\mu_t)\frac{\partial\eta_t}{\partial\mu_t} \\ &= \eta_t(\hat{\beta}^{(k)}) + (Y_t-\mu_t)\frac{\partial\eta_t}{\partial\mu_t} \end{aligned}$$

then the right-hand side of (1.44) is equal to $\mathbf{Z}'\mathbf{W}(\hat{\beta}^{(k)})\mathbf{q}^{(k)}$ according to (1.30) where the elements of the N-dimensional vector $\mathbf{q}^{(k)}$ are the $q_t^{(k)}$. With this and (1.30) applied to its left-hand side, expression (1.44) becomes

$$\mathbf{Z}'\mathbf{W}(\hat{\beta}^{(k)})\mathbf{Z}\hat{\beta}^{(k+1)} = \mathbf{Z}'\mathbf{W}(\hat{\beta}^{(k)})\mathbf{q}^{(k)}.$$

Therefore Fisher scoring method (1.43) simplifies to

$$\hat{\beta}^{(k+1)} = \left(\mathbf{Z}'\mathbf{W}(\hat{\beta}^{(k)})\mathbf{Z}\right)^{-1}\mathbf{Z}'\mathbf{W}(\hat{\beta}^{(k)})\mathbf{q}^{(k)}, \tag{1.45}$$

where $\mathbf{W}(\hat{\beta}^{(k)})$ and $\mathbf{q}^{(k)}$ are evaluated at $\hat{\beta}^{(k)}$. The limit of the recursion (1.45), as $k \to \infty$, is the maximum partial likelihood estimator $\hat{\beta}$. The iterative procedure (1.45) is called *iterative reweighted least squares* since each iteration has the same form as that of weighted least squares but with adjusted weight $\mathbf{W}(\hat{\beta}^{(k)})$ and adjusted dependent variable $\mathbf{q}^{(k)}$. Thus, maximizing the partial likelihood (1.23) reduces to an iterative weighted least squares procedure, valid for all generalized linear models regardless of the chosen link. The recursion is initialized by simply replacing the conditional means by the corresponding responses. This gives the first estimate for the weight matrix $\mathbf{W}$ and hence a starting point for $\boldsymbol{\beta}$. The iterations run until some criterion of convergence is satisfied.

1.4 ASYMPTOTIC THEORY

In the previous section we discussed partial likelihood estimation for time series following generalized linear models, and the computation of the maximum partial likelihood estimator (MPLE). In this section we discuss briefly the asymptotic behavior of the MPLE, postponing a more detailed treatment to Chapter 3. Under suitable regularity conditions the MPLE has some desirable properties similar to those enjoyed by maximum likelihood estimators. The presentation is along the lines of the well-established theory of maximum likelihood estimation, and likewise requires certain regularity conditions ([139]).

The following assumption is endemic to our presentation. It will be assumed throughout our treatment of time series following generalized linear models.

Assumption A

A1. The true parameter $\boldsymbol{\beta}$ belongs to an open set $B \subseteq R^p$.

A2. The covariate vector $\mathbf{Z}_{t-1}$ almost surely lies in a nonrandom compact subset Γ of R^p, such that $P[\sum_{t=1}^{N} \mathbf{Z}_{t-1}\mathbf{Z}'_{t-1} > \mathbf{0}] = 1$. In addition, $\mathbf{Z}'_{t-1}\boldsymbol{\beta}$ lies almost surely in the domain H of the inverse link function $h = g^{-1}$ for all $\mathbf{Z}_{t-1} \in \Gamma$ and $\boldsymbol{\beta} \in B$.

A3. The inverse link function h–defined in (A2)–is twice continuously differentiable and $|\partial h(\gamma)/\partial\gamma| \neq 0$.

A4. There is a probability measure ν on R^p such that $\int_{R^p} \mathbf{z}\mathbf{z}'\nu(d\mathbf{z})$ is positive definite, and such that under (1.6) and (1.7) for Borel sets $A \subset R^p$,

$$\frac{1}{N}\sum_{t=1}^{N} I_{[\mathbf{Z}_{t-1}\in A]} \rightarrow \nu(A)$$

in probability as $N \rightarrow \infty$, at the true value of $\boldsymbol{\beta}$.

Assumption A provides an efficacious mathematical framework that guarantees a well-defined model and consistent estimators under reasonable conditions. Specifically, A1 together with A3 guarantee that the second derivative of the partial log-likelihood function is continuous with respect to $\boldsymbol{\beta}$. In addition, the condition $|\partial h(\gamma)/\partial\gamma| \neq 0$ together with A2, assuming sufficiently large N, imply in particular that the conditional information matrix is positive definite with probability 1. Indeed, for any vector $\lambda \in R^p$, $\lambda \neq 0$,

$$\begin{aligned} \lambda' \mathbf{G}(\beta)_N \lambda &= \lambda' \left[\sum_{t=1}^{N} \mathbf{Z}_{t-1} \left(\frac{\partial \mu_t}{\partial \eta_t} \right)^2 \frac{1}{\sigma_t^2} \mathbf{Z}'_{t-1} \right] \lambda \\ &\geq \inf_t \left[\left(\frac{\partial \mu_t}{\partial \eta_t} \right)^2 \frac{1}{\sigma_t^2} \right] \left(\lambda' \sum_{t=1}^{N} \mathbf{Z}_{t-1} \mathbf{Z}'_{t-1} \lambda \right) \\ &> 0. \end{aligned}$$

Thus the unconditional information matrix is positive definite as well. The compactness stipulation in A2 is useful for deriving asymptotic bounds. Simulations as well as experience with real data indicate, however, that in applications the assumption can be softened and we may include covariates which take values in large bounded intervals with high probability. The last part of assumption A2 assures a mathematically tractable model. Assumption A4 calls for asymptotically "well behaved" covariates in the sense that if f is any continuous and bounded function on Γ taking values in R^p then,

$$\frac{\sum_{t=1}^{N} f(\mathbf{Z}_{t-1})}{N} \rightarrow \int_{R^p} f(\mathbf{z}) \nu(d\mathbf{z})$$

in probability as $N \rightarrow \infty$. Thus, the conditional information matrix $\mathbf{G}_N(\beta)$ has a non-random limit which we denote by $\mathbf{G}(\beta)$. That is, there exists a $p \times p$ limiting *information matrix per observation*, $\mathbf{G}(\beta)$, such that

$$\frac{\mathbf{G}_N(\beta)}{N} \rightarrow \mathbf{G}(\beta) \tag{1.46}$$

in probability, as $N \rightarrow \infty$. Explicit forms of the limiting information matrix will be given in later chapters. Notice that by A4, $\mathbf{G}(\beta)$ is a positive definite matrix at the true value of β and therefore its inverse exists. In general, $\mathbf{G}_N(\beta)/N$ could converge in probability to a random limit as in the *nonergodic* cases treated in [30].

1.4.1 Uniqueness and Existence

The score equation of a full likelihood may have multiple roots, or it may have no roots at all. Questions regarding existence and uniqueness of maximum likelihood estimators have been addressed by a number of authors including [11], [190], [247], [350], [389], and [423]. Regarding partial likelihood, the essential conditions needed for existence and uniqueness of MPLE's are similar to those of traditional maximum likelihood estimators, and asymptotic existence (as $N \rightarrow \infty$) is achieved in most interesting cases. The proof is postponed to the appendix of Chapter 3.

1.4.2 Large Sample Properties

The study of the asymptotic behavior of the maximum partial likelihood estimator is based on the stability of the conditional information matrix jointly with the central limit theorem for martingales. This is discussed next rather informally. A rigorous treatment is delayed to Chapter 3 on categorical time series.

The first step in establishing asymptotic properties of the maximum partial likelihood estimator concerns the asymptotic normality of the partial score (1.27). Recall that the partial score process is defined through the partial sums (1.28) evaluated at the true parameter. Then a straightforward calculation shows that

$$\mathrm{E}[\mathbf{S}_{t+1}(\boldsymbol{\beta})|\mathcal{F}_t] = \mathbf{S}_t(\boldsymbol{\beta})$$

since $\mathrm{E}[Y_{t+1} - \mu_{t+1}(\boldsymbol{\beta})|\mathcal{F}_t] = 0$. This shows that the process $\{\mathbf{S}_t(\boldsymbol{\beta})\}$ coupled with the filtration $\{\mathcal{F}_t\}$, $t = 1, \ldots, N$, forms a martingale. Using this fact, we can show that the asymptotic distribution of the partial score converges to a normal random variable. That is,

$$\frac{\mathbf{S}_N(\boldsymbol{\beta})}{\sqrt{N}} \to \mathcal{N}_p(\mathbf{0}, \mathbf{G}(\boldsymbol{\beta})), \tag{1.47}$$

in distribution, as $N \to \infty$, (see [152]). Here $\mathcal{N}_p$ stands for a p–dimensional normal distribution with covariance matrix $\mathbf{G}(\boldsymbol{\beta})$ defined in (1.46). The proof of this fact is based on verifying the conditions in [191, Corollary 3.1].

The second step in the quest for asymptotic properties of the maximum partial likelihood estimator is to show that the remainder term in equation (1.33) is small. It can be shown that

$$\frac{\mathbf{R}_N(\boldsymbol{\beta})}{N} \to \mathbf{0} \tag{1.48}$$

in probability as $N \to \infty$ (see [152]).

Next we observe that a Taylor series expansion of $\mathbf{S}_N(\hat{\boldsymbol{\beta}}) = \mathbf{0}$ to one term about $\boldsymbol{\beta}$ yields the following useful approximation up to terms asymptotically negligible in probability,

$$\sqrt{N}(\hat{\boldsymbol{\beta}} - \boldsymbol{\beta}) \approx \left(\frac{\mathbf{H}_N(\boldsymbol{\beta})}{N}\right)^{-1} \frac{1}{\sqrt{N}} \mathbf{S}_N(\boldsymbol{\beta}). \tag{1.49}$$

However, from (1.33), (1.46), and (1.48),

$$\frac{\mathbf{H}_N(\boldsymbol{\beta})}{N} = \frac{\mathbf{G}_N(\boldsymbol{\beta})}{N} - \frac{\mathbf{R}_N(\boldsymbol{\beta})}{N} \to \mathbf{G}(\boldsymbol{\beta}) \tag{1.50}$$

in probability, as $N \to \infty$. Therefore from (1.49) and (1.50) we obtain the useful approximation,

$$\sqrt{N}(\hat{\boldsymbol{\beta}} - \boldsymbol{\beta}) \approx \mathbf{G}^{-1}(\boldsymbol{\beta}) \frac{1}{\sqrt{N}} \mathbf{S}_N(\boldsymbol{\beta}),$$

which together with an appeal to Slutsky's theorem leads to the following fact.

Theorem 1.4.1 Under Assumption A the maximum partial likelihood estimator is almost surely unique for all sufficiently large N, and

1. The estimator is consistent and asymptotically normal,

$$\hat{\beta} \to \beta$$

in probability, and

$$\sqrt{N}(\hat{\beta} - \beta) \to \mathcal{N}_p(\mathbf{0}, \mathbf{G}^{-1}(\beta)),$$

in distribution, as $N \to \infty$.

2. The following holds in probability, as $N \to \infty$:

$$\sqrt{N}(\hat{\beta} - \beta) - \frac{1}{\sqrt{N}}\mathbf{G}^{-1}(\beta)\mathbf{S}_N(\beta) \to \mathbf{0}.$$

Theorem 1.4.1 shows that the maximum partial likelihood estimator enjoys the same useful asymptotic properties as those of the traditional maximum likelihood estimator. These large sample properties of the MPLE will be utilized in our discussion of hypothesis testing in the next section.

1.4.2.1 Example: Poisson Regression Let $\{Y_t\}$ be a time series of Poisson counts, and $\{\mathbf{Z}_{t-1}\}$ the corresponding covariate process, $t = 1, ..., N$. Then,

$$f(y_t|\mathbf{Z}_{t-1}) = \exp\{(y_t \log \mu_t - \mu_t) - \log y_t!\}, \quad y_t = 0, 1, 2, ...$$

and the conditional mean and variance are equal, $\mu_t = \sigma_t^2$. Assuming the canonical link,

$$\log \mu_t = \eta_t = \mathbf{Z}'_{t-1}\beta \tag{1.51}$$

or $\mu_t(\beta) = \exp(\mathbf{Z}'_{t-1}\beta)$. Some restrictions on β are discussed in Chapter 4. The log–partial likelihood is

$$\begin{aligned} l(\beta) &= \sum_{t=1}^{N} \{y_t \log \mu_t - \mu_t - \log y_t!\} \\ &= \sum_{t=1}^{N} \{y_t \mathbf{Z}'_{t-1}\beta - \exp(\mathbf{Z}'_{t-1}\beta) - \log y_t!\} \end{aligned}$$

and either by direct differentiation or from (1.39), the estimating equation is

$$\mathbf{S}_N(\beta) = \nabla l(\beta) = \sum_{t=1}^{N} \{Y_t - \exp(\mathbf{Z}'_{t-1}\beta)\} \mathbf{Z}_{t-1} = \mathbf{0}. \tag{1.52}$$

Again, either by direct differentiation or from (1.40), and (1.41),

$$\mathbf{H}_N(\beta) = \mathbf{G}_N(\beta) = -\nabla\nabla' l(\beta) = \sum_{t=1}^{N} \exp(\mathbf{Z}'_{t-1}\beta)\mathbf{Z}_{t-1}\mathbf{Z}'_{t-1}.$$

Therefore, from (1.46) and Assumption A4,

$$\mathbf{G}(\boldsymbol{\beta}) = \int_{R^p} \exp(\mathbf{z}'\boldsymbol{\beta})\mathbf{z}\mathbf{z}'\nu(d\mathbf{z}).$$

For large N, $\mathbf{G}_N(\boldsymbol{\beta})/N \approx \mathbf{G}(\boldsymbol{\beta})$, so that by Theorem 1.4.1, $\hat{\boldsymbol{\beta}}$ is approximately normal with mean $\boldsymbol{\beta}$ and approximate covariance matrix,

$$\mathbf{G}_N^{-1}(\boldsymbol{\beta}) = \left\{ \sum_{t=1}^{N} \exp(\mathbf{Z}_{t-1}'\boldsymbol{\beta})\mathbf{Z}_{t-1}\mathbf{Z}_{t-1}' \right\}^{-1}. \tag{1.53}$$

1.4.2.2 Prediction Intervals Let $h(x) = g^{-1}(x)$ denote the inverse link,

$$\mu_t = g^{-1}(\mathbf{Z}_{t-1}'\boldsymbol{\beta}) = h(\mathbf{Z}_{t-1}'\boldsymbol{\beta}).$$

Then from Theorem 1.4.1 and the delta method [366, p. 388], an asymptotic $100(1-\alpha)\%$ prediction interval is given by

$$\mu_t(\boldsymbol{\beta}) \doteq \mu_t(\hat{\boldsymbol{\beta}}) \pm z_{\alpha/2} \frac{|h'(\mathbf{Z}_{t-1}'\boldsymbol{\beta})|}{\sqrt{N}} \sqrt{\mathbf{Z}_{t-1}'\mathbf{G}^{-1}(\boldsymbol{\beta})\mathbf{Z}_{t-1}}, \tag{1.54}$$

where h' is the derivative of h, and $z_{\alpha/2}$ is the upper $\alpha/2$ point of the standard normal distribution.

1.5 TESTING HYPOTHESES

It is often desired to test the hypothesis that some of the regression parameters or functions thereof admit particular values. This in turn translates into appropriate statements regarding the covariates in the regression model. For example, if the hypothesis that the coefficient of a certain covariate is zero is accepted, it is reasonable to assume that covariate does not contribute much to the regression and it can be removed from the model.

In this section we will be concerned with the composite hypothesis

$$\mathrm{H}_0 : \mathbf{p}(\boldsymbol{\beta}) = \mathbf{0} \ \text{ against } \ \mathrm{H}_1 : \mathbf{p}(\boldsymbol{\beta}) \neq \mathbf{0}, \tag{1.55}$$

where $\mathbf{p}$ is a vector–valued function defined on R^p and taking values in R^r, $p > r$. We shall assume that the $p \times r$ matrix $\mathbf{P}(\boldsymbol{\beta}) = [\partial\mathbf{p}(\boldsymbol{\beta})/\partial\boldsymbol{\beta}]$ exists, is continuous with respect to $\boldsymbol{\beta}$, and its rank equals r.

As an example, consider the partition of the $p \times 1$ vector parameter $\boldsymbol{\beta} = (\boldsymbol{\beta}_1', \boldsymbol{\beta}_2')'$, where the dimension of $\boldsymbol{\beta}_2$ is $r \times 1$, $r < p$. If in (1.55) we let $\mathbf{p}(\boldsymbol{\beta}) = \boldsymbol{\beta}_2$, then the problem is to test the hypothesis that the subvector $\boldsymbol{\beta}_2$ is equal to $\mathbf{0}$.

An essential observation is that hypothesis (1.55) can equivalently be stated as

$$\mathrm{H}_0 : \boldsymbol{\beta} = \mathbf{m}(\boldsymbol{\theta}) \ \text{ against } \ \mathrm{H}_1 : \boldsymbol{\beta} \neq \mathbf{m}(\boldsymbol{\theta}), \tag{1.56}$$

for a continuously differentiable vector–valued function $\mathbf{m} : R^{p-r} \to R^p$. Here β is expressed in terms of a parameter $\boldsymbol{\theta}$ of a lower dimension,

$$\beta_j = m_j(\theta_1, ..., \theta_{p-r}), \quad j = 1, ..., p.$$

We denote by $\mathbf{M}$ the $p \times (p-r)$ matrix $[\partial \mathbf{m}(\boldsymbol{\theta})/\partial \boldsymbol{\theta}]$ and assume that its rank is equal to $p - r$.

To illustrate the equivalence of (1.55) and (1.56), suppose $p = 5$ so that $\beta = (\beta_1, \beta_2, \beta_3, \beta_4, \beta_5)'$ and for $r = 3$,

$$\mathbf{p}(\beta) = \begin{pmatrix} \beta_1 - \beta_2 \\ \beta_3 - \beta_4 \\ \beta_5 \end{pmatrix}.$$

Then, hypothesis (1.55) is restated as (1.56) by letting $\boldsymbol{\theta} = (\theta_1, \theta_2)'$ and

$$\mathbf{m}(\boldsymbol{\theta}) = (\ \theta_1, \theta_1, \theta_2, \theta_2, 0 \ )'$$

in which case $\beta_j = m_j(\theta_1, \theta_2), j = 1, ..., 5$.

Let $\tilde{\beta}$ be the MPLE of β obtained from the estimator of $\boldsymbol{\theta}$ under hypothesis (1.55), and let $\hat{\beta}$ be the unrestricted MPLE. Then, the following test statistics are commonly used for testing of (1.55):

- The log-partial likelihood ratio statistic

$$\lambda_N = 2\left\{l(\hat{\beta}) - l(\tilde{\beta})\right\}. \tag{1.57}$$

- The Wald statistic

$$w_N = N\mathbf{p}'(\hat{\beta})\left[\mathbf{P}'(\hat{\beta})\mathbf{G}^{-1}(\hat{\beta})\mathbf{P}(\hat{\beta})\right]^{-1}\mathbf{p}(\hat{\beta}). \tag{1.58}$$

- The score statistic

$$c_N = \frac{1}{N}\mathbf{S}'_N(\tilde{\beta})\mathbf{G}^{-1}(\tilde{\beta})\mathbf{S}_N(\tilde{\beta}). \tag{1.59}$$

In what follows, we shall show that the distribution of all these test statistics converges to a chi–square random variable when the hypothesis (1.55) is true. The method for deriving the asymptotic distribution is entirely analogous to the case of independent data discussed, for example, in [366, pp. 415–419].

Consider first the Wald statistic which is clearly the simplest to deal with. Theorem 1.4.1 and expression (1.55) imply that

$$\sqrt{N}\mathbf{p}(\hat{\beta}) \to \mathcal{N}_r\left(\mathbf{0}, \mathbf{P}'(\beta)\mathbf{G}^{-1}(\beta)\mathbf{P}(\beta)\right)$$

in distribution, as $N \to \infty$. Thus

$$N\mathbf{p}'(\hat{\beta})\left[\mathbf{P}'(\beta)\mathbf{G}^{-1}(\beta)\mathbf{P}(\beta)\right]^{-1}\mathbf{p}(\hat{\beta}) \to \chi^2_r$$

in distribution, as $N \to \infty$. The continuity of $\mathbf{P}$ together with the consistency of $\hat{\beta}$ imply

$$w_N \to \chi_r^2 \tag{1.60}$$

in distribution as $N \to \infty$.

We derive next the asymptotic distribution of the log–partial likelihood ratio test statistic. From the expansion of $l(\beta)$ to two terms about $\hat{\beta}$ and using the fact that $\mathbf{H}_N(\beta)/N \approx \mathbf{G}(\beta)$ we have

$$\begin{aligned} 2\{l(\hat{\beta}) - l(\beta)\} &\approx \sqrt{N}(\hat{\beta} - \beta)' \frac{\mathbf{G}_N(\beta)}{N} \sqrt{N}(\hat{\beta} - \beta) \\ &\approx \frac{1}{\sqrt{N}} \mathbf{S}_N'(\beta) \mathbf{G}^{-1}(\beta) \frac{1}{\sqrt{N}} \mathbf{S}_N(\beta). \end{aligned}$$

Similarly we can prove under (1.55) that

$$2\{l(\tilde{\beta}) - l(\beta)\} \approx \frac{1}{\sqrt{N}} \mathbf{S}_N'(\beta) \mathbf{M} \mathbf{G}_R^{-1}(\beta) \mathbf{M}' \frac{1}{\sqrt{N}} \mathbf{S}_N(\beta)$$

with $\mathbf{M} = [\partial \mathbf{m}(\boldsymbol{\theta})/\partial \boldsymbol{\theta}]$, and $\mathbf{G}_R(\beta) = \mathbf{M}'\mathbf{G}(\beta)\mathbf{M}$. Then by subtraction, we can see that when (1.55) holds,

$$\begin{aligned} 2\{l(\hat{\beta}) - l(\beta)\} &- 2\{l(\tilde{\beta}) - l(\beta)\} \\ = 2\{l(\hat{\beta}) &- l(\tilde{\beta})\} \\ \approx \frac{1}{\sqrt{N}} \mathbf{S}_N'(\beta)[\mathbf{G}^{-1}(\beta) &- \mathbf{M}\mathbf{G}_R^{-1}(\beta)\mathbf{M}'] \frac{1}{\sqrt{N}} \mathbf{S}_N(\beta). \end{aligned}$$

The last expression is a quadratic form in an asymptotically normal vector with mean $\mathbf{0}$ and covariance matrix $\mathbf{G}(\beta)$ corresponding to the $p \times p$ matrix $\mathbf{A} \equiv [\mathbf{G}^{-1}(\beta) - \mathbf{M}\mathbf{G}_R^{-1}(\beta)\mathbf{M}']$. It is easily seen that $\mathbf{A}\mathbf{G}(\beta)\mathbf{A} = \mathbf{A}$, and that, since the trace is invariant under cyclical permutations, meaning that $tr(\mathbf{AB}) = tr(\mathbf{BA})$, we have

$$tr(\mathbf{A}\mathbf{G}(\beta)) = tr(\mathbf{I}_{p \times p}) - tr(\mathbf{I}_{(p-r)\times(p-r)}) = p - (p - r) = r.$$

This implies that the asymptotic distribution of the log–partial likelihood ratio λ_N under the null hypothesis is χ_r^2. That is,

$$\lambda_N \to \chi_r^2 \tag{1.61}$$

in distribution as $N \to \infty$. Here we have used the fact that since $\mathbf{A}\mathbf{G}(\beta)$ is idempotent, $(\mathbf{A}\mathbf{G}(\beta))^2 = \mathbf{A}\mathbf{G}(\beta)$, its rank is equal to its trace.

The asymptotic distribution of the score statistic can be worked out along the same lines as well. Thus, note that the restricted maximum partial likelihood estimator satisfies

$$\frac{\partial l(\beta)}{\partial \beta} + \mathbf{P}(\beta)\gamma = \mathbf{0}, \quad \mathbf{p}(\beta) = \mathbf{0} \tag{1.62}$$

with γ an r–dimensional vector of Lagrange multipliers. Expanding the partial score about the true value and using (1.62), we can show that

$$\frac{1}{\sqrt{N}}\tilde{\gamma} \rightarrow \mathcal{N}_r\left(\mathbf{0}, \left[\mathbf{P}^{'}(\beta)\mathbf{G}^{-1}(\beta)\mathbf{P}(\beta)\right]^{-1}\right)$$

as $N \rightarrow \infty$, in distribution, where $\tilde{\gamma}$ is defined from $\mathbf{S}_N(\tilde{\beta}) = -\mathbf{P}(\tilde{\beta})\tilde{\gamma}$. Thus, we obtain

$$\begin{aligned} c_N &= \frac{1}{N}\mathbf{S}_N^{'}(\tilde{\beta})\mathbf{G}^{-1}(\tilde{\beta})\mathbf{S}_N(\tilde{\beta}) \\ &= \frac{1}{N}\tilde{\gamma}^{'}\mathbf{P}^{'}(\tilde{\beta})\mathbf{G}^{-1}(\tilde{\beta})\mathbf{P}(\tilde{\beta})\tilde{\gamma}. \end{aligned}$$

Therefore,

$$c_N \rightarrow \chi^2_r \tag{1.63}$$

in distribution as $N \rightarrow \infty$.

The previous discussion is summarized in the following theorem.

Theorem 1.5.1 Under Assumption A and hypothesis (1.55), the three test statistics λ_N, w_N and c_N have the same asymptotic distribution. Their asymptotic distribution is chi–square with r degrees of freedom.

A particular form of hypothesis (1.55) which arises frequently in applications is the following general linear hypothesis.

$$\text{H}_0 : \mathbf{C}\beta = \beta_0 \text{ against } \text{H}_1 : \mathbf{C}\beta \neq \beta_0, \tag{1.64}$$

where $\mathbf{C}$ is an appropriate known matrix with full rank r, $r < p$. It is straightforward to verify that the Wald test becomes

$$w_N = \{\mathbf{C}\hat{\beta} - \beta_0\}'\{\mathbf{C}\mathbf{G}^{-1}(\hat{\beta})\mathbf{C}'\}^{-1}\{\mathbf{C}\hat{\beta} - \beta_0\},$$

while the functional form of the likelihood ratio and score tests remain unchanged.

1.6 DIAGNOSTICS

Diagnostics in regression analysis consists of procedures for exploring and testing the adequacy and goodness of fit of fitted models. In the context of generalized linear models this refers in particular to the examination of several types of residuals and *deviance analysis*. Deviance analysis is carried out routinely through a statistic called the *scaled deviance* and the closely related information criteria *AIC* and *BIC*. The *deviance* itself is a constant multiple of the scaled deviance and it is important to make a distinction between the two when the constant differs from 1. We shall see a more detailed analysis of goodness of fit in later chapters.

1.6.1 Deviance

To understand the notion of deviance in generalized linear models it is necessary to revisit briefly the log–partial likelihood ratio test. In (1.56), the null hypothesis states that $\boldsymbol{\beta}$ with dimension p is expressed in terms of a lower dimensional vector $\boldsymbol{\theta}$ with dimension $p-r$, in which case the asymptotic distribution of the log–partial likelihood ratio (1.57) is chi–square with degrees of freedom equal to $p-(p-r)=r$. For a generalized linear model each $\mu_t, t=1,...,N$, is a function of a p–dimensional vector of regression parameters $\boldsymbol{\beta}=(\beta_1,...,\beta_p)'$, and the maximum partial likelihood estimates $\hat{\mu}_t$ are expressed in terms of $\hat{\boldsymbol{\beta}}$, where $p<N$. This is an example of a reduced model where each of N parameters is expressed in terms of a smaller set of $p<N$ parameters; the same holds for the corresponding estimates. But we also can entertain a *saturated* model with as many parameters as observations where each μ_t is estimated directly from the data $Y_1,...,Y_N$ disregarding $\boldsymbol{\beta}$. In this latter case there are N free parameters $\mu_1,...,\mu_N$, and the maximum partial likelihood estimate of μ_t is $\tilde{\mu}_t=Y_t$. Denote by $l(\hat{\boldsymbol{\mu}};\mathbf{y})$ the maximum log partial likelihood from the reduced model, and by $l(\mathbf{y};\mathbf{y})$ the maximum log partial likelihood corresponding to the saturated model. Since for exponential family models $l(\mathbf{y};\mathbf{y})\geq l(\hat{\boldsymbol{\mu}};\mathbf{y})$, the statistic,

$$D\equiv 2\{l(\mathbf{y};\mathbf{y})-l(\hat{\boldsymbol{\mu}};\mathbf{y})\} \tag{1.65}$$

has been suggested as a measure of goodness of fit [313]. D is referred to as the *scaled deviance*. The scaled deviance depends on a known or consistently estimated scale parameter ϕ, and in either case the preceding discussion suggests that to some degree, $D\sim\chi^2_{N-p}$. Indeed, for independent normal data with the identity link and a known variance, $D\sim\chi^2_{N-p}$ exactly, but in general the chi-square approximation may not be effective due to a variable number of parameters. Notice that in the preceding discussion the number of parameters in the unrestricted model was fixed whereas in the saturated model the corresponding number N varies, suggesting that the asymptotic argument need not apply. Furthermore, even when the fit is good and the χ^2_{N-p} approximation holds, D can still be large with an expected value of $N-p$ approximately.

The *deviance* itself is defined as ϕD, and is free of ϕ. From this we obtain an approximately unbiased moment estimator for ϕ,

$$\hat{\phi}=\frac{\text{Deviance}}{N-p}.$$

The so called analysis of deviance of nested models is a reiteration of the log partial likelihood ratio test procedure. Let D_0 correspond to $\mathrm{H}_0:\ \boldsymbol{\beta}=\boldsymbol{\beta}_0$, and D_1 to $\mathrm{H}_1:\ \boldsymbol{\beta}=\boldsymbol{\beta}_1$, where $\boldsymbol{\beta}_1$ is a p-dimensional vector and $\boldsymbol{\beta}_0$ is a subvector of $\boldsymbol{\beta}_1$ with dimension $q\times 1$, $q<p$. Then, approximately for large N, $D_0\sim\chi^2_{N-q}$ and $D_1\sim\chi^2_{N-p}$, and from (1.57), (1.61), and the definition (1.65), under H_0

$$D_0-D_1\sim\chi^2_{p-q} \tag{1.66}$$

The null hypothesis H_0 is rejected for large values of $D_0 - D_1$, greater than the upper α point of the χ^2_{p-q} distribution. When ϕ is unknown, in analogy with the normal case, an alternative test statistic for testing H_0 is

$$\frac{(D_0 - D_1)/(p-q)}{D_1/(N-p)}.$$

It is free of ϕ and distributed under H_0, roughly, as $F_{p-q,N-p}$.

1.6.2 Model Selection Criteria

Evaluation and selection among several competing models may be based on Akaike's information criterion (AIC) introduced by Akaike [6], [7] and used widely in applications. The AIC criterion is defined as a function of the number of independent model parameters,

$$\text{AIC}(p) = -2\log \text{PL}(\hat{\beta}) + 2p, \tag{1.67}$$

where $\hat{\beta}$ is the maximum partial likelihood estimator of β and p is the "model order", $p = \dim(\beta)$. We choose the model corresponding to p that minimizes (1.67). Evidently, the AIC criterion penalizes against models with unrealistically large p.

To improve the large sample performance of the AIC criterion in the sense of obtaining consistent estimates of the order p, several modifications of the AIC criterion have been suggested in the literature. In [371], [380] the $2p$ term in (1.67) is replaced by $p\log N$, and in [196] it is replaced by $2p\log\log N$, where N denotes the time series length. The so called Bayesian information criterion (BIC) formulated by Schwarz [380] is defined as the following function of p,

$$\text{BIC}(p) = -2\log \text{PL}(\hat{\beta}) + p\log N. \tag{1.68}$$

The BIC estimates the model order consistently in most cases. Further discussion and properties of the various information criteria can be found in [99].

1.6.3 Residuals

By "residual" is meant a certain deviation of a fitted from an observed value. Residual analysis is important in assessing the goodness of fit–how well the fitted model explains the observed data–of a regression model, and in judging the impact and significance of covariates on the response.

Let $\hat{\mu}_t = \mu_t(\hat{\beta})$. There are several ways to define residuals in the context of time series following generalized linear models. The most obvious definition is that of the so called *raw* or *response* residuals

$$\hat{e}_t = Y_t - \hat{\mu}_t, \quad t = 1, \ldots, N. \tag{1.69}$$

Three popular additional types of residuals, *Pearson, working,* and *deviance*, are defined in terms of the raw residuals as follows. The Pearson residuals are the standardized version,

$$\hat{r}_t = \frac{Y_t - \hat{\mu}_t}{\sqrt{V(\hat{\mu}_t)}}, \quad t = 1, \ldots, N. \tag{1.70}$$

The working residuals are a different standardized version,

$$\hat{wr}_t = \frac{Y_t - \hat{\mu}_t}{\partial \mu_t / \partial \eta_t}, \quad t = 1, \ldots, N, \tag{1.71}$$

where $\partial \mu_t / \partial \eta_t$ is evaluated at $\hat{\beta}$. The deviance residuals are given by

$$\hat{d}_t = \text{sign}(Y_t - \hat{\mu}_t)\sqrt{2\left[l_t(Y_t) - l_t(\hat{\mu}_t)\right]}, \quad t = 1, \ldots, N, \tag{1.72}$$

where the sum of squares of deviance residuals is equal to the deviance statistic (1.65). In the so called *Anscombe residuals*, Y_t and $\hat{\mu}_t$ are replaced by $A(Y_t)$ and $A(\hat{\mu}_t)$ where $A(\cdot)$ is a normalizing transformation given by

$$A(\cdot) = \int \frac{d\mu}{V^{1/3}(\mu)}.$$

As an example, for Poisson regression the residuals take on the form

$$\hat{r}_{A,t} = \frac{\frac{2}{3}(Y_t^{2/3} - \hat{\mu}_t^{2/3})}{\hat{\mu}_t^{1/6}}, \quad t = 1, \ldots, N. \tag{1.73}$$

For a discussion of this and more examples of Anscombe residuals see [313]. Certain generalizations are defined and discussed in [108]. In the context of generalized linear models see [286], [348], and [352], for example.

1.6.3.1 Autocorrelation of Pearson Residuals

Define the Pearson deviation as

$$r_t = \frac{Y_t - \mu_t}{\sqrt{V(\mu_t)}}, \quad t = 1, \ldots, N, \tag{1.74}$$

and assume without loss of generality that $\text{Var}\left[Y_t \mid \mathcal{F}_{t-1}\right] = \phi V(\mu_t)$. That is, $\alpha_t(\phi) = \phi$ and $\omega_t = 1$ in equation (1.13). Then $\{r_t\}$ is a martingale difference with finite variance since,

$$\text{E}\left[r_t \mid \mathcal{F}_{t-1}\right] = 0, \tag{1.75}$$

$$\text{E}\left[r_t\right] = \text{E}\left\{\text{E}\left[r_t \mid \mathcal{F}_{t-1}\right]\right\} = 0 \tag{1.76}$$

and

$$\text{E}\left[r_t^2\right] = \text{E}\left\{\text{E}\left[r_t^2 \mid \mathcal{F}_{t-1}\right]\right\} = \phi. \tag{1.77}$$

Considering (1.77), the lag k innovation autocorrelation function of the random sequence $\{r_t\}, t = 1, \ldots, N$ is defined by the form

$$\rho_r(k) = \frac{1}{N} \frac{\sum_{t=k+1}^{N} r_t r_{t-k}}{\phi}, \quad k = 1, .., m. \tag{1.78}$$

The large sample properties of the MPLE established in Theorem 1.4.1 imply that $\hat{r}_t$ in (1.70) estimates r_t consistently, and hence $\rho_r(k)$ can be estimated from

$$\hat{\rho}_r(k) = \frac{1}{N} \frac{\sum_{t=k+1}^{N} \hat{r}_t \hat{r}_{t-k}}{\hat{\phi}}, \quad k = 1, .., m, \tag{1.79}$$

where $\hat{\phi}$ is a consistent estimator of ϕ given by

$$\hat{\phi} = \frac{1}{N-p}\sum_{t=1}^{N}\frac{(Y_t - \hat{\mu}_t)^2}{V(\hat{\mu}_t)} = \frac{1}{N-p}\sum_{t=1}^{N}\hat{r}_t^2. \tag{1.80}$$

The following theorem gives the asymptotic distribution of $\hat{\rho}_r(k)$ under the hypothesis that the model is correct [298].

Theorem 1.6.1 When the model is correct, the vector

$$\frac{1}{\sqrt{N}}\hat{\boldsymbol{\rho}}_r = \left(\frac{\hat{\rho}_r(1)}{\sqrt{N}}, \frac{\hat{\rho}_r(2)}{\sqrt{N}}, \ldots, \frac{\hat{\rho}_r(m)}{\sqrt{N}}\right)'$$

for some $m > 0$ is asymptotically normally distributed with mean $\mathbf{0}$ and covariance matrix

$$\mathbf{I}_m - \phi^{-1}\mathbf{K}'(\boldsymbol{\beta})\mathbf{G}^{-1}(\boldsymbol{\beta})\mathbf{K}(\boldsymbol{\beta}),$$

where the mth column of the $p \times m$ matrix $\mathbf{K}$ is given by

$$\lim_{N\to\infty}\frac{1}{N}\sum_t\left[\mathbf{Z}_{t-1}V^{-1/2}(\mu_t)\frac{\partial\mu_t}{\partial\eta_t}r_{t-i}\right], \quad i = 1, \ldots, m.$$

The basic idea of the proof is to show that the vector $\sqrt{N}\boldsymbol{\rho}_r = \sqrt{N}\,(\rho_r(1), \ldots, \rho_r(m))'$ is a zero mean square integrable martingale whose asymptotic distribution is m–dimensional standard normal, and then use a first-order Taylor expansion of $\sqrt{N}\hat{\rho}_r(k)$ about $(\boldsymbol{\beta}', \phi)'$. For details see [298].

Estimated standard errors for $\hat{\boldsymbol{\rho}}_r$ are obtained by substituting $\hat{\boldsymbol{\beta}}$ and $\hat{\phi}$ for $\boldsymbol{\beta}$ and ϕ appearing in the covariance matrix in Theorem 1.6.1 leads. Moreover, an overall significance test of the residual autocorrelation can be based on the quadratic form

$$n\hat{\boldsymbol{\rho}}_r'\left(\mathbf{I}_m - \hat{\phi}^{-1}\hat{\mathbf{K}}'\hat{\mathbf{G}}^{-1}\hat{\mathbf{K}}\right)^{-1}\hat{\boldsymbol{\rho}}_r, \tag{1.81}$$

which is asymptotically distributed as a chi–square random variable with m degrees of freedom. When the response process is a stationary autoregressive process such that $V(\mu_t) = 1$ and $\phi = \sigma^2$, then (1.81) reduces to the so called Box–Pierce portmanteau statistic [61, p. 291].

1.6.3.2 White Noise Test

In an ideal situation the residuals should resemble a sequence of uncorrelated random variable with mean 0 and a fixed variance, referred to as *white noise*. In general, if $Y_1, ..., Y_N$ is a Gaussian white noise sequence with sample mean $\bar{Y} = \sum_{t=1}^{N} Y_t/N$ and sample autocorrelation function (acf)

$$\hat{\rho}(k) = \frac{\sum_{t=1}^{N-k}(Y_t - \bar{Y})(Y_{t+k} - \bar{Y})}{\sum_{t=1}^{N}(Y_t - \bar{Y})^2}, \quad k = 0, 1, ...,$$

then $\hat{\rho}(0) = 1$, and for each $k \geq 1$ and sufficiently large N, $\hat{\rho}(k)$ is approximately normal with mean 0 and variance $1/N$ [19]. Because of this fact plots of the sample acf of the residuals with horizontal lines at $\pm 1.96/\sqrt{N}$ are used routinely in time series regression to get a sense of how close the residual sequence is to white noise. We shall see examples of this graphical tool throughout the book.

1.7 QUASI-PARTIAL LIKELIHOOD

A careful examination of the partial score (1.27) reveals that it depends only on the specification of the conditional mean in terms of the inverse link $g^{-1}(\cdot)$ applied to $\eta_t = \mathbf{Z}_{t-1}'\boldsymbol{\beta}$, and the relationship between the conditional mean and variance. This suggests an extension of the class of generalized linear models by introducing a partial score function that depends only on the link and the relationship between the conditional mean and variance, disregarding all other distributional assumptions. The generalized score shares many of the properties of the partial score obtained from a partial likelihood, but is no longer the result of any partial likelihood per se. Its integral with respect to $\boldsymbol{\beta}$, however, when it exists, resembles a partial likelihood and is referred to as *quasi-partial likelihood*. Quasi-partial likelihood, or more precisely the corresponding partial score, enables consistent inference for a wide class of models that satisfy only first- and second-moment assumptions, and may be used in situations when the statistician is unable or does not wish to make distributional assumptions leading to partial likelihood.

Let $\{Y_t\}$, $t = 1, \ldots, N$, be a time series with conditional mean

$$\mu_t(\boldsymbol{\beta}) = \mathrm{E}\left[Y_t \mid \mathcal{F}_{t-1}\right], \tag{1.82}$$

and *true* conditional variance

$$\sigma_t^2(\boldsymbol{\beta}) = \mathrm{Var}[Y_t \mid \mathcal{F}_{t-1}]. \tag{1.83}$$

Furthermore, suppose there exists a function g such that

$$g(\mu_t) = \mathbf{Z}_{t-1}'\boldsymbol{\beta},$$

where $\{\mathbf{Z}_{t-1}\}$, $t = 1, \ldots, N$, denotes a covariate vector process. We now assume a conditional *working variance* of Y_t given $\mathcal{F}_{t-1}$ that depends on the conditional mean,

$$\phi V(\mu_t), \tag{1.84}$$

where ϕ is an unknown dispersion parameter and $V(\cdot)$ is a variance function that depends on μ_t. A variance model of the form (1.84) is $\phi\mu_t$ with $\phi > 1$, which may hold for count data where the conditional variance exceeds the conditional mean. Alternative models are suggested by the variance of mixtures of the Poisson distribution such as Poisson-gamma. The working variance serves as a sensible model and may be different than the true variance $\sigma_t^2(\boldsymbol{\beta})$.

The *quasi-partial score* for the estimation of the unknown regression coefficients is defined by the sum,

$$\mathbf{S}_N(\beta) = \sum_{t=1}^{N} \mathbf{Z}_{t-1} \frac{\partial \mu_t}{\partial \eta_t} \frac{(Y_t - \mu_t(\beta))}{\phi V(\mu_t(\beta))}, \tag{1.85}$$

with $\eta_t = \mathbf{Z}'_{t-1}\beta$. Observe that (1.85) has the same form as (1.27) with the true conditional variance replaced by the working variance. However only first and second moments are needed to define (1.85) whereas (1.27) is calculated through the properties of the exponential family of distributions. As mentioned, the quasi-partial score may be thought of as the derivative of some quasi-partial likelihood when the latter exists.

Under regularity conditions, including a correctly specified link and that Assumption A holds, a consistent (local) root of the equation $\mathbf{S}_N(\beta) = \mathbf{0}$, denoted by $\hat{\beta}_{\mathrm{q}}$, exists and is asymptotically normal.

From (1.82) we have

$$\mathrm{E}\left[\mathbf{S}_N(\beta)\right] = \mathbf{0}$$

and

$$\mathrm{E}\left[\mathbf{S}_{t+1}(\beta) \mid \mathcal{F}_t\right] = \mathbf{S}_t(\beta),$$

so that the process $\{\mathbf{S}_t(\beta), \mathcal{F}_t\}$, $t = 1, \ldots, N$, forms a martingale. The asymptotic distribution of this martingale is normal:

$$\frac{\mathbf{S}_N(\beta)}{\sqrt{N}} \to \mathcal{N}_p(\mathbf{0}, \mathbf{G}_1(\beta)),$$

in distribution as $N \to \infty$, where the $p \times p$ matrix $\mathbf{G}_1(\beta)$ is defined from the following probability limit as $N \to \infty$,

$$\frac{1}{N} \sum_{t=1}^{N} \mathbf{Z}_{t-1} \left(\frac{\partial \mu_t}{\partial \eta_t}\right)^2 \frac{\sigma_t^2(\beta)}{\phi^2 V^2(\mu_t(\beta))} \mathbf{Z}'_{t-1} \to \mathbf{G}_1(\beta). \tag{1.86}$$

Following the proof of Theorem 1.4.1, the quasi maximum partial likelihood estimator $\hat{\beta}_{\mathrm{q}}$ is asymptotically normally distributed,

$$\sqrt{N}\left(\hat{\beta}_{\mathrm{q}} - \beta\right) \to \mathcal{N}_p\left(\mathbf{0}, \mathbf{G}^{-1}(\beta)\mathbf{G}_1(\beta)\mathbf{G}^{-1}(\beta)\right) \tag{1.87}$$

with $\mathbf{G}(\beta)$ being the probability limit of

$$\frac{1}{N} \sum_{t=1}^{N} \mathbf{Z}_{t-1} \left(\frac{\partial \mu_t}{\partial \eta_t}\right)^2 \frac{1}{\phi V(\mu_t(\beta))} \mathbf{Z}'_{t-1}, \tag{1.88}$$

and $\mathbf{G}_1(\beta)$ is given by (1.86). Apparently the efficiency of $\hat{\beta}_{\mathrm{q}}$ increases as the working variance approaches the true variance and when equality is reached, that is, when $\sigma_t^2 = \phi V(\mu_t)$, then $\mathbf{G}^{-1}(\beta)\mathbf{G}_1(\beta)\mathbf{G}^{-1}(\beta) = \mathbf{G}^{-1}(\beta)$, where $\mathbf{G}(\beta)$ is the information matrix given in (1.46). Note that the asymptotic covariance matrix in (1.87) is independent of ϕ unless $\sigma_t^2(\boldsymbol{\beta})$ is a function of ϕ.

In practice, the covariance matrix of $\hat{\boldsymbol{\beta}}_{\mathrm{q}}$ is estimated by replacing the parameters $\phi, \boldsymbol{\beta}, \sigma_t^2(\boldsymbol{\beta})$ by their respective estimates. The true conditional variance $\sigma_t^2(\boldsymbol{\beta})$ needed for (1.87) is replaced by $(Y_t - \mu_t(\hat{\boldsymbol{\beta}}_{\mathrm{q}}))^2$. The parameter ϕ is estimated from

$$\hat{\phi} = \frac{1}{N-p} \sum_{t=1}^{N} \hat{r}_t^2,$$

where $\hat{r}_t$ is the Pearson residual,

$$\hat{r}_t = \frac{(Y_t - \mu_t(\hat{\boldsymbol{\beta}}_{\mathrm{q}}))}{\sqrt{V(\mu_t(\hat{\boldsymbol{\beta}}_{\mathrm{q}}))}},$$

and $V(\cdot)$ is the variance function.

The quasi-partial likelihood theory also covers problems of testing hypotheses by using the test statistics (1.57), (1.58) and (1.59) replacing $\hat{\boldsymbol{\beta}}$ by $\hat{\boldsymbol{\beta}}_{\mathrm{q}}$ (see [209, Ch. 9] and [282]).

The idea of quasi likelihood was originated by Wedderburn [422] in the context of generalized linear models for independent data, and it was further developed in [312] and [313]. Wedderburn was the first to realize that (1.85) provides estimators of the regression parameter $\boldsymbol{\beta}$ for any choice of link and variance functions without reference to any member of the exponential family of distributions. It should be noted that the quasi likelihood is a special case of the methodology of estimating functions. See, for example, the articles [175], [177] and the texts [176], [209].

1.7.0.1 Example: Nonlinear Least Squares

With link g, $\mu_t = g^{-1}(\eta_t)$, and $\eta_t = \mathbf{Z}_{t-1}'\boldsymbol{\beta}$, suppose we wish to estimate $\boldsymbol{\beta}$ by minimizing the sum of squares

$$\sum_{t=1}^{N} \{Y_t - \mu_t(\boldsymbol{\beta})\}^2 = \sum_{t=1}^{N} \{Y_t - g^{-1}(\mathbf{Z}_{t-1}'\boldsymbol{\beta})\}^2.$$

Differentiating with respect to $\boldsymbol{\beta}$ and setting the result equal to $\mathbf{0}$ we obtain,

$$\nabla \sum_{t=1}^{N} \{Y_t - \mu_t(\boldsymbol{\beta})\}^2 = \sum_{t=1}^{N} \mathbf{Z}_{t-1} \frac{\partial \mu_t}{\partial \eta_t} (Y_t - \mu_t(\boldsymbol{\beta})) = \mathbf{0},$$

and we see that nonlinear least squares correspond to a quasi-partial likelihood where $\phi V(\mu_t(\boldsymbol{\beta}))$ is constant.

1.7.0.2 Example: Over-Dispersion in Count Data

As in Poisson regression with canonical link discussed in Example 1.4.2.1, assume that $\log \mu_t = \eta_t = \mathbf{Z}_{t-1}'\boldsymbol{\beta}$, or equivalently $\mu_t = \exp(\eta_t)$. However, unlike in Poisson regression, assume that

$\sigma_t^2 = \phi V(\mu_t) = \phi\mu_t$, where $\phi > 1$, a case referred to as *over-dispersion*. In this case the quasi-partial score (1.85) reduces to

$$\mathbf{S}_N(\beta) = \sum_{t=1}^{N} \mathbf{Z}_{t-1} \frac{(Y_t - \mu_t(\beta))}{\phi}, \tag{1.89}$$

yielding the estimating equation,

$$\sum_{t=1}^{N} \mathbf{Z}_{t-1} \left(Y_t - \exp(\mathbf{Z}'_{t-1}\beta)\right) = \mathbf{0}. \tag{1.90}$$

Therefore, either directly from (1.90) or from (1.87), remembering that $\partial\mu_t/\partial\eta_t = \mu_t$, for large N,

$$\text{Cov}(\hat{\beta}_q) \approx \phi \left\{ \sum_{t=1}^{N} \exp(\mathbf{Z}'_{t-1}\beta)\mathbf{Z}_{t-1}\mathbf{Z}'_{t-1} \right\}^{-1}. \tag{1.91}$$

We see that relative to (1.53) in Poisson regression, the variability of the regression coefficients increases since $\phi > 1$. Interestingly, the estimating equation (1.90) is identical to (1.52) in Poisson regression, however, the true conditional variance σ_t^2 in the present example is inflated by ϕ and hence the covariance matrices in (1.53) and (1.91) are not the same.

1.7.1 Generalized Estimating Equations

The generalized estimating equations (GEE) method is a multivariate version of quasi-likelihood that was developed in [300], [354], [446], primarily in the context of longitudinal data analysis, where periodic measurements are made on a process and on its associated covariates. Typically, in longitudinal studies there are multiple short independent time series. For example, in child development studies, annual medical and social data are collected from children until they reach adulthood to identify risk factors over time. It is very plausible that time series obtained from an individual exhibit internal as well as cross correlation, but that series obtained from different individuals are independent. A standard reference on the analysis of longitudinal data is [123].

The basic framework of generalized estimating equations methodology is as follows. Let $\mathbf{Y}_i = (Y_{i1}, \ldots, Y_{in_i})'$ be the $n_i \times 1$ vector representing the response and $\mathbf{Z}_i = (\mathbf{Z}_{i1}, \ldots, \mathbf{Z}_{in_i})'$ an $n_i \times p$ matrix of covariate values for the i'th subject, $i = 1, \ldots, K$. Here $\mathbf{Z}_{ij}$ corresponds to Y_{ij} and both are observed at time t_{ij}, $i = 1, \ldots, K, j = 1, ..., n_i$. Let μ_{ij} denote the conditional expectation of the response Y_{ij} given the corresponding covariate matrix $\mathbf{Z}_i$, $\mu_{ij} = \text{E}[Y_{ij} \mid \mathbf{Z}_i]$, and assume the relationship $g(\mu_{ij}) = \mathbf{Z}'_{ij}\beta$, with g a known link function. As in quasi-partial likelihood, we model the conditional working variance of Y_{ij} given $\mathbf{Z}_i$ as $\phi V(\mu_{ij})$, where V is the variance function and ϕ is an unknown dispersion parameter. Conditional on the covariates, assume that $\text{Corr}\,[Y_{ij}, Y_{ik}] = \rho(\mu_{ij}, \mu_{ik}, \boldsymbol{\alpha})$, where $\rho(\cdot)$ is a known function and $\boldsymbol{\alpha}$ is a vector of unknown parameters. It follows that the conditional *working covariance matrix* of $\mathbf{Y}_i$ given $\mathbf{Z}_i$ is equal to

$$\mathbf{V}_i = \mathbf{A}_i^{1/2}\mathbf{R}_i\mathbf{A}_i^{1/2}, \quad i = 1, \ldots, K,$$

where

$$\mathbf{A}_i = \text{diag}\,(\phi V(\mu_{ij}(\boldsymbol{\beta})))$$

and the (j, k) element of $\mathbf{R}_i$ is $\text{Corr}\,[Y_{ij}, Y_{ik}]$, conditional on $\mathbf{Z}_i$. Notice that the working covariance matrix $\mathbf{V}_i$ depends on $\boldsymbol{\alpha}$ through $\mathbf{R}_i$, and that it may be different than the true conditional covariance matrix of $\mathbf{Y}_i$ given $\mathbf{Z}_i$.

GEE is defined by the quasi–score equation

$$\mathbf{U}(\boldsymbol{\beta}, \boldsymbol{\alpha}) = \sum_{i=1}^{K} \left(\frac{\partial \boldsymbol{\mu}_i}{\partial \boldsymbol{\beta}}\right)' \mathbf{V}_i^{-1}\left(\mathbf{Y}_i - \boldsymbol{\mu}_i\right) = \mathbf{0}, \tag{1.92}$$

where $\boldsymbol{\mu}_i = (\mu_{i1}, \ldots, \mu_{in_i})'$. Notice that in (1.92) the derivative is with respect to $\boldsymbol{\beta}$ whereas in (1.85) the derivative is with respect to η, resulting in the explicit manifestation of the covariates in the expression of the quasi-partial score.

In practice, $\boldsymbol{\alpha}$ is replaced by any $K^{1/2}$ consistent estimator $\hat{\boldsymbol{\alpha}}$, and an estimator for $\boldsymbol{\beta}$, denoted by $\hat{\boldsymbol{\beta}}_{\text{gee}}$, is obtained as a solution of the revised estimating equations,

$$\mathbf{U}(\boldsymbol{\beta}, \hat{\boldsymbol{\alpha}}) = \sum_{i=1}^{K} \left(\frac{\partial \boldsymbol{\mu}_i}{\partial \boldsymbol{\beta}}\right)' \mathbf{V}_i^{-1}(\hat{\boldsymbol{\alpha}})\left(\mathbf{Y}_i - \boldsymbol{\mu}_i\right) = \mathbf{0}. \tag{1.93}$$

It can be shown that $\hat{\boldsymbol{\beta}}_{\text{gee}}$ is asymptotically as efficient as if $\boldsymbol{\alpha}$ were known, and that under mild regularity conditions,

$$\sqrt{K}\left(\hat{\boldsymbol{\beta}}_{\text{gee}} - \boldsymbol{\beta}\right) \to \mathcal{N}_p(\mathbf{0}, \mathbf{V}_{\text{gee}}) \tag{1.94}$$

in distribution as $K \to \infty$, where

$$\begin{aligned}
\mathbf{V}_{\text{gee}} &= \lim_{K\to\infty} \left\{\sum_{i=1}^{K} \left(\frac{\partial \boldsymbol{\mu}_i}{\partial \boldsymbol{\beta}}\right)' \mathbf{V}_i^{-1} \left(\frac{\partial \boldsymbol{\mu}_i}{\partial \boldsymbol{\beta}}\right)\right\}^{-1} \\
&\times \left\{\sum_{i=1}^{K} \left(\frac{\partial \boldsymbol{\mu}_i}{\partial \boldsymbol{\beta}}\right)' \mathbf{V}_i^{-1} \text{Var}\,[\mathbf{Y}_i]\, \mathbf{V}_i^{-1} \left(\frac{\partial \boldsymbol{\mu}_i}{\partial \boldsymbol{\beta}}\right)\right\} \\
&\times \left\{\sum_{i=1}^{K} \left(\frac{\partial \boldsymbol{\mu}_i}{\partial \boldsymbol{\beta}}\right)' \mathbf{V}_i^{-1} \left(\frac{\partial \boldsymbol{\mu}_i}{\partial \boldsymbol{\beta}}\right)\right\}^{-1}
\end{aligned} \tag{1.95}$$

and $\text{Var}\,[\mathbf{Y}_i]$ is the true (conditional) variance. See [183], [300].

The GEE methodology leads to consistent estimates of the regression coefficients provided the link function $g(\cdot)$ has been correctly specified. A robust estimator of $\mathbf{V}_{\text{gee}}$ is obtained by evaluating the matrices $(\partial \boldsymbol{\mu}_i / \partial \boldsymbol{\beta})$ and $\mathbf{V}_i$ at the estimated parameters and substituting $\left(\mathbf{Y}_i - \boldsymbol{\mu}_i(\hat{\boldsymbol{\beta}}_{\text{gee}})\right)\left(\mathbf{Y}_i - \boldsymbol{\mu}_i(\hat{\boldsymbol{\beta}}_{\text{gee}})\right)'$ for $\text{Var}\left[\mathbf{Y}_i\right]$. The estimates for the regression parameters and the asymptotic covariance matrix are consistent regardless of the choice of the correlation structure.

Recent efforts directed at extending GEE for the purpose of more efficient estimators of both regression and association (correlation) parameters have been focused on the so called second-order generalized estimating equations (GEE2), based on a joint estimating equation for regression and association parameters. Simulation studies indicate that when the regression parameters are of interest, GEE2 results in more efficient estimators than those from GEE. However, for wrongly specified link function for association, the regression estimates may be asymptotically biased even when the link function for the mean has been correctly specified. Moreover, the choice of a sensible working correlation matrix for GEE2 involves the third and fourth moment of the response variable, which may result in less efficient regression estimates. See [80], [302], [303], [354], [356], [450], and the review article [149].

In summary, the generalized estimating equations method, although it was originated from likelihood considerations, is not likelihood based as it does not require the complete specification of joint or conditional distributions. Under regularity conditions it produces consistent estimators for the regression parameters when the link function for the response mean is correctly specified regardless of the choice of the working covariance matrix. Moreover, the resulting estimators are nearly efficient relative to maximum likelihood estimators given a sensible choice of a working covariance matrix.

1.8 REAL DATA EXAMPLES

1.8.1 A Note on Computation

Partial likelihood estimation of time series following generalized linear models can be carried out by standard statistical software packages such as S-PLUS or SAS. This follows from the fact that the score equation (1.27) is the same as the score equation obtained from independent data. In particular, this means that the standard errors–approximated by inverting (1.30) and taking square roots of the diagonal elements–can be derived from standard software except that the results should be interpreted conditionally.

Regarding S-PLUS, the computation can be carried out by the function `glm()` and the library `MASS` described in [418]. More information about GLM can be found at `http://www.maths.uq.oz.au/~gks/research/glm`.

1.8.2 A Note on Model Building

Quite generally, the process of model building in modern time series analysis consists of the entertainment of several possible sensible models, parameter estimation in these models, hypothesis testing and deviance analysis, examination of residuals, graphical considerations including displays of the observed and fitted values, and model selection by means of information criteria. This procedure does not guarantee the identification of the "true" model, but it often leads to the selection of reasonable and useful regression models. This process is illustrated in this section in terms of two examples of environmental time series.

1.8.3 Analysis of Mortality Count Data

The preceding development is applied in this section to sampled smoothed mortality count data and related covariate series. The data set in question refers to total mortality and associated covariates in Los Angeles County during a period of 10 years from January 1, 1970, to December 31, 1979. The original data consisted of three daily mortality time series including total mortality (Y), two weather series, temperature (T) and relative humidity (RH), and six pollution series, carbon monoxide (CO), sulfur dioxide (SO_2), nitrogen dioxide (NO_2), hydrocarbons (HC), ozone (OZ), and particulates (KM). However, from the experience gained in [387], we shall concentrate on the relationship between *smoothed or filtered* versions of total mortality, temperature (weather), and carbon monoxide (pollutant). The data and their sources are described and analyzed in detail in [387] by linear regression in terms of functions of the covariates and second-order autoregressive noise. Other regression models are employed in [388]. The ensuing analysis uses Poisson regression with canonical link (1.51) and is of a somewhat different nature.

From spectral coherency considerations (e.g., see [388]), the original 11 daily series were first low-filtered (or smoothed) as to remove frequencies above 0.1 cycles per day, using a specially designed linear filter of the form $\sum_j a_j x_{t-j}$. This means that the smoothed data consist mainly of periodic components with periods longer than 10 days. The smoothed series were then sampled weekly, resulting in shorter 11 series each of length $N = 508$ observations[1]. The weekly sampled smoothed data of total mortality (expressed in death per day) and temperature, and the logarithm of the weekly sampled filtered carbon monoxide series are plotted in Figure 1.1. The three series display a very similar oscillatory pattern, a fact echoed by the estimated autocorrelations. For simplicity, we shall still refer to these weekly smoothed series as mortality (Y_t), temperature (T_t), and log-carbon ($\log(CO_t)$), respectively.

The fact that the count data were first smoothed does not preclude the application of Poisson regression with the corresponding canonical link. This nonstandard use of the Poisson model becomes more palatable if we think of it against the backdrop of estimating functions. In the present example, the Poisson model provides a convenient as well as useful estimating function, the justification of which emanates from the quality of the fit shown in Figure 1.2 and the residuals in Figure 1.3. Another way to justify the procedure is to think of the smoothed data as approximate integer–valued data.

[1]Thanks are due to Robert H. Shumway for kindly supplying the sampled LA mortality data.

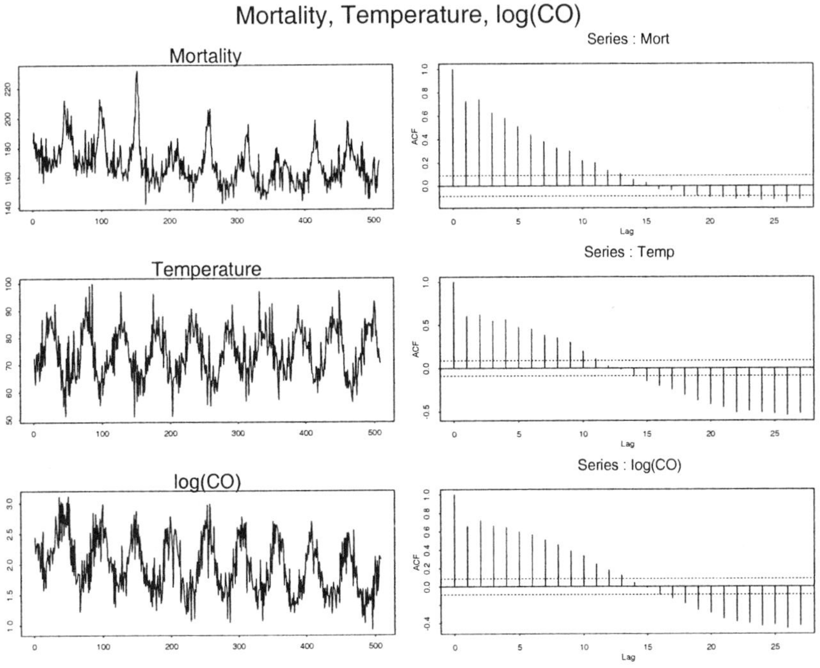

Fig. 1.1 Weekly data of filtered total mortality and temperature, and log-filtered CO, and the corresponding estimated autocorrelation functions. $N = 508$.

Table 1.1 Covariates and η_t used in Poisson regression. $S = \mathrm{SO}_2$, $N = \mathrm{NO}_2$. To recover η_t, insert the β's. For Model 2, $\eta_t = \beta_0 + \beta_1 Y_{t-1} + \beta_2 Y_{t-2}$, etc.

Model 0	$T_t + RH_t + CO_t + S_t + N_t + HC_t + OZ_t + KM_t$
Model 1	Y_{t-1}
Model 2	$Y_{t-1} + Y_{t-2}$
Model 3	$Y_{t-1} + Y_{t-2} + T_{t-1}$
Model 4	$Y_{t-1} + Y_{t-2} + T_{t-1} + \log(CO_t)$
Model 5	$Y_{t-1} + Y_{t-2} + T_{t-1} + T_{t-2} + \log(CO_t)$
Model 6	$Y_{t-1} + Y_{t-2} + T_t + T_{t-1} + \log(CO_t)$

Table 1.1 lists seven models, where Model 0 is recorded for the sake of comparison, and Table 1.2 provides diagnostics statistics pertaining to these models. These include the sum of squares of the working residuals (WR), $(Y_t - \hat{\mu}_t)/(\partial\mu_t/\partial\eta_t)$, evaluated at $\hat{\beta}$, the mean sum of squares (MSE) of the response or raw residuals, $Y_t - \hat{\mu}_t$, the sum of squares of Pearson residuals χ^2, the scaled deviance D with df $= N - p$ degrees of freedom, and the AIC and BIC computed up to an additive constant from the deviance as $D + 2p$ and $D + p\log(N)$, respectively. We see that the AIC is minimized at Model 6 while the BIC is minimized at Model 4. To resolve this, in testing the significance of T_t from the scaled deviances of Models 4 and 6, the log-partial likelihood test (1.66)

gives a p-value of 0.076, meaning that T_t is not quite significant. Thus, we opt for Model 4 as suggested by the BIC. Likewise, from the scaled deviances of Models 4 and 5, T_{t-2} is apparently not significant with a p-value of 0.873. By contrast, from Models 3 and 4, $\log(CO_t)$ is very significant with p-value on the order of 10^{-8}, and in testing the significance of the covariates T_{t-1} and $\log(CO_t)$ using the scaled deviances of Models 2 and 4, we obtain a p-value on the order of 10^{-11}. Thus, T_{t-1} and $\log(CO_t)$ are very significant and are therefore included in the chosen Model 4.

The fitted Model 4 is given by, $\hat{\mu}_t = \mu_t(\hat{\beta}) = \exp(\mathbf{Z}'_{t-1}\hat{\beta})$, where

$$\hat{\mu}_t = \exp\left\{\hat{\beta}_0 + \hat{\beta}_1 Y_{t-1} + \hat{\beta}_2 Y_{t-2} + \hat{\beta}_3 T_{t-1} + \hat{\beta}_4 \log(CO_t)\right\} \tag{1.96}$$

where $\hat{\beta} = (\hat{\beta}_0, \hat{\beta}_1, \hat{\beta}_2, \hat{\beta}_3, \hat{\beta}_4)' = (4.50513, 0.00189, 0.00184, -0.00133, 0.04683)'$, and $\mathbf{Z}_{t-1} = (1, Y_{t-1}, Y_{t-2}, T_{t-1}, \log(CO_t))'$. The corresponding estimated standard errors obtained from $\mathbf{G}_N^{-1}(\hat{\beta})$ in (1.53) are $(0.06944, 0.00036, 0.00037, 0.00044, 0.00871)'$. Figures 1.2 and 1.3 indicate that the fit is quite reasonable, and that the residuals of Model 4 are appreciably whiter than those from Model 0. Figure 1.4 depicts 95% prediction intervals approximated by $Y_t = \hat{\mu}_t \pm 1.96\sqrt{\hat{\mu}_t}$.

By comparison, the model suggested in [387] has the form

$$Y_t = \alpha_0 + \alpha_1 T_t + \alpha_2 T_t^2 + \beta_1 \log(P_t) + X_t,$$

where P_t is a pollution series such as CO_t, and $X_t = \phi_1 X_{t-1} + \phi_2 X_{t-2} + W_t$ is an autoregressive series of order two with W_t a sequence of uncorrelated zero-mean normal random variables (Gaussian white noise). Interestingly, if T_t^2 is added to Model 4, there is further reduction in the AIC (181.67) and in the residuals sums of squares (WR=0.988, MSE=57.87, $\chi^2 = 169.95$). However, the BIC (207.05) is slightly larger than that of Model 4, and the resulting standard errors are relatively large, having the same order of magnitude as that of the respective estimated regression coefficients.

Table 1.2 Comparison of seven Poisson regression models. $N = 508$.

Model	p	WR	MSE	χ^2	D	df	AIC	BIC
0	9	1.858	108.99	320.17	315.69	499	333.69	371.76
1	2	1.620	93.02	276.11	276.07	506	280.07	288.53
2	3	1.295	75.52	222.32	222.23	505	228.23	240.92
3	4	1.191	69.04	203.78	203.52	504	211.52	228.44
4	5	1.020	59.38	174.91	174.55	503	184.55	205.71
5	6	1.020	59.38	174.89	174.53	502	186.53	211.91
6	6	0.999	58.44	171.72	171.41	502	183.41	208.79

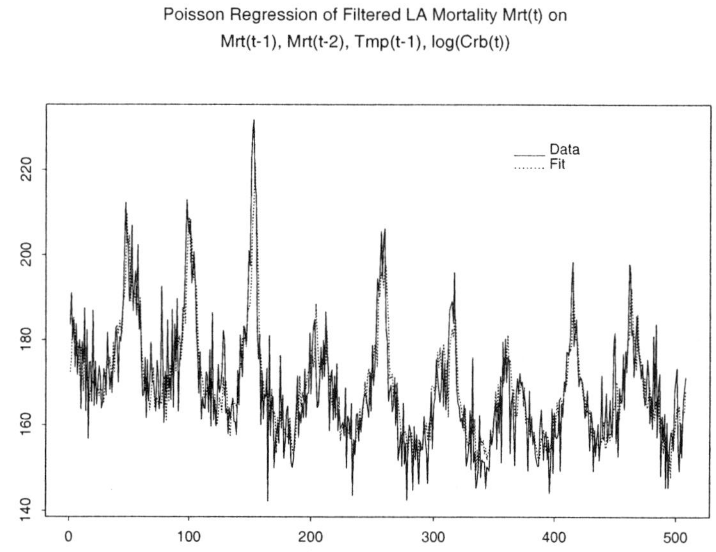

Fig. 1.2 Observed and predicted weekly filtered total mortality from Model 4.

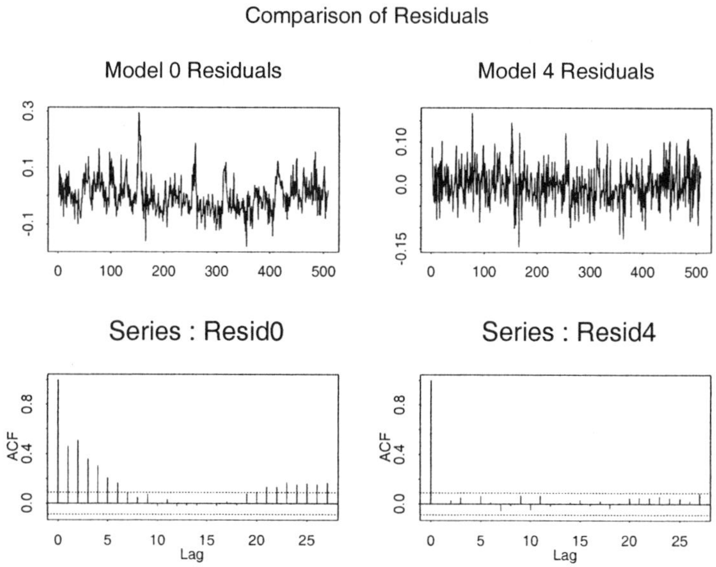

Fig. 1.3 Working residuals from Models 0 and 4, and their respective estimated autocorrelation.

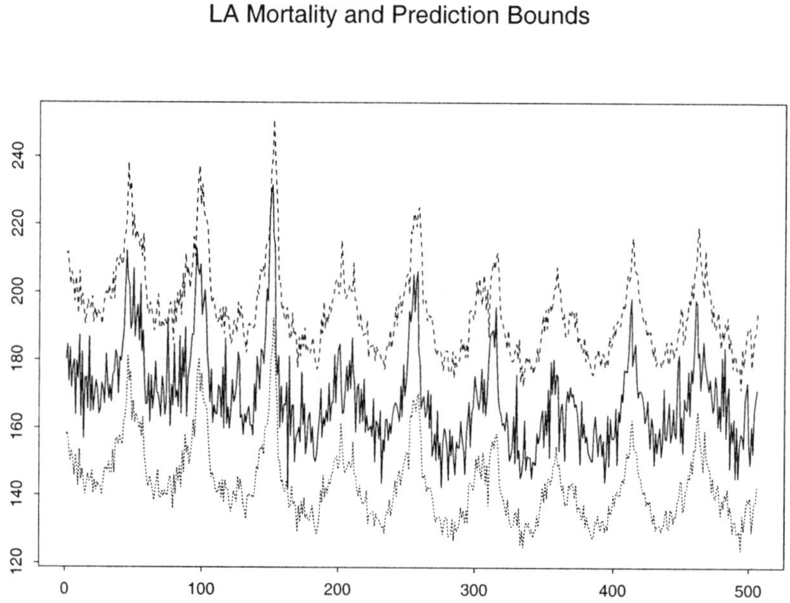

Fig. 1.4 95% prediction intervals from Model 4 for weekly smoothed LA mortality.

1.8.4 Application to Evapotranspiration

Evapotranspiration is the water lost to the atmosphere by the processes of evaporation and transpiration. Evaporation is the water loss from sources such as lakes, reservoirs, marshes, soil moisture, and glaciers. Transpiration is the loss from plant surfaces such as grain fields and rain forests.

We consider a time series of monthly evapotranspiration rates (mm/day), X_t, $t = 1, ..., 96$, obtained during the period 1960 to 1967 at Yotvata, southern Israel, by the Hydrological Service. The data are provided and analyzed in [250] by clipping the seasonal difference $W_t = X_t - X_{t-12}$,

$$Y_{t-12} = \begin{cases} 1, & \text{if } W_t \geq \bar{W} \\ 0, & \text{if } W_t < \bar{W} \end{cases}$$

$t = 13, \ldots, 96$. The clipped series $Y_t, t = 1, \ldots, 84$, is given by the binary sequence,

111111110001111100000001111000100000111111

110011000000011111100000000011111110000000

In fitting logistic (1.20) and probit (1.21) regression models, using the binary series $Y_t, t = 5, ..., 84$ ($N = 80$), we entertain the models listed in Table 1.3.

Table 1.3 Candidate models η_t

Model 1	$\beta_0 + \beta_1 Y_{t-1}$
Model 2	$\beta_0 + \beta_1 Y_{t-1} + \beta_2 Y_{t-2}$
Model 3	$\beta_0 + \beta_1 Y_{t-1} + \beta_2 Y_{t-2} + \beta_3 Y_{t-3}$
Model 4	$\beta_0 + \beta_1 Y_{t-1} + \beta_2 Y_{t-2} + \beta_3 Y_{t-3} + \beta_4 Y_{t-4}$

Tables 1.4 and 1.5 report the regression results following the format of Table 1.2. Notice that the working residuals in the present case are computed from $(Y_t - \hat{\pi}_t)/(\partial \pi_t / \partial \eta_t)$, so that the division is by $\hat{\pi}_t(1 - \hat{\pi}_t)$, a very small quantity when $\hat{\pi}_t$ is close to 0 or 1. This explains the relatively large WR values in the tables. The AIC and BIC are defined up to an additive constant as $D + 2p$ and $D + p \log N$, respectively. The tables show that both regression models yield very similar results consistently.

Based on the BIC criterion, Model 1, a first-order model, emerges as the best model in both cases. The fitted logistic model is

$$\hat{\pi}_t = \pi_t(\hat{\boldsymbol{\beta}}) = \frac{1}{1 + \exp\left\{-(\hat{\beta}_0 + \hat{\beta}_1 Y_{t-1})\right\}}$$

with $\hat{\boldsymbol{\beta}} = (\hat{\beta}_0, \hat{\beta}_1)' = (-1.61, 2.93)'$ and corresponding standard errors $(0.41, 0.57)$. Similarly the fitted probit model is

$$\hat{\pi}_t = \pi_t(\hat{\boldsymbol{\beta}}) = \Phi(\hat{\beta}_0 + \hat{\beta}_1 Y_{t-1})$$

Table 1.4 Comparison of the models from Table 1.3 using the logistic link

Model	p	WR	MSE	χ^2	D	df	AIC	BIC
1	2	520.45	0.15	79.23	76.96	78	80.96	85.72
2	3	573.37	0.14	78.35	74.83	77	80.83	87.97
3	4	594.09	0.14	78.59	74.35	76	82.35	91.88
4	5	616.87	0.14	79.30	74.21	75	84.21	96.12

Table 1.5 Comparison of models from Table 1.3 using the probit link

Model	p	WR	MSE	χ^2	D	df	AIC	BIC
1	2	167.66	0.15	79.54	76.96	78	80.96	85.72
2	3	173.97	0.14	78.31	74.77	77	80.77	87.92
3	4	177.61	0.14	78.55	74.31	76	82.31	91.83
4	5	181.57	0.14	79.17	74.21	75	84.21	96.11

with estimated coefficients $(-0.96, 1.77)'$ and corresponding standard errors $(0.22, 0.32)$.

The AIC, barely minimized in both cases at Model 2, points to a second-order model as a possibility. In this case, the fitted second-order logistic model is

$$\hat{\pi}_t = \pi_t(\hat{\beta}) = \frac{1}{1 + \exp\left\{-(\hat{\beta}_0 + \hat{\beta}_1 Y_{t-1} + \hat{\beta}_2 Y_{t-2})\right\}}$$

with $\hat{\beta} = (\hat{\beta}_0, \hat{\beta}_1, \hat{\beta}_2)' = (-1.43, 3.93, -1.34)'$ and corresponding standard errors $(0.42, 1.06, 1.06)$. The fitted second-order probit model is

$$\hat{\pi}_t = \pi_t(\hat{\beta}) = \Phi(\hat{\beta}_0 + \hat{\beta}_1 Y_{t-1} + \hat{\beta}_2 Y_{t-2})$$

with estimated coefficients $(-0.87, 2.29, -0.72)'$ and corresponding standard errors $(0.24, 0.53, 0.53)$.

To resolve the problem of first- versus second-order model, we test the significance of the covariate Y_{t-2} using the log-partial likelihood ratio test. The test gives p-values of 0.144 (logistic) and 0.139 (probit), neither terribly small nor substantially large. The uncertainty incurs as a result of somewhat indecisive p-values suggests that perhaps further analysis is needed to be able to choose between Models 1 and 2. However, an augmented model is a more prudent choice when there is appreciable probability that the lesser model may result in an underfit, and we opt for Model 2. Figures 1.5 and 1.6 give credence to this conclusion as does goodness of fit based on *power divergence*, a topic we shall take up in Section 3.4.2 (Table 1.6).

Figure 1.5 shows the autocorrelation plots of the Pearson residuals for the second-order model, Model 2, for both the logistic and probit links. As expected, both Figures 1.5 (a) and (b) are very similar, both pointing to reasonably white residuals. Figure 1.6 depicts both observed versus predicted values–that is, estimated success probabilities $\hat{\pi}_t$–for the logistic model. The second-order model fits the data reasonably well. The problem of prediction becomes more challenging at the points where the observed process makes a transition from 0 to 1 and vice versa.

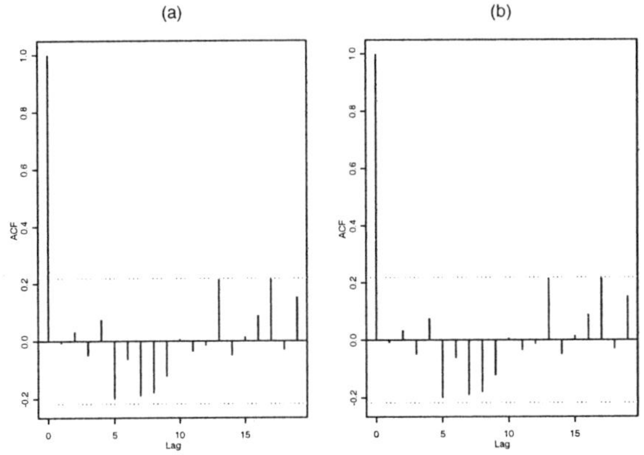

Fig. 1.5 Residual analysis for Model 2 of Table 1.3 for the evapotranspiration data. (a) Estimated autocorrelation of Pearson residuals from the logistic model. (b) Estimated autocorrelation of Pearson residuals from the probit model.

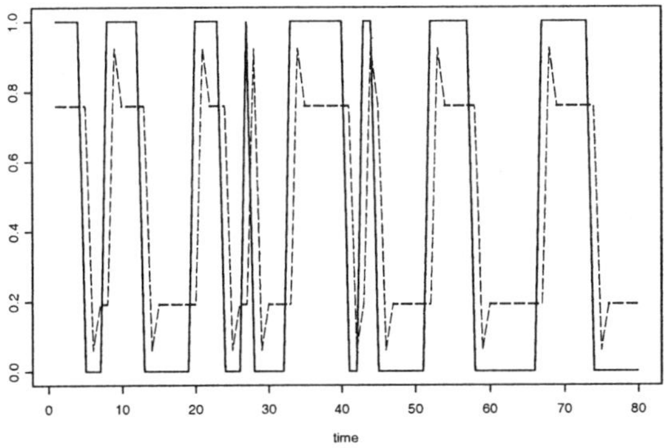

Fig. 1.6 Observed (solid line) and predicted (dash line) for the evatranspiration data using Model 2 of Table 1.3 with the logistic link

Interestingly, in [250] and [306, Ch. 4.7] it is assumed that Y_t is a Markov chain of some order, and the problem is to estimate the order. In [250] the author does not rule out a second-order chain despite of a relatively high p-value when testing "first-order" versus "second-order," while in [306] it is concluded the chain is of first-order based on a BIC value of 88.73. From [306, Table 4.22] the BIC value corresponding to a second-order Markov chain is 94.13, second smallest out of six different models.

Table 1.6 Values of the power divergence statistic for the second-order model applied to evapotranspiration data.

λ	-0.5	-0.2	0.5	0.66
Logistic	-0.161	-0.078	-0.281	-0.380
Probit	-0.388	-0.271	-0.341	-0.496
λ	1	2.5	4	5
Logistic	-0.570	-0.907	-0.964	-0.977
Probit	-0.730	-0.957	-0.978	-0.982

1.9 PROBLEMS AND COMPLEMENTS

1. *Properties of conditional expectation.* For jointly distributed random variables X, Y, use the property $\mathrm{E}[g(X)Y] = \mathrm{E}[g(X)\mathrm{E}(Y|X)]$, for all g such that $\mathrm{E}[g(X)Y]$ is finite, to show the following.

   a. $\mathrm{E}(a_1Y_1 + a_2Y_2|X) = a_1\mathrm{E}(Y_1|X) + a_2\mathrm{E}(Y_2|X)$
   b. $\mathrm{E}[\mathrm{E}(Y|X)] = \mathrm{E}(Y)$
   c. $\mathrm{Var}(Y) = \mathrm{Var}[\mathrm{E}(Y|X)] + \mathrm{E}[\mathrm{Var}(Y|X)]$
   d. $\mathrm{Corr}^2[Y, \mathrm{E}(Y|X)] = \mathrm{Var}[\mathrm{E}(Y|X)]/\mathrm{Var}(Y)$
   e. $\mathrm{E}[Y|\mathrm{E}(Y|X)] = \mathrm{E}(Y|X)$
   f. $\mathrm{E}[\mathrm{E}(Y|X_1, X_2)|X_1] = \mathrm{E}(Y|X_1)$
   g. $\mathrm{E}[Y|B] = \mathrm{E}[YI_B]/P(B)$, where I_B is the indicator of B.

2. *Construction of natural exponential families* [234],[342]. Let Y be a nondegenerate random variable with density $f_0(y)$ with respect to measure ν supported on $\mathcal{Y}$. Here ν is either Lebesgue measure (continuous case) or a counting measure (discrete case) or a mixture of the two (mixed case). Let $M_0(\theta)$ be the moment generating function of Y,

$$M_0(\theta) = \int_{\mathcal{Y}} e^{\theta y} f_0(y)\nu(dy),$$

   and define the *natural parameter space*, $\Theta = \{\theta \in R^1 : M_0(\theta) < \infty\}$. A natural exponential (parametric) family with respect to ν, $\{f(y;\theta), y \in \mathcal{Y}, \theta \in \Theta\}$, can be constructed by normalizing the "exponential tilt" $e^{\theta y} f_0(y)$,

$$f(y;\theta) = \frac{e^{\theta y} f_0(y)}{M_0(\theta)}, \; y \in \mathcal{Y}, \; \theta \in \Theta.$$

   Clearly, $f(y; 0) = f_0(y)$ is a member of the family, and all members have the same support $\mathcal{Y}$, regardless of θ. Construct natural exponential families by tilting:
   a. The standard normal density (get $\mathcal{N}(\theta, 1)$.)

b. The Poisson(1) density (get Poisson($\exp(\theta)$).)
c. A density $f_0(y)$ with respect to a mixed measure with an atom at zero.

3. Show that the gamma($\nu, \nu/\mu$) pdf

$$f(y;\theta,\phi) = \frac{(\nu/\mu)^\nu}{\Gamma(\nu)} y^{\nu-1} \exp\{-(\nu/\mu)y\},\ y > 0,$$

has the form (1.6) with mean μ, $\theta(\mu) = -1/\mu$, $b(\theta) = \log(\mu) = -\log(-\theta)$, $V(\mu) = \mu^2$, $\phi = 1/\nu$.

Discuss the danger in using the canonical link and how it can be overcome by using the log link which guarantees nonnegative estimated responses [334].

4. Show that the inverse Gaussian pdf IG($\mu, 1/\sigma^2$)

$$f(y;\theta,\phi) = \left\{\frac{1}{2\pi\sigma^2 y^3}\right\}^{1/2} \exp\left\{-\frac{(y-\mu)^2}{2\mu^2\sigma^2 y}\right\},\ y > 0,$$

has the form (1.6) with mean μ, $\theta(\mu) = -1/2\mu^2$, $b(\theta) = -1/\mu = -(-2\theta)^{1/2}$, $V(\mu) = \mu^3$, $\phi = \sigma^2$.

5. *Mean parametrization* [78]. Suppose $\{Y_t\}$ is a time series with a covariate vector process $\{\mathbf{Z}_{t-1}\}$, $t = 1, \ldots, N$. Conditionally Y_t has a distribution from an exponential family parametrized by its conditional mean,

$$f(y_t; \mu_t \mid \mathcal{F}_t) = \exp\{a(\mu_t) + b(y_t) + c(\mu_t) y_t\},$$

where $\mu_t = \mu(\mathbf{Z}_{t-1}, \boldsymbol{\beta})$ and $a(\cdot), b(\cdot), c(\cdot)$ are differentiable functions.

a. Show that the Poisson family with parameter $\mu_t = \mu(\mathbf{Z}_{t-1}, \boldsymbol{\beta})$ is a special case.
b. Show that

$$\mu_t = \mathrm{E}[Y_t \mid \mathcal{F}_{t-1}] = -[c'(\mu_t)]^{-1} a'(\mu_t),$$

$$\sigma_t^2 = \mathrm{Var}[Y_t \mid \mathcal{F}_{t-1}] = [c'(\mu_t)]^{-1}.$$

c. Show that the partial likelihood score

$$\mathbf{S}_N(\boldsymbol{\beta}) = \nabla \log \mathrm{PL}(\boldsymbol{\beta}) = \sum_{t=1}^{N} \frac{1}{\sigma_t^2}(Y_t - \mu_t)\nabla \mu_t$$

is a zero-mean martingale.
d. Define the cumulative conditional information matrix, $\mathbf{G}_N(\boldsymbol{\beta})$, by

$$\mathbf{G}_N(\boldsymbol{\beta}) = \sum_{t=1}^{N} \mathrm{Cov}\left[\frac{1}{\sigma_t^2}(Y_t - \mu_t)\nabla \mu_t \mid \mathcal{F}_{t-1}\right]$$

and let $\mathbf{H}_N(\beta) \equiv -\nabla\nabla\,'l(\beta)$. Let $\mathbf{G}, \mathbf{H}$ be the respective probability limits of $\mathbf{G}_N(\beta)/N$ and $\mathbf{H}_N(\beta)/N$. Argue that there are conditions such that

$$\sqrt{N}(\hat{\beta} - \beta) \rightarrow \mathcal{N}(\mathbf{0}, \mathbf{H}^{-1}(\beta)\mathbf{G}(\beta)\mathbf{H}^{-1}(\beta)),$$

in distribution, as $N \rightarrow \infty$.

6. *Properties of the GARMA(p,q) model* [35], [36]. Recall equation (1.9) which defines the GARMA(p,q) process. Denote by $\Phi(\mathcal{B}) = 1 - \phi_1\mathcal{B} - \ldots - \phi_p\mathcal{B}^p$, $\Theta(\mathcal{B}) = 1 + \theta_1\mathcal{B} + \ldots + \theta_q\mathcal{B}^q$, $\Psi(\mathcal{B}) = \Phi^{-1}(\mathcal{B})\Theta(\mathcal{B}) = 1 + \psi_1\mathcal{B} + \psi_2\mathcal{B}^2 + \ldots$, $\Psi^{(2)}(\mathcal{B}) = 1 + \psi_1^2\mathcal{B} + \psi_2^2\mathcal{B}^2 + \ldots$, where $\mathcal{B}$ is the backward shift operator and $\Phi(\mathcal{B})$ is assumed invertible. Show that when g is the identity function, then
a. The marginal mean of Y_t is equal to $\mathbf{X}_t'\gamma$.
b. The marginal variance of Y_t is given by $\mathrm{E}\left[\Psi^{(2)}(\mathcal{B})\mathrm{Var}[Y_t \mid \mathcal{F}_{t-1}]\right]$.

7. *Partial likelihood inference for the GARMA(p,q) model* [35], [36]. Suppose that the response process $\{Y_t\}$, $t = 1, \ldots, N$ follows a GARMA(p,q) model and set $\beta = (\gamma', \phi', \theta')'$ where $\phi = (\phi_1, \ldots, \phi_p)'$ and $\theta = (\theta_1, \ldots, \theta_q)'$. Show that partial likelihood estimation of the regression coefficients β is based on the following partial score

$$\sum_{t=m+1}^{N} \frac{\partial\eta_t}{\partial\beta}\frac{\partial\mu_t}{\partial\eta_t}\frac{(Y_t - \mu_t)}{\mathrm{Var}[Y_t \mid \mathcal{F}_{t-1}]},$$

where $m > j$, $j = \max(p, q)$, and that the calculation of $\partial\eta_t/\partial\beta$ is carried out by the following recursive equations for $t > j$:

$$\frac{\partial\eta_t}{\partial\gamma_l} = X_{t;l} - \sum_{i=1}^{p}\phi_i X_{t-i;l} - \sum_{i=1}^{q}\theta_i\frac{\partial\eta_{t-i}}{\partial\gamma_l}, l = 1, \ldots, s$$

$$\frac{\partial\eta_t}{\partial\phi_l} = (g(y_{t-l}) - \mathbf{X}_{t-l}'\gamma) - \sum_{i=1}^{q}\theta_i\frac{\partial\eta_{t-i}}{\partial\phi_l}, l = 1, \ldots, p$$

$$\frac{\partial\eta_t}{\partial\theta_l} = (g(y_{t-l}) - \eta_{t-l}) - \sum_{i=1}^{p}\theta_i\frac{\partial\eta_{t-i}}{\partial\theta_l}, l = 1, \ldots, q,$$

where $X_{t;l}$ is the l´th component of the s–dimensional vector $\mathbf{X}_t$.

8. Show that when $\alpha_t(\phi) = 1$ for all t, and with the canonical link, the partial score equation (1.29) for β simplifies to

$$\mathbf{S}_N(\beta) = \nabla \log \mathrm{PL}(\beta) = \sum_{t=1}^{N}\mathbf{Z}_{t-1}\{Y_t - \mu_t(\beta)\} = \mathbf{0}.$$

9. Suppose $Y_1, \ldots, Y_N$, are independent $\mathcal{N}(\mu_i, \sigma^2)$ such that $\mu_i = \mathbf{x}'_i\beta = x_{i1}\beta_1 + \cdots + x_{ip}\beta_p$, and $\mathbf{X} = (\mathbf{x}_1, \ldots, \mathbf{x}_N)'$ is of full rank p. Show that the scaled

deviance is the scaled sum of squares of the residuals, $D = \sum_{i=1}^{N}(Y_i - \hat{\mu}_i)^2/\sigma^2$, and that $D \sim \chi^2_{N-p}$ exactly. Hence deduce that for $\mu_i = \beta$, all i, $D \sim \chi^2_{N-1}$.

10. For independent $Y_i \sim$ Poisson(β), $i = 1, ..., N$, argue that, approximately,

$$D = 2\sum_{i=1}^{N} Y_i \log(Y_i/\bar{Y}) \sim \chi^2{}_{N-1}.$$

More generally, with $\mu_i = \mu_i(\boldsymbol{\beta})$, where $\boldsymbol{\beta}$ is a $p \times 1$ vector, show that the scaled deviance is given by

$$D = 2\left\{\sum_{i=1}^{N} Y_i \log(Y_i/\hat{\mu}_i) - \sum_{i=1}^{N} Y_i + \sum_{i=1}^{N} \hat{\mu}_i\right\},$$

where $\hat{\mu}_i = \mu_i(\hat{\boldsymbol{\beta}})$ is obtained under the Poisson GLM model, and argue that its approximate distribution is χ^2_{N-p}.

11. For independent binomial random variables $Y_i \sim \mathrm{b}(m_i, \pi_i), \pi_i = \pi_i(\beta_1, ..., \beta_p)$, show that

$$D = 2\sum_{i=1}^{N}\left\{Y_i \log\left(\frac{Y_i}{\hat{\mu}_i}\right) + (m_i - Y_i)\log\left(\frac{m_i - Y_i}{m_i - \hat{\mu}_i}\right)\right\}.$$

Argue that for sufficiently large m_i, $D \sim \chi^2{}_{N-p}$ approximately [313, p. 118]. Note that

$$D = 2\sum o \log\left(\frac{o}{e}\right),$$

where o, e are observed and estimated expected frequencies, respectively, in tabular data, and the summation is over all possible $2N$ cells [124, Ch. 8].

12. Discuss regularity conditions under which the generalized Pearson goodness of fit statistic with a known scale parameter ϕ,

$$\chi^2 = \frac{1}{\phi}\sum_{i=1}^{N} \frac{\omega_i(Y_i - \hat{\mu}_i)^2}{V(\hat{\mu}_i)}$$

is asymptotically χ^2_{N-p} approximately [148], [313].

13. *The general linear model for longitudinal data* [123]. Suppose that Y_{ij} denotes the observed data for the i'th unit, $i = 1, \ldots, m$ taken at time j, $j = 1, \ldots, n$. Assume further $\mathbf{X}_{ij}$ denotes a p–dimensional vector of covariates associated with each of the Y_{ij} values. The random variables Y_{ij} follow the linear model

$$Y_{ij} = \beta_1 X_{ij1} + \ldots + \beta_p X_{ijp} + \epsilon_{ij} \tag{1.97}$$

for $i = 1, \ldots, m$, $j = 1, \ldots, n$, where the regression coefficients β_i are unknown and ϵ_{ij} are normal random variables. For the classical linear model theory the random variables ϵ_{ij} are assumed mutually independent. However in the context of longitudinal data analysis, ϵ_{ij} are correlated within subject so that $\boldsymbol{\epsilon}_i = (\epsilon_{i1}, \ldots, \epsilon_{in})'$–an $n \times 1$ vector–is assumed to follow the multivariate normal distribution with mean $\mathbf{0}$ and covariance matrix $\sigma^2 \mathbf{V}_0$. Common choices for $\mathbf{V}_0$ include the *uniform correlation* model

$$\mathbf{V}_0 = (1 - \rho)\mathbf{I} + \rho\mathbf{J},$$

where $\mathbf{I}$ is the $n \times n$ identity matrix and all the elements of $\mathbf{J}$ are equal to 1, or the so called *exponential correlation* model

$$v_{jk}^0 = \rho^{|j-k|},$$

where v_{jk}^0 is the jk'th element of $\mathbf{V}_0$, and $| \rho | \leq 1$.

(a) Show that model (1.97) is equivalent to

$$\mathbf{Y} = \mathbf{X}\boldsymbol{\beta} + \boldsymbol{\epsilon}, \tag{1.98}$$

where $\mathbf{Y}$ consists of all the nm observations, $\mathbf{X}$ is an $(nm) \times p$ matrix of the covariates whose $n(i-1)+j$'th row is given by $\mathbf{X}_{ij}$ and $\boldsymbol{\epsilon}$ follows the nm–dimensional normal distribution with mean $\mathbf{0}$ and covariance matrix $\sigma^2\mathbf{V}$ with $\mathbf{V}$a block–diagonal matrix with non–zero blocks $\mathbf{V}_0$.

(b) The *weighted least squares* estimator of $\boldsymbol{\beta}$ is the value that minimizes

$$(\mathbf{Y} - \mathbf{X}\boldsymbol{\beta})' \mathbf{W} (\mathbf{Y} - \mathbf{X}\boldsymbol{\beta}),$$

where $\mathbf{W}$ is a suitable *weight* matrix. Show that if $\mathbf{W} = \mathbf{V} = \mathbf{I}$, then the weighted least squares estimator of $\boldsymbol{\beta}$ reduces to the ordinary least squares estimator. Moreover, if $\mathbf{W} = \mathbf{V}^{-1}$ then the weighted least squares estimator of $\boldsymbol{\beta}$ is equal to the maximum likelihood estimator.

(c) Discuss maximum likelihood estimation of σ^2 and $\mathbf{V}_0$.

It turns out that maximum likelihood estimation for σ^2 and $\mathbf{V}_0$ is problematic due to the fact that consistent estimators of the covariance structure are obtained by $\mathbf{X}$ with large number of columns whereas design matrices with small number of columns lead to unbiased estimators. An alternative estimation method is that of *restricted maximum likelihood estimation* [123].

14. *Generalized Additive Models*. Generalized additive models (GAM), introduced in [202] and detailed further in [203], extend the class of generalized linear models in the sense that the linear component of the model is replaced by an additive component. That is, if $Y_1, \ldots, Y_N$ are independent random variables from the exponential family of distributions observed jointly with covariates $\mathbf{Z}_1, \ldots, \mathbf{Z}_N$, then the linear predictor $\sum_j Z_{ij}\beta_j$ is replaced by

$$\sum_j s_j(Z_{ij}),$$

where s_j is an unspecified function. For example, if $Y_1, \ldots, Y_N$ are Bernoulli random variables, then the logistic model can be generalized as

$$\log\left[\frac{\mathrm{P}[Y_i = 1|\mathbf{Z}_i]}{1 - \mathrm{P}[Y_i = 1|\mathbf{Z}_i]}\right] = \beta_0 + s_1(Z_{i1}) + s_1(Z_{i2}) + \ldots + s_p(Z_{ip}).$$

The function $s_1(.), \ldots, s_p(.)$ are estimated by a backfitting procedure in connection with the maximum likelihood algorithm (see [203, Ch.6]). Modeling of time series data through generalized additive models has been considered in [219]. Topics related to GAM include partially linear model, time varying coefficients, projection pursuit regression and basis function approach [144, Ch. 5].

15. Consider a time series following generalized linear models and assume the canonical link. Suppose that the MPLE of the regression parameter $\boldsymbol{\beta}$, at time t, is $\hat{\boldsymbol{\beta}}$. Assume that a new observation, say $(Y_{t+1}, \mathbf{Z}_t)$ becomes available.
a. Prove the following recursive algorithm that expresses $\hat{\boldsymbol{\beta}}_{t+1}$ in terms of $\hat{\boldsymbol{\beta}}$ (see [153]).

$$\hat{\boldsymbol{\beta}}_{t+1} = \hat{\boldsymbol{\beta}} + \mathbf{G}_{t+1|t}^{-1}\mathbf{Z}_t(Y_{t+1} - \mu_{t+1|t})$$

with $\mathbf{G}_{t+1|t} = \mathbf{G}_{t+1}(\hat{\boldsymbol{\beta}})$ and $\mu_{t+1|t} = \mu_{t+1}(\hat{\boldsymbol{\beta}})$.
b. By writing

$$\mathbf{G}_{t+1|t} = \mathbf{G}_{t|t} + \mathbf{Z}_t\mathbf{Z}_t'\sigma_{t+1|t}^2,$$

where $\sigma_{t+1|t}^2 = \sigma_{t+1}^2(\hat{\boldsymbol{\beta}})$, and applying the matrix inversion lemma in Problem 5, Chapter 6, show that

$$\begin{aligned}
\hat{\boldsymbol{\beta}}_{t+1} &= \hat{\boldsymbol{\beta}} + \frac{\mathbf{D}_{t|t}\mathbf{Z}_t}{1 + \mathbf{Z}_t'\mathbf{D}_{t|t}\mathbf{Z}_t(\sigma_{t+1|t}^2)^{-1}}(Y_{t+1} - \mu_{t+1|t}) \\
\mathbf{D}_{t+1|t} &= \mathbf{D}_{t|t} - \frac{\mathbf{D}_{t|t}\mathbf{Z}_t\mathbf{Z}_t'\mathbf{D}_{t|t}}{(\sigma_{t+1|t}^2)^{-1} + \mathbf{Z}_t\mathbf{D}_{t|t}\mathbf{Z}_t'},
\end{aligned}$$

where $\mathbf{D}_{t|t} = \mathbf{G}_{t|t}^{-1}$.

2

Regression Models for Binary Time Series

Consider a binary time series $\{Y_t\}$ taking the values 0 or 1, and related covariate or auxiliary stochastic data represented by a $p \times 1$ column vector $\{\mathbf{Z}_{t-1}\}, t = 1, 2, 3, \cdots$. Against the backdrop of the general framework presented in Chapter 1, we wish to study a special case: the regression problem of estimating the conditional success probability

$$P_{\boldsymbol{\beta}}(Y_t = 1 | \mathcal{F}_{t-1}), \tag{2.1}$$

where $\boldsymbol{\beta}$ is a p-dimensional parameter vector, and $\mathcal{F}_{t-1}$ represents all that is known to the observer at time $t-1$ about the time series and the covariate information. More precisely, the problem is to model the conditional probability (2.1) by a regression model depending on $\boldsymbol{\beta}$, and then estimate the latter given a binary time series and its time dependent random covariates.

As before, $\mathcal{F}_{t-1}$ stands for the σ-field,

$$\mathcal{F}_{t-1} = \sigma(Y_{t-1}, Y_{t-2}, \ldots, \mathbf{Z}_{t-1}, \mathbf{Z}_{t-2}, \cdots)$$

generated by the past values of the series and of the covariate data. No harm is caused, however, if $\mathcal{F}_{t-1}$ is thought of as the past "history" or "information" as embodied in the past data

$$Y_{t-1}, Y_{t-2}, \ldots, \mathbf{Z}_{t-1}, \mathbf{Z}_{t-2}, \ldots$$

It follows that $\mathcal{F}_{t-1} \subset \mathcal{F}_t$ because more is known at time t than at time $t-1$.

Following the general formulation outlined in Chapter 1, the response process $\{Y_t\}$ may be stationary or nonstationary, and the *time dependent random covariate* vector process $\{\mathbf{Z}_{t-1}\}$ may represent one or more time series and functions thereof that influence the evolution of the primary series of interest $\{Y_t\}$. The covariate vector process $\{\mathbf{Z}_{t-1}\}$ need not be stationary per se, however, it is required to possess the "nice" long-term behavior described by Assumption A in Chapter 1. Conveniently, $\mathbf{Z}_{t-1}$ may contain past values of Y_t and/or past values of an underlying process that produces Y_t.

Quintessential examples of binary time series are those obtained by dichotomies such as

$$Y_t \equiv I_{[X_t \in C]} = \begin{cases} 1, & \text{if} \quad X_t \in C \\ 0, & \text{if} \quad X_t \in \bar{C}, \end{cases} \tag{2.2}$$

where $\bar{C}$ is the complement of the set C. Important special cases of (2.2) are binary time series obtained by clipping an underlying process X_t at a fixed level r,

$$Y_t \equiv I_{[X_t \geq r]} = \begin{cases} 1, & \text{if} \quad X_t \geq r \\ 0, & \text{if} \quad X_t < r. \end{cases} \tag{2.3}$$

In these cases we interpret (2.1) as prediction of the future events $\{X_t \in C\}$ and $\{X_t \geq r\}$ from past covariate information $\mathbf{Z}_{t-1}$ where the latter may include past values of both X_t and Y_t.

Typical examples of $\mathbf{Z}_{t-1}$ are

$$\mathbf{Z}_{t-1} = (1, X_{t-12}, Y_{t-1}, Y_{t-2}, \log(W_{t-1}))'$$

or, when interactions are entertained,

$$\mathbf{Z}_{t-1} = (1, W_t^3, X_{t-1}, X_{t-1}W_t, W_{t-2}Y_{t-3})'$$

and so on, given an underlying one-dimensional process $\{X_t\}$ and an auxiliary process $\{W_t\}$. Since $\mathbf{Z}_{t-1}$ may contain functions of past data, it may also contain past durations observed in the primary or covariate data such as the most recent length of 1-run in Y_t prior to time t. Moreover, in order to characterize an observed oscillation in $\{Y_t\}$ or in any of its covariates, $\mathbf{Z}_{t-1}$ may incorporate random as well as deterministic periodic functions. Notice that in the second case the auxiliary W is assumed known at time t but not the underlying process X. This occurs, for example, in daily rainfall-runoff data when the amount of rainfall W is known at time t but the runoff information X is delayed. Other examples include the cases when W is a deterministic function, or when it represents a time-shifted process.

2.1 LINK FUNCTIONS FOR BINARY TIME SERIES

To guarantee that a given model for (2.1) yields proper probability estimates, we choose suitable inverse links $h \equiv F$ that map the real line onto the interval $[0, 1]$. A logical first-order model then is [258],

$$P(Y_t = 1 | X_{t-1} = x_{t-1}) = F(x_{t-1}), \tag{2.4}$$

where X_t is an underlying process and F is some known cumulative distribution function. Various parametric generalizations of the form

$$P(Y_t = 1|\mathcal{F}_{t-1}) = F(\beta'\mathbf{W}_{t-1} + \theta_1 Y_{t-1} + \cdots \theta_q Y_{t-q}), \tag{2.5}$$

where $\mathbf{W}_t$ is a covariate vector have been considered by quite a few authors including [58], [87], [102], [107], [140], [182], [208], [254] [274], [301], [331], [395], [447]. In this chapter we consider regression models of the form (2.5) and the associated question of what constitutes a reasonable choice for F. Nonparametric prediction methods for ergodic and stationary binary time series have been considered in [189], [330], [329], and [443].

2.1.1 The Logistic Regression Model

We argue next that for a binary time series, when conditioning only on past values of the series itself, the cumulative distribution function (cdf) of the standard logistic distribution,

$$F_l(x) = \frac{e^x}{1+e^x} = \frac{1}{1+e^{-x}}, \quad -\infty < x < \infty \tag{2.6}$$

is the natural choice for F. In other words, $F_l^{-1}(x) = \log(x/(1-x))$ is the natural link under some conditions.

Let F_l be given by (2.6). For a binary time series $Y_1, Y_2, ...,$ consider the conditional probability of Y_t given all its past values,

$$P_{y_t y_{t-1}\cdots y_1} \equiv P(Y_t = y_t | Y_{t-1} = y_{t-1}, ..., Y_1 = y_1). \tag{2.7}$$

Then we can see that [251, p. 9],

$$P(Y_2 = y_2|Y_1 = y_1) = P_{00}\left(\frac{P_{01}}{P_{00}}\right)^{y_1}\left(\frac{P_{10}}{P_{00}}\right)^{y_2}\left(\frac{P_{11}P_{00}}{P_{01}P_{10}}\right)^{y_2 y_1}$$

or,

$$\log\left\{\frac{P(Y_2 = 1|Y_1 = y_1)}{P(Y_2 = 0|Y_1 = y_1)}\right\} = \theta_2 + \theta_{12} y_1,$$

where

$$\theta_2 = \log\left(\frac{P_{10}}{P_{00}}\right), \quad \theta_{12} = \log\left(\frac{P_{11}P_{00}}{P_{01}P_{10}}\right)$$

and so,

$$P(Y_2 = 1|Y_1 = y_1) = F_l(\theta_2 + \theta_{12} y_1).$$

Likewise,

$$\begin{aligned} P(Y_3 = y_3 \quad | \quad & Y_2 = y_2, Y_1 = y_1) = P_{000} \\ & \times \left(\frac{P_{001}}{P_{000}}\right)^{y_1}\left(\frac{P_{010}}{P_{000}}\right)^{y_2}\left(\frac{P_{100}}{P_{000}}\right)^{y_3} \end{aligned}$$

$$\times \left(\frac{P_{011}P_{000}}{P_{010}P_{001}} \right)^{y_2 y_1} \left(\frac{P_{101}P_{000}}{P_{001}P_{100}} \right)^{y_3 y_1} \left(\frac{P_{110}P_{000}}{P_{010}P_{100}} \right)^{y_3 y_2}$$
$$\times \left(\frac{P_{111}P_{010}P_{001}P_{100}}{P_{011}P_{110}P_{101}P_{000}} \right)^{y_3 y_2 y_1}$$

or, with obvious notation,

$$\log \left\{ \frac{P(Y_3 = 1|Y_2 = y_2, Y_1 = y_1)}{P(Y_3 = 0|Y_2 = y_2, Y_1 = y_1)} \right\} = \theta_3 + \theta_{13}y_1 + \theta_{23}y_2 + \theta_{123}y_1y_2$$

or, equivalently,

$$P(Y_3 = 1|Y_2 = y_2, Y_1 = y_1) = F_l(\theta_3 + \theta_{13}y_1 + \theta_{23}y_2 + \theta_{123}y_1y_2).$$

In general, for any binary time series $Y_1, Y_2, ...$, stationary or nonstationary,

$$\log \left\{ \frac{P(Y_t = 1|Y_{t-1} = y_{t-1}, ..., Y_1 = y_1)}{P(Y_t = 0|Y_{t-1} = y_{t-1}, ..., Y_1 = y_1)} \right\}$$
$$= \theta_t + \sum_{i<t} \theta_{it}y_i + \sum_{i<j<t} \theta_{ijt}y_iy_j + \cdots + \theta_{1\ldots t}y_1 \cdots y_{t-1} \tag{2.8}$$

or, equivalently,

$$P(Y_t = 1|Y_{t-1} = y_{t-1}, ..., Y_1 = y_1) \tag{2.9}$$
$$= F_l(\theta_t + \sum_{i<t} \theta_{it}y_i + \sum_{i<j<t} \theta_{ijt}y_iy_j + \cdots + \theta_{1\ldots t}y_1 \cdots y_{t-1}).$$

This is the saturated case of *logistic regression* where the explanatory variables are all the past values of the series.

The number of parameters in the conditional log-odds (2.8), 2^{t-1}, decreases as the degree of independence increases. Under complete independence, all interactions vanish and (2.8) reduces to the marginal log-odds of Y_t, and when second and higher interaction terms vanish we obtain a curtailed model,

$$\log \left\{ \frac{P(Y_t = 1|Y_{t-1} = y_{t-1}, ..., Y_1 = y_1)}{P(Y_t = 0|Y_{t-1} = y_{t-1}, ..., Y_1 = y_1)} \right\} = \theta_t + \sum_{i<t} \theta_{it}y_i. \tag{2.10}$$

Note that in both (2.8) and (2.10), the parameters and their number change with t so that the standard notion of statistical consistency in parameter estimation does not apply. This is rectified by fixing the parameters and their number so that more and more information is obtained about a fixed set of parameters. As a result, we obtain more practical autoregressive models for binary time series a special case of which is the autoregressive model of order p,

$$\log \left\{ \frac{P(Y_t = 1|Y_{t-1} = y_{t-1}, ..., Y_1 = y_1)}{P(Y_t = 0|Y_{t-1} = y_{t-1}, ..., Y_1 = y_1)} \right\}$$
$$= \theta_0 + \theta_1 y_{t-1} + \cdots + \theta_p y_{t-p}. \tag{2.11}$$

The previous discussion shows that for binary time series Y_t, whatever the degree and type of dependence of Y_t on the past values $Y_{t-1}, Y_{t-2}, \ldots$, logistic regression as expressed in (2.8)–(2.11) occurs naturally. There are other instances where logistic regression emerges naturally as a consequence of certain conditions. For example, if Z is a random variable whose conditional distribution given a binary random variable Y is a member of the exponential family, then logistic regression of Y on Z is the true model (see Problem 1). Another example is provided in terms of a clipped autoregressive process with logistic noise discussed in Section 2.1.1.1 below. All this points to the standard logistic cdf as a plausible candidate for (2.4) and (2.5), and also motivates the general *logistic regression model* for (2.1),

$$\pi_t(\boldsymbol{\beta}) \equiv P_{\boldsymbol{\beta}}(Y_t = 1|\mathcal{F}_{t-1}) = F_l(\boldsymbol{\beta}'\mathbf{Z}_{t-1}) = \frac{1}{1+\exp[-\boldsymbol{\beta}'\mathbf{Z}_{t-1}]}, \tag{2.12}$$

where $\boldsymbol{\beta}$ is a column vector parameter of the same dimension p as $\mathbf{Z}_{t-1}$, and $\mathbf{Z}_{t-1}$ is a covariate vector process as described above. The inverse transformation of (2.12) is the canonical link for binary data referred to as *logit*,

$$\text{logit}(\pi_t(\boldsymbol{\beta})) \equiv \log\left\{\frac{\pi_t(\boldsymbol{\beta})}{1-\pi_t(\boldsymbol{\beta})}\right\} = \boldsymbol{\beta}'\mathbf{Z}_{t-1}. \tag{2.13}$$

It is emphasized that (2.13) is an assumed model while (2.8) is a general fact regarding binary time series; (2.8) promotes (2.13).

Interestingly, the logistic model (2.12) possesses a certain invariance property with respect to prospective/retrospective sampling not shared by other models (links) for binary data. The logistic model is the only one among GLM's for binary data for which the retrospective and prospective likelihoods differ by an intercept [239], [355] (Problem 2).

Starting from the 1940s on, there is a very large body of work on logistic regression and its applications throughout the life, physical, and social sciences. An early reference is [43], while many more recent references describing a wide range of applications include [2], [107], [123], [273], [313], and [376]. For example, applications of weighted logistic regression in health surveys taking into account inclusion probabilities are described in [273].

2.1.1.1 Logistic Autoregression To show that the model (2.12) is not vacuous, consider this (see [395]). Let $\{X_t\}, t = 0, 1, 2, \ldots$, be an autoregressive process of order p,

$$X_t = \gamma_0 + \gamma_1 X_{t-1} + \ldots + \gamma_p X_{t-p} + \lambda\epsilon_t, \tag{2.14}$$

where λ is a constant, and the ϵ_t are i.i.d. random variables logistically distributed, $\epsilon_t \sim f_l(x) = e^x/(1+e^x)^2$, with mean 0 and variance $\pi^2/3$. This is the probability density corresponding to the logistic distribution function (2.6). Now fix a threshold $r \in (-\infty, \infty)$, and define a binary time series by clipping X_t at r as in (2.3). Then $Y_t = I_{[X_t \geq r]}$ satisfies (2.12) for each fixed r with

$$\mathbf{Z}_{t-1} = (1, X_{t-1}, X_{t-2}, \ldots, X_{t-p})'$$

and

$$\beta = \frac{1}{\lambda}(\gamma_0 - r, \gamma_1, \ldots, \gamma_p)'.$$

That is,

$$\pi_t(\beta) = \frac{1}{1 + \exp[-(\gamma_0 - r + \gamma_1 X_{t-1} + \cdots + \gamma_p X_{t-p})/\lambda]}. \tag{2.15}$$

2.1.2 Probit and Other Links

We have modeled $\pi_t(\beta)$ in (2.12) in terms of $F(\beta'\mathbf{Z}_{t-1})$, where F is the logistic cdf (2.6), however, there are numerous other possibilities, some motivated by convenience while others by hazard or other considerations [144, p. 86].

Parametric hazard models are a source for useful link functions. As an example, consider Cox's proportional hazards [326],

$$\lambda(t; \mathbf{z}) = \lambda_0(t) \exp(\beta'\mathbf{z}),$$

where $\mathbf{z}$ is the value of a covariate vector, and $\lambda_0(t)$ is a baseline hazard function. Here t is continuous and in principle we could replace $\beta'\mathbf{z}$ by $v(\beta'\mathbf{z})$ to obtain different models for different choices of v. The corresponding cdf admits the representation,

$$F(t; \mathbf{z}) = 1 - \exp(-\exp(\beta'\mathbf{z} + \log w_0(t))),$$

where $w_0(t) = \int_0^t \lambda_0(u)du$. Switching back to our notation and discrete time, this motivates a regression model in terms of the extreme value distribution $F(x) = 1 - \exp(-\exp(x))$,

$$\pi_t(\beta) \equiv P_{\beta}(X_t = 1|\mathcal{F}_{t-1}) = 1 - \exp(-\exp(\beta'\mathbf{Z}_{t-1}))$$

with the so called *complementary log-log* (cloglog) corresponding link,

$$\beta'\mathbf{Z}_{t-1} = \log\{-\log(1 - \pi_t(\beta))\}. \tag{2.16}$$

Another useful inverse link is defined by $F \equiv \Phi$, where Φ is the standard normal distribution function. In this case we obtain what is known as the *probit* model [147],

$$\pi_t(\beta) \equiv P_{\beta}(X_t = 1|\mathcal{F}_{t-1}) = \Phi(\beta'\mathbf{Z}_{t-1}). \tag{2.17}$$

The four most common link functions $g(\pi_t(\beta))$ for binary data are listed in Table 2.1, and their graphs are plotted in Figure 2.1. Practical differences between these links occur for probabilities closer to 0 or 1. Observe that the four links appear to be approximately linear for values of success probabilities around 0.5. Figure 2.2 conveys the same impression in terms of the distribution functions of the uniform, normal, and logistic distributions where all three have mean 0 and variance $\pi^2/3$. This suggests that under appropriate restrictions on the parameters, the identity link can be used as well. Assuming the identity link may be helpful in establishing

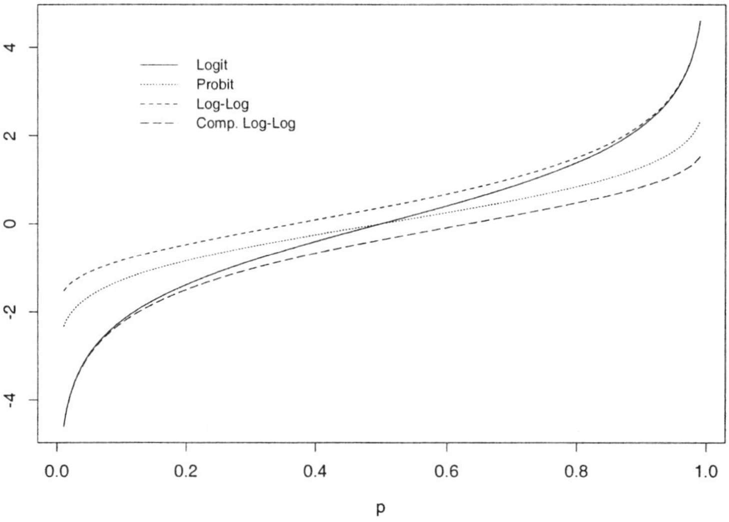

Fig. 2.1 Common link functions for binary time series.

autocorrelation and spectral properties for stationary binary time series (see Problem 5).

Several other parametric link functions have been proposed in the literature. In particular, [22], [186], [328], [430] propose one–parameter families and [93], [112], [353], [351], [407] consider two–parameter families.

Table 2.1 Link functions for binary time series.

logit (canonical)	$\beta' \mathbf{Z}_{t-1} = \log\{\pi_t(\beta)/(1 - \pi_t(\beta))\}$
probit	$\beta' \mathbf{Z}_{t-1} = \Phi^{-1}\{\pi_t(\beta)\}$
log-log	$\beta' \mathbf{Z}_{t-1} = -\log\{-\log(\pi_t(\beta))\}$
Complementary log-log	$\beta' \mathbf{Z}_{t-1} = \log\{-\log(1 - \pi_t(\beta))\}$

In the next section we consider statistical inference about β assuming quite generally that the inverse link function is a continuous cdf possessing a density, and in a subsequent chapter we show how the general results are readily reduced in the important special case of logistic regression. The same reduction is achieved when employing other link functions; see Problem 12 regarding probit regression.

For a general (inverse) link, $\pi_t(\beta) \equiv P_{\beta}(Y_t = 1|\mathcal{F}_{t-1}) = F(\beta' \mathbf{Z}_{t-1})$, Problem 4 lists some helpful facts that exemplify certain calculations in the following sections.

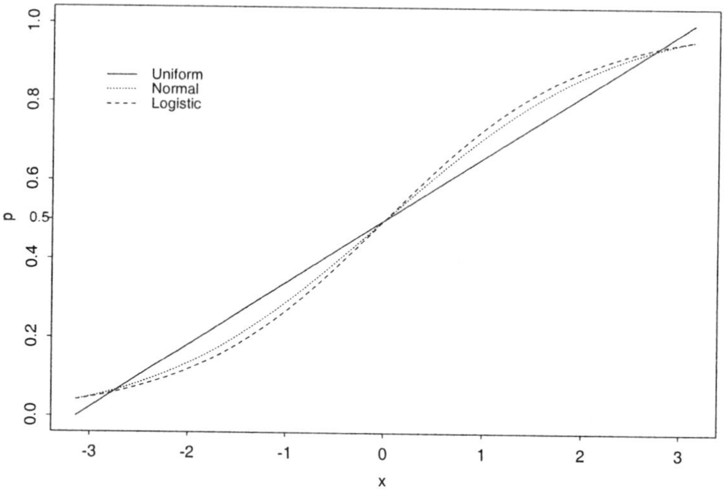

Fig. 2.2 Distribution functions of the uniform, normal, and logistic distributions all having mean 0 and variance $\pi^2/3$.

2.2 PARTIAL LIKELIHOOD ESTIMATION

The results and calculations of this section can be deduced quite readily from the general theory presented in Chapter 1. However, given the peculiarity and special nature of binary time series, it is beneficial on practical and pedagogical grounds to rederive some of the results anew in this important special case. This also serves as an illustration as well as validation to some extent of the general theory.

Let $\{Y_t\}$ be a binary time series with covariate vector series $\{\mathbf{Z}_{t-1}\}, t = 1, ..., N$, and consider the general regression model,

$$\pi_t(\boldsymbol{\beta}) \equiv P_{\boldsymbol{\beta}}(Y_t = 1|\mathcal{F}_{t-1}) = F(\boldsymbol{\beta}'\mathbf{Z}_{t-1}). \tag{2.18}$$

Any suitable inverse link $h \equiv F$ that maps the real line onto the interval $[0, 1]$ will do, but we shall assume without loss of too much generality that F is a differentiable cumulative distribution function (cdf) with probability density function (pdf) $f = F'$. Notice that all the four cases in Table 2.1 are links obtained by inverting cdf's.

In terms of the notation in Chapter 1,

$$\pi_t(\boldsymbol{\beta}) = \mu_t(\boldsymbol{\beta}) = F(\boldsymbol{\beta}'\mathbf{Z}_{t-1}),$$

or,

$$\mu_t = F(\eta_t), \quad \eta_t = \boldsymbol{\beta}'\mathbf{Z}_{t-1}, \quad \frac{\partial \mu_t}{\partial \eta_t} = f(\eta_t),$$

and

$$\sigma_t^2(\beta) = \mathrm{Var}[Y_t \mid \mathcal{F}_{t-1}] = F(\beta'\mathbf{Z}_{t-1})[1 - F(\beta'\mathbf{Z}_{t-1})].$$

We have

$$P_{\beta}(Y_t = y_t|\mathcal{F}_{t-1}) = [\pi_t(\beta)]^{y_t}[1 - \pi_t(\beta)]^{1-y_t}$$

and the partial likelihood of β takes on the simple product form,

$$\begin{aligned}\mathrm{PL}(\beta) &= \prod_{t=1}^{N}[\pi_t(\beta)]^{y_t}[1 - \pi_t(\beta)]^{1-y_t} \\ &= \prod_{t=1}^{N}[F(\beta'\mathbf{Z}_{t-1})]^{y_t}[1 - F(\beta'\mathbf{Z}_{t-1})]^{1-y_t}. \end{aligned} \quad (2.19)$$

In this section we study the maximizer $\hat{\beta}$ of $\mathrm{PL}(\beta)$, the maximum partial likelihood estimator (MPLE) of β, by maximizing with respect to β the log-partial likelihood,

$$\begin{aligned} l(\beta) &= \log \mathrm{PL}(\beta) \\ &= \sum_{t=1}^{N}\{y_t \log F(\beta'\mathbf{z}_{t-1}) + (1 - y_t)\log(1 - F(\beta'\mathbf{z}_{t-1}))\}. \end{aligned} \quad (2.20)$$

Assuming differentiability, when $\hat{\beta}$ exists it can be obtained from an estimating equation referred to as the partial likelihood equation,

$$\nabla \log \mathrm{PL}(\beta) = \mathbf{0}. \quad (2.21)$$

From (2.20), the score vector for binary time series reduces to

$$\mathbf{S}_N(\beta) \equiv \nabla \log \mathrm{PL}(\beta) = \sum_{t=1}^{N} \mathbf{Z}_{t-1} D(\beta'\mathbf{Z}_{t-1})(Y_t - \pi_t(\beta)), \quad (2.22)$$

where

$$D(x) \equiv \frac{f(x)}{F(x)(1 - F(x))}$$

and the partial sums,

$$\mathbf{S}_t(\beta) \equiv \sum_{s=1}^{t} \mathbf{Z}_{s-1} D(\beta'\mathbf{Z}_{s-1})(Y_s - \pi_s(\beta))$$

define the score vector process, $\{\mathbf{S}_t(\beta)\}$, $t = 1, \cdots, N$. Observe that the score process, being the sum of martingale differences, is a martingale with respect to the filtration $\mathcal{F}_0 \subset \mathcal{F}_1 \subset \mathcal{F}_2 \subset \cdots$. That is, $E[\mathbf{S}_t(\beta)|\mathcal{F}_{t-1}] = \mathbf{S}_{t-1}(\beta)$. Clearly, $E[\mathbf{S}_t(\beta)] = \mathbf{0}$.

For binary time series, the cumulative conditional information matrix $\mathbf{G}_N(\beta)$ and the residual matrix $\mathbf{R}_N(\beta)$, reduce to

$$\begin{aligned}\mathbf{G}_N(\beta) &= \sum_{t=1}^{N} \mathbf{Z}_{t-1}\mathbf{Z}'_{t-1} D(\beta'\mathbf{Z}_{t-1}) f(\beta'\mathbf{Z}_{t-1}) \\ &= \sum_{t=1}^{N} \mathbf{Z}_{t-1}\mathbf{Z}'_{t-1} \frac{f^2(\beta'\mathbf{Z}_{t-1})}{F(\beta'\mathbf{Z}_{t-1})(1-F(\beta'\mathbf{Z}_{t-1}))}\end{aligned} \tag{2.23}$$

and

$$\mathbf{R}_N(\beta) = \sum_{t=1}^{N} \mathbf{Z}_{t-1}\mathbf{Z}'_{t-1} W(\beta'\mathbf{Z}_{t-1})(Y_t - \pi_t(\beta)), \tag{2.24}$$

where

$$W(x) = \frac{d}{dx} D(x).$$

Then the *observed information matrix*, $\mathbf{H}_N(\beta) \equiv \nabla\nabla'(-\log \mathrm{PL}(\beta))$, satisfies

$$\mathbf{H}_N(\beta) \equiv \nabla\nabla'(-\log \mathrm{PL}(\beta)) = \mathbf{G}_N(\beta) - \mathbf{R}_N(\beta). \tag{2.25}$$

By Assumption A in Chapter 1, $\mathbf{G}_N(\beta)/N$ has a limit in probability which we can now compute directly as

$$\frac{\mathbf{G}_N(\beta)}{N} \to \mathbf{G}(\beta) = \int_{\mathcal{R}^p} \mathbf{z}\mathbf{z}' \frac{f^2(\beta'\mathbf{z})}{F(\beta'\mathbf{z})(1-F(\beta'\mathbf{z}))} \nu(d\mathbf{z}), \tag{2.26}$$

as $N \to \infty$. Since

$$\frac{\mathbf{R}_N(\beta)}{N} \to \mathbf{0},$$

in probability, as $N \to \infty$, it follows from (2.25) that

$$\frac{\mathbf{H}_N(\beta)}{N} \to \mathbf{G}(\beta), \tag{2.27}$$

in probability, as $N \to \infty$, where $\mathbf{G}(\beta)$ is the information matrix per single observation for estimating β. By Assumption A it is positive definite and hence also nonsingular for every β.

For the sake of clarity and completeness we repeat a classical argument presented already in Chapter 1. Accordingly, by expanding $\mathbf{0} = \nabla \log \mathrm{PL}(\hat{\beta})$ using Taylor series to one term about β, we obtain the useful approximation up to terms asymptotically negligible in probability,

$$\begin{aligned}\sqrt{N}(\hat{\beta} - \beta) &\approx \left(-\frac{1}{N}\nabla\nabla' \log \mathrm{PL}(\beta))^{-1} \frac{1}{\sqrt{N}} \nabla \log \mathrm{PL}(\beta\right) \\ &\approx \mathbf{G}^{-1}(\beta) \frac{1}{\sqrt{N}} \mathbf{S}_N(\beta).\end{aligned} \tag{2.28}$$

Moreover,

$$\frac{1}{\sqrt{N}}\mathbf{S}_N(\beta) \rightarrow \mathcal{N}(\mathbf{0}, \mathbf{G}(\beta)),$$

in distribution, as $N \rightarrow \infty$. Thus, from the large sample theory in Chapter 1 we have the following result.

Theorem 2.2.1 [152], [254], [395]. The MPLE $\hat{\beta}$ is almost surely unique for all sufficiently large N, and as $N \rightarrow \infty$,

(i)

$$\hat{\beta} \rightarrow \beta,$$

in probability,

(ii)

$$\sqrt{N}(\hat{\beta} - \beta) \rightarrow \mathcal{N}_p(\mathbf{0}, \mathbf{G}^{-1}(\beta)),$$

in distribution, and

(iii)

$$\sqrt{N}(\hat{\beta} - \beta) - \frac{1}{\sqrt{N}}\mathbf{G}^{-1}(\beta)\mathbf{S}_N(\beta) \rightarrow \mathbf{0},$$

in probability.

An immediate application of Theorem 2.2.1 is in constructing prediction intervals for $\pi_t(\beta)$ from $\mathbf{Z}_{t-1}$. By the delta method (see [366, p. 388]), (ii) in Theorem 2.2.1 implies that

$$\sqrt{N}(\pi_t(\hat{\beta}) - \pi_t(\beta)) \rightarrow \mathcal{N}(0, \gamma'\mathbf{G}^{-1}(\beta)\gamma)$$

in distribution, as $N \rightarrow \infty$, where

$$\gamma = \nabla \pi_t(\beta) = \nabla F(\beta'\mathbf{Z}_{t-1}) = f(\beta'\mathbf{Z}_{t-1})\mathbf{Z}_{t-1}.$$

Therefore, an asymptotic $100(1-\alpha)\%$ prediction interval is given by

$$\pi_t(\beta) \doteq \pi_t(\hat{\beta}) \pm z_{\alpha/2}\frac{f(\beta'\mathbf{Z}_{t-1})}{\sqrt{N}}\sqrt{\mathbf{Z}'_{t-1}\mathbf{G}^{-1}(\beta)\mathbf{Z}_{t-1}}. \tag{2.29}$$

2.3 INFERENCE FOR LOGISTIC REGRESSION

The results of the previous section can be illustrated quite clearly by treating the important special case of logistic regression (2.12), where $F = F_l$ given in (2.6) is the inverse of the canonical link associated with binary data. In this canonical link case, all the previous calculations are greatly simplified due to the property of the (standard) logistic distribution that for all x,

$$D(x) \equiv \frac{f_l(x)}{F_l(x)(1 - F_l(x))} = 1.$$

For this case $W(x) = 0$ and therefore also $\mathbf{R}_N(\boldsymbol{\beta}) = \mathbf{0}$ as expected, since in general $\mathbf{R}_N(\boldsymbol{\beta})$ vanishes whenever the link is canonical.

Thus, the score vector has the simplified form,

$$\mathbf{S}_N(\boldsymbol{\beta}) \equiv \nabla \log \mathrm{PL}(\boldsymbol{\beta}) = \sum_{t=1}^{N} \mathbf{Z}_{t-1}(Y_t - \pi_t(\boldsymbol{\beta})) \tag{2.30}$$

and the sample information matrix reduces to

$$\begin{aligned}
\mathbf{H}_N(\boldsymbol{\beta}) = \mathbf{G}_N(\boldsymbol{\beta}) &\equiv \nabla\nabla'(-\log \mathrm{PL}(\boldsymbol{\beta})) \\
&= \sum_{t=1}^{N} \mathbf{Z}_{t-1}\mathbf{Z}'_{t-1}\pi_t(\boldsymbol{\beta})(1 - \pi_t(\boldsymbol{\beta})) \\
&= \sum_{t=1}^{N} \mathbf{Z}_{t-1}\mathbf{Z}'_{t-1}\frac{\exp(\boldsymbol{\beta}'\mathbf{Z}_{t-1})}{[1 + \exp(\boldsymbol{\beta}'\mathbf{Z}_{t-1})]^2}.
\end{aligned} \tag{2.31}$$

It is easily seen that $\mathbf{G}_N(\boldsymbol{\beta})$ is the sum of conditional covariance matrices,

$$\mathbf{G}_N(\boldsymbol{\beta}) = \sum_{t=1}^{N} \mathrm{Var}_{\boldsymbol{\beta}}[\mathbf{Z}_{t-1}(Y_t - \pi_t(\boldsymbol{\beta}))|\mathcal{F}_{t-1}]$$

and that the sample information matrix per single observation $\mathbf{G}_N(\boldsymbol{\beta})/N$ converges in probability to a special case of the limit (2.26),

$$\frac{\mathbf{G}_N(\boldsymbol{\beta})}{N} \to \mathbf{G}(\boldsymbol{\beta}) = \int_{\mathcal{R}^p} \frac{e^{\boldsymbol{\beta}'\mathbf{z}}}{(1 + e^{\boldsymbol{\beta}'\mathbf{z}})^2}\mathbf{z}\mathbf{z}'\nu(d\mathbf{z}). \tag{2.32}$$

The next example illustrates the large sample behavior of $\hat{\boldsymbol{\beta}}$ described by Theorem 2.2.1.

2.3.0.1 Example: Incorporation of a Periodic Component Consider a binary time series $\{Y_t\}$, $t = 1, ..., N$, obeying a logistic autoregression model containing a deterministic periodic component,

$$\mathrm{logit}(\pi_t(\boldsymbol{\beta})) = \beta_1 + \beta_2 \cos\left(\frac{2\pi t}{12}\right) + \beta_3 Y_{t-1} \tag{2.33}$$

so that $\mathbf{Z}_{t-1} = (1, \cos(2\pi t/12), Y_{t-1})'$. Two time series from the model and their corresponding success probabilities, $\pi_t(\boldsymbol{\beta})$, are plotted in Figures 2.3 and 2.4.

The process described by (2.33) is not stationary per se, but it is still instructive to examine the sample autocorrelation (acf) of time series from this process. This, however, must be done with caution as the following special cases indicate. In general, when the periodic component is sufficiently dominant the sample acf is likewise periodic. Figure 2.5 depicts the sample acf corresponding to the series from Figures 2.3 and 2.4 in addition to two other cases. The acf suggests, correctly, that the first and

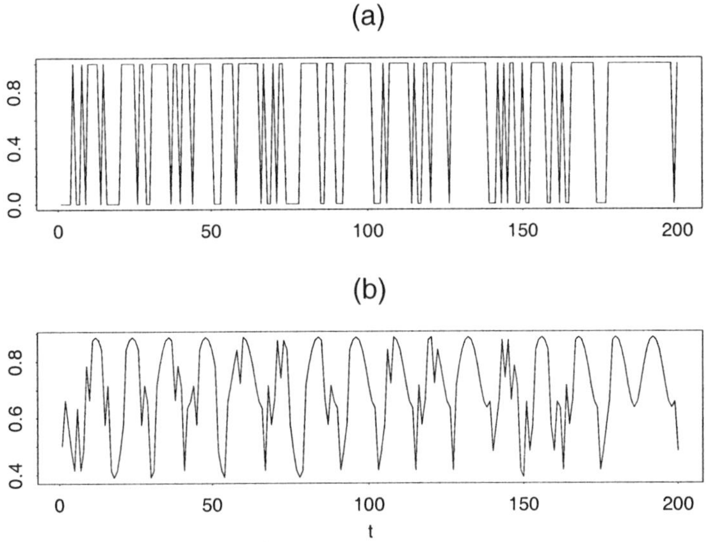

Fig. 2.3 Logistic autoregression with a sinusoidal component. a. Y_t. b. $\pi_t(\boldsymbol{\beta})$ where $\text{logit}(\pi_t(\boldsymbol{\beta})) = 0.3 + 0.75\cos(2\pi t/12) + y_{t-1}$.

fourth cases pertain to a sinusoidal component, and that the third case is dominated by a negative AR component. The second case where the acf displays insignificant correlation when the series is not white noise is quite elusive and can be misleading.

To illustrate the asymptotic normality result (ii) in Theorem 2.2.1, the model was simulated 1000 times for different N. In each run, the partial likelihood estimates of the β_j were obtained by maximizing (2.19). This gives 1000 estimates $\hat{\beta}_j, j = 1, 2, 3$, from which sample means and variances were computed. The theoretical variances of the estimators were approximated by inverting $\mathbf{G}_N(\boldsymbol{\beta})$ in (2.31). The results are summarized in Table 2.2.

In addition, for the case $\boldsymbol{\beta} = (0.3, 0.75, 1)'$, $N = 200$, the histograms–each made of 1000 values–of the normalized $\hat{\beta}_j$ are compared graphically with the probability density of the standard normal distribution in Figure 2.6.

A graphical illustration of the prediction limits (2.29) is given in Figure 2.7 where we can see that $\pi_t(\boldsymbol{\beta})$ is nestled quite comfortably within the prediction limits. Again we inverted (2.31) for the approximation $\mathbf{G}_N^{-1}(\boldsymbol{\beta}) \approx \mathbf{G}^{-1}(\boldsymbol{\beta})/N$.

Evidently, Table 2.2, Figure 2.6, and Figure 2.7, point to a match between the theory and the simulation results.

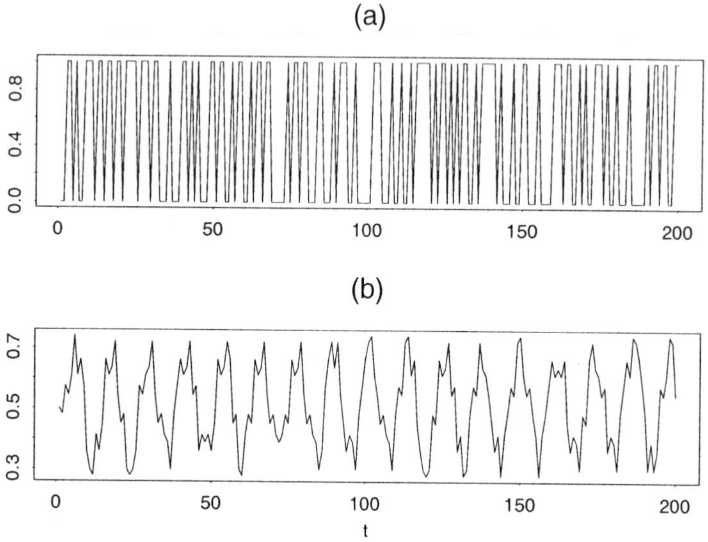

Fig. 2.4 Logistic autoregression with a sinusoidal component. a. Y_t. b. $\pi_t(\beta)$ where $\text{logit}(\pi_t(\beta)) = 0.3 - 0.75\cos(2\pi t/12) - 0.5y_{t-1}$.

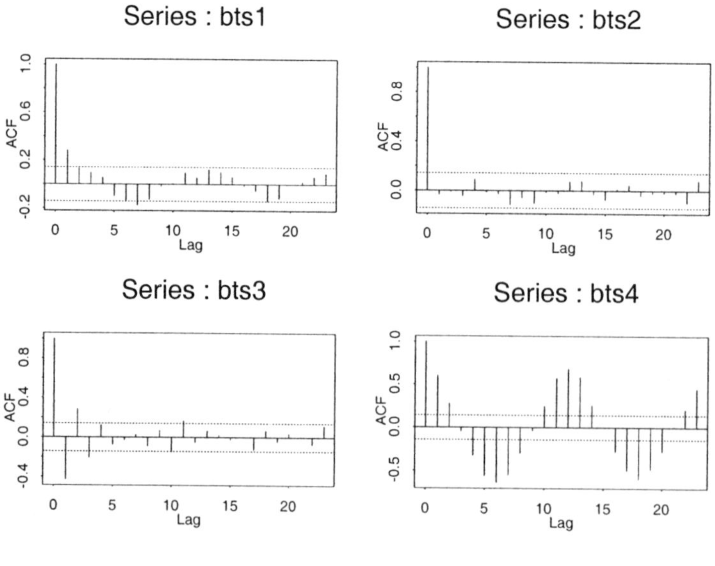

Fig. 2.5 Sample autocorrelation of logistic autoregression with a sinusoidal component. 1. $\beta = (.3, .75, 1)'$. 2. $\beta = (.3, -.75, -0.5)'$. 3. $\beta = (0.5, 0.8, -2)'$. 4. $\beta = (0.5, 5, 1)'$.

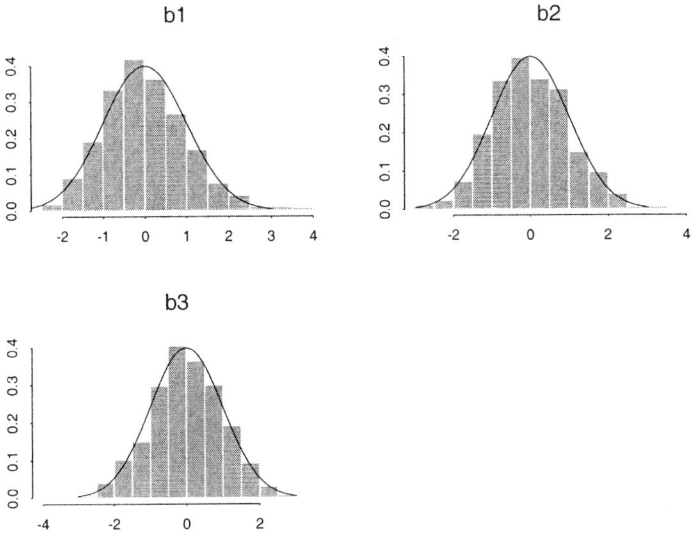

Fig. 2.6 Histograms of normalized MPLE's for parameters from the model (2.33) with $\beta = (0.3, 0.75, 1)'$, $N = 200$. Each histogram consists of 1000 estimates.

Table 2.2 Theoretical versus experimental results for logistic autoregression plus a cosine component. The theoretical and estimated SE's are close. The row-wise respective values of N are 200,500,1000,250,500,200,300,1000. Thus, for the fourth row $N = 250$.

β			$\hat{\beta}$			$\mathbf{G}_N^{-1}$			$\hat{\text{SE}}$		
β_1	β_2	β_3	$\hat{\beta}_1$	$\hat{\beta}_2$	$\hat{\beta}_3$	$\hat{\beta}_1$	$\hat{\beta}_2$	$\hat{\beta}_3$	$\hat{\beta}_1$	$\hat{\beta}_2$	$\hat{\beta}_3$
.3	$-.75$	$-.5$	.33	$-.78$	$-.56$	.22	.22	.30	.23	.23	.30
.3	$-.75$	$-.5$	.31	$-.77$	$-.52$	.14	.14	.19	.14	.14	.20
.3	$-.75$	$-.5$	.31	$-.76$	$-.51$	.10	.10	.14	.10	.10	.14
$-.4$	.80	3.0	$-.38$	.75	2.95	.31	.31	.42	.35	.30	.43
$-.4$	.80	3.0	$-.39$	.79	2.97	.23	.22	.30	.24	.22	.30
.3	.75	1.0	.30	.73	.97	.25	.22	.31	.28	.25	.35
.3	.75	1.0	.31	.73	.97	.23	.20	.28	.23	.20	.29
.3	.75	1.0	.31	.75	.99	.12	.11	.15	.13	.12	.16

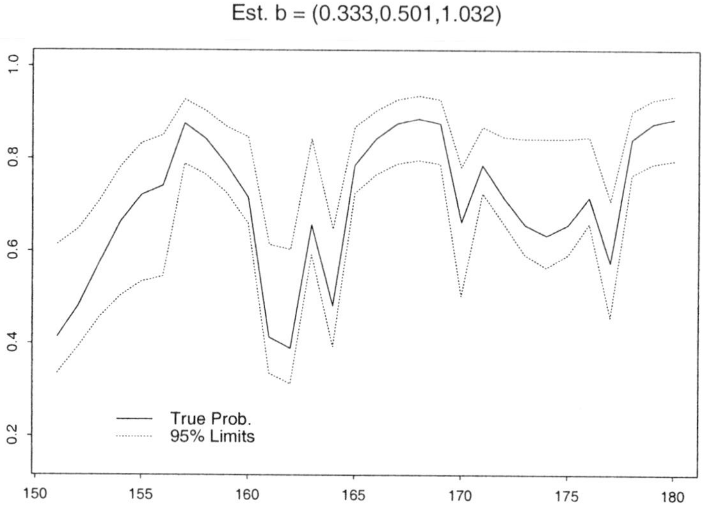

Fig. 2.7 Probability limits (2.29) for $\pi_t(\boldsymbol{\beta}), t = 151, ..., 180$, in (2.33). $\boldsymbol{\beta} = (0.3, 0.75, 1)'$, $\hat{\boldsymbol{\beta}} = (0.333, 0.501, 1.032)'$, $N = 200$.

2.3.1 Asymptotic Relative Efficiency

How efficient are maximum partial likelihood estimates compared with the usual maximum likelihood estimates ? While no general answer to this question is available, an idea can be obtained from special cases. In point of fact, we can give a partial answer in the case of an $AR(p)$ process as discussed in Section 2.1.1.1. In this special case, a fully specified model for $\{Y_t, \mathbf{Z}_{t-1}\}$ is readily available, and a comparison is possible via the information matrices corresponding to partial and full likelihoods [395].

Specifically, set $r = 0, \lambda = 1, \gamma_0 = 0$, and consider the stationary $AR(p)$ process, $X_t = \beta_1 X_{t-1} + \cdots + \beta_p X_{t-p} + \epsilon_t$. Here, $\boldsymbol{\beta} = (\beta_1, \cdots, \beta_p)'$, $Y_t = I_{[X_t \geq 0]}$, $\mathbf{Z}_{t-1} = (X_{t-1}, X_{t-2}, \cdots, X_{t-p})'$, and

$$\epsilon_t = X_t - \boldsymbol{\beta}' \mathbf{Z}_{t-1} \tag{2.34}$$

are i.i.d. logistic random variables with density, $f_l(x) = e^x/(1+e^x)^2$. The variable ϵ_t is independent of $\mathcal{F}_{t-1} = \sigma(\mathbf{Z}_s, s < t)$.

Under the assumption of stationarity, let $\mathbf{Z}$ be distributed as $\mathbf{Z}_{t-1}$. Also let ϵ be distributed as ϵ_t, independently of $\mathbf{Z}$. The notation $\mathrm{E}_{\boldsymbol{\beta}}$ is used to denote mathematical expectation when the true parameter is $\boldsymbol{\beta}$.

Corresponding to (2.12) and (2.19), the partial likelihood information matrix defined in (2.32) is given by

$$\begin{aligned} \mathbf{G}^{PL}(\boldsymbol{\beta}) \equiv \mathbf{G}(\boldsymbol{\beta}) &= \mathrm{E}_{\boldsymbol{\beta}}\left[\frac{e^{\boldsymbol{\beta}'\mathbf{Z}}}{(1+e^{\boldsymbol{\beta}'\mathbf{Z}})^2}\mathbf{Z}\mathbf{Z}'\right] \\ &= \mathrm{E}_{\boldsymbol{\beta}}[f_l(\boldsymbol{\beta}'\mathbf{Z})\mathbf{Z}\mathbf{Z}']. \end{aligned} \tag{2.35}$$

By fixing $\mathbf{Z}_0 = (X_1, \cdots, X_p)'$, the full likelihood $\mathrm{L}(\boldsymbol{\beta})$ based on the observations $X_{p+1}, \cdots, X_N$, is given by

$$\mathrm{L}(\boldsymbol{\beta}) = \prod_{t=p+1}^{N} f_l(\epsilon_t), \tag{2.36}$$

where ϵ_t is given in (2.34). The information matrix is obtained in exactly the same manner as in the partial likelihood case, as the limit of the sample information per observation about $\boldsymbol{\beta}$. This information matrix, equal to the inverse of the asymptotic covariance matrix for the maximum likelihood estimator of $\boldsymbol{\beta}$ when the true parameter value is $\boldsymbol{\beta}$, is given by

$$\begin{aligned} \mathbf{G}^{L}(\boldsymbol{\beta}) \equiv 2\mathrm{E}_{\boldsymbol{\beta}}\left[\frac{e^{\epsilon}}{(1+e^{\epsilon})^2}\mathbf{Z}\mathbf{Z}'\right] &= 2\mathrm{E}_{\boldsymbol{\beta}}[f_l(\epsilon)\mathbf{Z}\mathbf{Z}'] \\ &= \frac{1}{3}\mathrm{E}_{\boldsymbol{\beta}}[\mathbf{Z}\mathbf{Z}'] \end{aligned} \tag{2.37}$$

upon noting that ϵ and $\mathbf{Z}$ are independent and $\int_{-\infty}^{\infty} f_l(x)f_l(x)dx = 1/6$.

Since $f_l(x) \leq 1/4$, it follows immediately from (2.35) and (2.37) that for every vector $\mathbf{b} \in \mathcal{R}^p$ we have

$$\mathbf{b}'\mathbf{G}^{PL}(\boldsymbol{\beta})\mathbf{b} \leq \frac{3}{4}\mathbf{b}'\mathbf{G}^{L}(\boldsymbol{\beta})\mathbf{b}. \tag{2.38}$$

The inequality (2.38) implies that each component β_j from $\boldsymbol{\beta}$ can be estimated with *asymptotic relative efficiency* (ARE) at best $3/4$ via the partial likelihood logistic regression method as compared with a complete maximum likelihood $AR(p)$ analysis (see Problem 10). From $f_l(\boldsymbol{\beta}'\mathbf{Z}) = \pi_t(\boldsymbol{\beta})(1-\pi_t(\boldsymbol{\beta}))$ in (2.35), the worst ARE is obtained for those $\pi_t(\boldsymbol{\beta})$ that are close to 1 or 0, that is, the case where prediction is very good!

A similar result is obtained in the probit regression model (2.17) where it can be checked that the upper bound for the asymptotic relative efficiency is $2/\pi$, and again the ARE is much worse when $\tilde{\pi}_t(\boldsymbol{\beta})$ values fall frequently near 1 and 0 (see Problem 12).

2.4 GOODNESS OF FIT

There are several ways to measure the goodness of fit of regression models for binary time series. We first present a general method based on the (scaled) *deviance*, and then describe an alternative illustrated in terms of logistic regression.

2.4.1 Deviance

One way to measure goodness of fit is by means of the scaled deviance

$$D \equiv 2\{l(\mathbf{y};\mathbf{y}) - l(\hat{\boldsymbol{\mu}};\mathbf{y})\}$$

discussed in Section 1.6.1. Recall that the scaled deviance depends on a scale parameter ϕ, and that to some degree, $D \sim \chi^2_{N-p}$.

Concerning binary time series, there is a complication that hinders the use of the deviance [148], [313, p. 121]. To make the point, consider logistic regression applied to independent 0-1 Bernoulli data $Y_t \sim \mathrm{b}(1, \pi_t)$ with $p \times 1$ covariate vectors $\mathbf{x}_t$. Then, since $y_t \log(1 - y_t) = (1 - y_t)\log(1 - y_t) = 0$, and $\log(\hat{\pi}_t/(1 - \hat{\pi}_t)) = \mathbf{x}'_t\hat{\boldsymbol{\beta}}$,

$$\begin{aligned} D &= 2\sum_{t=1}^{N}\left\{-y_t \log\left(\frac{\hat{\pi}_t}{1-\hat{\pi}_t}\right) - \log(1-\hat{\pi}_t)\right\} \\ &= 2\sum_{t=1}^{N}\{-y_t\mathbf{x}'_t\hat{\boldsymbol{\beta}} - \log(1-\hat{\pi}_t)\}. \end{aligned}$$

As N increases this tends to a degenerate distribution depending on $\boldsymbol{\beta}$ a fact that renders the deviance problematic for goodness of fit.

2.4.2 Goodness of Fit Based on Response Classification

The limited usefulness of the deviance for binary data prompts the search for alternatives where the number of degrees of freedom is fixed. In analogy with the classical Pearson goodness of fit statistic, a useful way to test goodness of fit is to *classify* the binary responses Y_t according to a *finite* number of mutually exclusive events defined in terms of the covariates $\mathbf{Z}_{t-1}$, and then check for each category the deviation of the number of positive responses from its conditional expected value [378]. This is illustrated for the general regression model (2.18).

Let $C_1, \cdots, C_k$ constitute a partition of $\mathcal{R}^p$. For $j = 1, \cdots, k$ define

$$M_j \equiv \sum_{t=1}^{N} I_{[\mathbf{Z}_{t-1} \in C_j]} Y_t$$

and

$$E_j(\boldsymbol{\beta}) \equiv \sum_{t=1}^{N} I_{[\mathbf{Z}_{t-1} \in C_j]} \pi_t(\boldsymbol{\beta}).$$

Put $\mathbf{M} \equiv (M_1, \cdots, M_k)'$, $\mathbf{E}(\boldsymbol{\beta}) \equiv (E_1(\boldsymbol{\beta}), \cdots, E_k(\boldsymbol{\beta}))'$. The goodness of fit can be tested with the help of quadratic forms of the type

$$(\mathbf{M} - \mathbf{E}(\hat{\boldsymbol{\beta}}))'\mathbf{V}(\mathbf{M} - \mathbf{E}(\hat{\boldsymbol{\beta}})),$$

where $\mathbf{V}$ is a suitable $k \times k$ matrix.

For testing the hypothesis that $\beta = \beta_0$, we can use statistics of the form

$$\sum_{j=1}^{k}(M_j - E_j(\beta_0))^2 W_j$$

using appropriate normalizations and weights W. A useful special case is the goodness of fit statistic (2.39) defined below.

The next result follows readily from Theorem 2.2.1, and several applications of the multivariate Martingale Central Limit Theorem (see Section 3.4.1). Assume that the true parameter is β.

Theorem 2.4.1 [254], [395]. Consider the general regression model (2.18) where F is a cdf with density f. Let $C_1, \cdots, C_k$, be a partition of $\mathcal{R}^p$. Then we have as $N \to \infty$.

(i)
$$\sqrt{N}((\mathbf{M} - \mathbf{E}(\beta))'/N, (\hat{\beta} - \beta)')' \to \mathcal{N}_{p+k}(\mathbf{0}, \mathbf{\Sigma}),$$

in distribution, where $\mathbf{\Sigma}$ is a square matrix of dimension $p + k$,

$$\mathbf{\Sigma} = \begin{pmatrix} \mathbf{A} & \mathbf{B}' \\ \mathbf{B} & \mathbf{G}^{-1}(\beta) \end{pmatrix}.$$

Here $\mathbf{A}$ is a diagonal $k \times k$ matrix with the jth diagonal element given by

$$\sigma_j^2 \equiv \int_{C_j} F(\beta'\mathbf{z})(1 - F(\beta'\mathbf{z}))\nu(d\mathbf{z}).$$

The matrix $\mathbf{G}^{-1}(\beta)$ is the limiting $p \times p$ inverse of the information matrix, and the jth column of $\mathbf{B}$ is given by

$$\mathbf{b}_j \equiv \mathbf{G}^{-1}(\beta) \int_{C_j} \mathbf{z} D(\beta'\mathbf{z}) F(\beta'\mathbf{z})(1 - F(\beta'\mathbf{z}))\nu(d\mathbf{z}).$$

(ii) As $N \to \infty$,

$$\frac{(\mathbf{E}(\hat{\beta}) - \mathbf{E}(\beta))}{\sqrt{N}} - \sqrt{N}\mathbf{B}'\mathbf{G}(\beta)(\hat{\beta} - \beta) \to \mathbf{0},$$

in probability.

(iii) As $N \to \infty$, the asymptotic distribution of the statistic

$$\chi^2(\beta) \equiv \frac{1}{N}\sum_{j=1}^{k}(M_j - E_j(\beta))^2/\sigma_j^2 \tag{2.39}$$

is χ_k^2.

Corollary 2.4.1 Under the assumption of logistic regression (2.12), Theorem 2.4.1 holds with

$$\sigma_j^2 = \int_{C_j} \frac{e^{\beta' \mathbf{z}}}{(1+e^{\beta' \mathbf{z}})^2} \nu(d\mathbf{z})$$

and

$$\mathbf{b}_j = \mathbf{G}^{-1}(\beta) \int_{C_j} \frac{e^{\beta' \mathbf{z}}}{(1+e^{\beta' \mathbf{z}})^2} \mathbf{z}\nu(d\mathbf{z}).$$

To verify the structure of $\mathbf{\Sigma}$ see Problem 13. In verifying (2.39) in Theorem 2.4.1 it is helpful to note that $M_j - E_j(\beta)$ is a zero-mean martingale with asymptotic variance $N\sigma_j^2$ and that for $j \neq j'$, $M_j - E_j(\beta)$ and $M_{j'} - E_{j'}(\beta)$ are orthogonal.

2.4.2.1 Model Adequacy: Training Versus Testing Data To assess model adequacy in applications, we often use the goodness of fit statistic $\chi^2(\hat{\beta})$, replacing β by its estimator $\hat{\beta}$ in (2.39). In this case $\chi^2(\hat{\beta})$ is stochastically smaller than $\chi^2(\beta)$ when $\hat{\beta}$ and $(\mathbf{M}-\mathbf{E}(\beta))$ are obtained from the same (training) data, and stochastically larger when $\hat{\beta}$ is obtained from one data set (training data) but $(\mathbf{M} - \mathbf{E}(\beta))$ comes from a different independent data set (testing data). To see this, write

$$\chi^2(\hat{\beta}) = \frac{1}{N}(\mathbf{M} - \mathbf{E}(\hat{\beta}))'\mathbf{A}^{-1}(\mathbf{M} - \mathbf{E}(\hat{\beta})).$$

But for N sufficiently large and a one-term Taylor series expansion,

$$\begin{aligned} \frac{1}{\sqrt{N}}\{\mathbf{M} - \mathbf{E}(\hat{\beta})\} &= \frac{1}{\sqrt{N}}\{(\mathbf{M} - \mathbf{E}(\beta)) - (\mathbf{E}(\hat{\beta}) - \mathbf{E}(\beta))\} \\ &\approx \frac{1}{\sqrt{N}}\{\mathbf{M} - \mathbf{E}(\beta)\} - \sqrt{N}\mathbf{B}'\mathbf{G}(\beta)(\hat{\beta} - \beta) \end{aligned}$$

and therefore (see Problem 14),

$$\mathrm{E}(\chi^2(\hat{\beta})) \approx tr\{(\mathbf{A} - \mathbf{B}'\mathbf{G}(\beta)\mathbf{B})\mathbf{A}^{-1}\} = k - \sum_{j=1}^{k} (\mathbf{B}'\mathbf{G}(\beta)\mathbf{B})_{jj}/\sigma_j^2.$$

On the other hand, when $\hat{\beta}$ and $(\mathbf{M} - \mathbf{E}(\beta))$ are obtained from independent data sets,

$$\mathrm{E}(\chi^2(\hat{\beta})) \approx tr\{(\mathbf{A} + \mathbf{B}'\mathbf{G}(\beta)\mathbf{B})\mathbf{A}^{-1}\} = k + \sum_{j=1}^{k} (\mathbf{B}'\mathbf{G}(\beta)\mathbf{B})_{jj}/\sigma_j^2.$$

Appropriate modification of the degrees of freedom must be made to accommodate the two cases.

2.4.2.2 Example: Illustration of Theorem 2.4.1 To demonstrate the tendency of the goodness of fit statistic (2.39) toward a chi-square distribution with the indicated number of degrees of freedom, consider again the logistic regression model

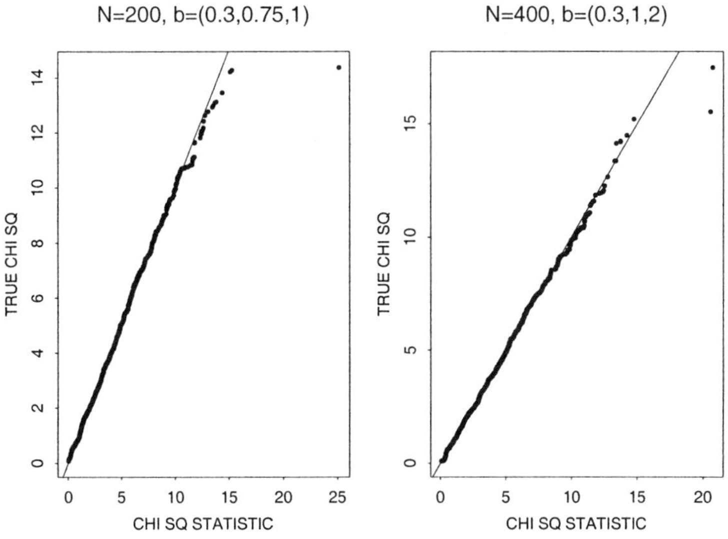

Fig. 2.8 Q-Q plots from 1000 independent goodness of fit statistics (2.39), and 1000 independent χ_4^2 random variables.

(2.33) in Example 2.3.0.1. It is sufficient to partition the set of values of $\mathbf{Z}_{t-1}$ into disjoint sets. Let

$$C_1 = \{\mathbf{Z} : Z_1 = 1, -1 \leq Z_2 < 0, Z_3 = 0\}$$

$$C_2 = \{\mathbf{Z} : Z_1 = 1, -1 \leq Z_2 < 0, Z_3 = 1\}$$

$$C_3 = \{\mathbf{Z} : Z_1 = 1, 0 \leq Z_2 \leq 1, Z_3 = 0\}$$

$$C_4 = \{\mathbf{Z} : Z_1 = 1, 0 \leq Z_2 \leq 1, Z_3 = 1\}.$$

Then, $k = 4$, M_j is the sum of those Y_t's for which $\mathbf{Z}_{t-1}$ is in C_j, $j = 1, 2, 3, 4$, and the $E_j(\boldsymbol{\beta})$ are obtained similarly. In forming (2.39), we replace σ_j^2 by its estimator,

$$\tilde{\sigma}_j^2 = \frac{1}{N} \sum_{t=1}^{N} I_{[\mathbf{Z}_{t-1} \in C_j]} \pi_t(\boldsymbol{\beta})(1 - \pi_t(\boldsymbol{\beta})), \tag{2.40}$$

where $\pi_t(\boldsymbol{\beta})$ is given in (2.12). The Q-Q plots in Figure 2.8 were obtained from 1000 independent time series (2.33) of length $N = 200$ ($\boldsymbol{\beta} = (0.3, 0.75, 1)'$) and $N = 400$ ($\boldsymbol{\beta} = (0.3, 1, 2)'$), and 1000 independent χ_4^2 random variables. Except for a few outliers, the χ_4^2 approximation is quite good.

2.5 REAL DATA EXAMPLES

2.5.1 Rainfall Prediction

Daily precipitation in inches and hundredths at Mount Washington, NH, over a period of 107 days ($N = 107$) from 6.1.1971 to 9.15.1971, is displayed in Figure 2.9 together with the associated binary time series Y_t. Here 1 stands for rain and 0 for no rain. The numerical data are given in Table 2.5.

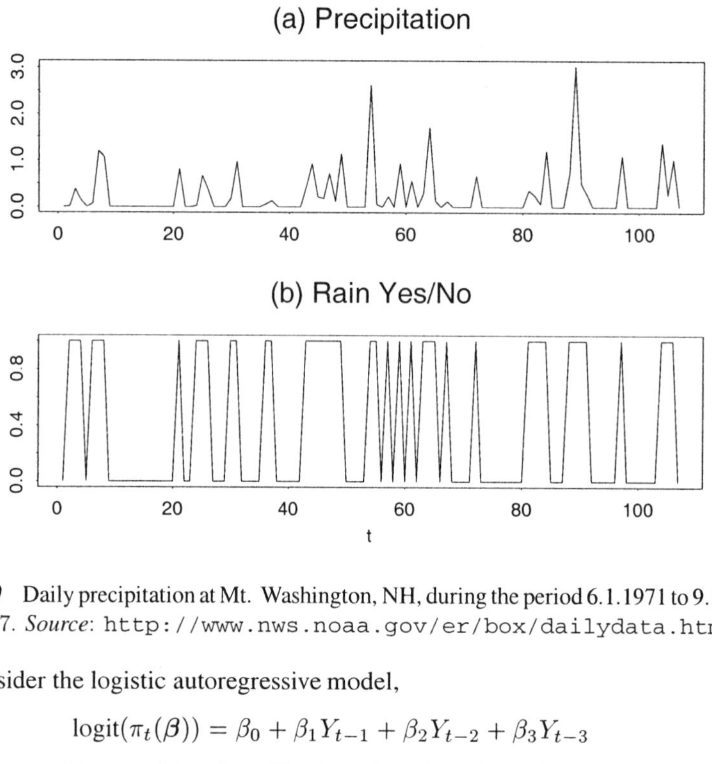

Fig. 2.9 Daily precipitation at Mt. Washington, NH, during the period 6.1.1971 to 9.15.1971. $N = 107$. *Source*: `http://www.nws.noaa.gov/er/box/dailydata.htm`

Consider the logistic autoregressive model,

$$\text{logit}(\pi_t(\boldsymbol{\beta})) = \beta_0 + \beta_1 Y_{t-1} + \beta_2 Y_{t-2} + \beta_3 Y_{t-3} \tag{2.41}$$

with $\mathbf{Z}_{t-1} = (1, Y_{t-1}, Y_{t-2}, Y_{t-3})'$. The value of the MPLE is

$$\hat{\boldsymbol{\beta}} = (-0.737, 1.337, 0.512, -1.176)'$$

and its approximate covariance matrix, obtained by inverting $\mathbf{G}_N$, is given by

$$\mathbf{G}_N^{-1} = \begin{pmatrix} 0.105 & -0.058 & -0.048 & -0.037 \\ -0.058 & 0.218 & -0.052 & -0.046 \\ -0.048 & -0.052 & 0.236 & -0.075 \\ -0.037 & -0.046 & -0.075 & 0.246 \end{pmatrix}.$$

From this, taking the square root along the diagonal, the corresponding standard errors of the $\hat{\beta}$'s are

$$0.324, 0.467, 0.486, 0.495.$$

To test the hypothesis

$$\mathrm{H}_0 : \beta_3 = 0 \text{ against } \mathrm{H}_1 : \beta_3 \neq 0$$

we use the log–partial likelihood ratio statistic (see Section 1.5). Its value is $\lambda_N = 2\left\{l(\hat{\beta}) - l(\tilde{\beta})\right\} = 6.16$ with one degree of freedom. The p-value is 0.013 and the null hypothesis is rejected at all levels greater than 0.013. Likewise, the hypothesis $\mathrm{H}_0 : \beta_1 = 0$ is rejected as $\lambda_N(\beta) = 8.95$ with a single degree of freedom gives a p-value of 0.0028. On the other hand in testing,

$$\mathrm{H}_0 : \beta_2 = 0 \text{ against } \mathrm{H}_1 : \beta_2 \neq 0$$

the log–partial likelihood ratio statistic is relatively small, $\lambda_N(\beta) = 1.13$, again with a single degree of freedom. The p-value is 0.288 and the null hypothesis is not rejected.

We are thus led to entertaining the curtailed autoregressive model

$$\mathrm{logit}(\pi_t(\beta)) = \beta_0 + \beta_1 Y_{t-1} + \beta_3 Y_{t-3} \tag{2.42}$$

with $\mathbf{Z}_{t-1} = (1, Y_{t-1}, Y_{t-3})'$. In this case the MPLE is

$$\hat{\beta} = (-0.638, 1.459, -1.028)'$$

and the corresponding SE's are 0.308,0.452,0.468.

Goodness of fit of (2.42) can be checked by means of the goodness of fit statistic $\chi^2(\hat{\beta})$ replacing β by its estimator $\hat{\beta}$ in (2.39). This also means that σ_j^2 is estimated by

$$\hat{\sigma}_j^2 = \frac{1}{N}\sum_{t=1}^{N} I_{[\mathbf{Z}_{t-1}\in C_j]}\pi_t(\hat{\beta})(1-\pi_t(\hat{\beta})). \tag{2.43}$$

Note that in (2.40) β is known, unlike the case here where the estimator $\hat{\beta}$ is used instead. Let

$$C_1 = \{\mathbf{Z} : Z_1 = 1, Z_2 = 0, Z_3 = 0\}$$
$$C_2 = \{\mathbf{Z} : Z_1 = 1, Z_2 = 0, Z_3 = 1\}$$
$$C_3 = \{\mathbf{Z} : Z_1 = 1, Z_2 = 1, Z_3 = 0\}$$
$$C_4 = \{\mathbf{Z} : Z_1 = 1, Z_2 = 1, Z_3 = 1\}.$$

As in Example 2.4.2.2, $k = 4$, M_j is the sum of those Y_t's for which $\mathbf{Z}_{t-1}$ is in C_j, $j = 1, 2, 3, 4$, and the $E_j(\hat{\beta})$ are obtained similarly. We have $\chi^2(\hat{\beta}) = 0.606$ and this, in light of our earlier discussion, should be compared with critical values from a chi-square distribution with degree of freedom less than 4. For 4 degrees of freedom the p-value is 0.962 while a single degree of freedom gives a p-value of 0.436. The true p-value is most likely between these two numbers, which makes the model difficult to reject based on $\chi^2(\hat{\beta})$.

Table 2.3 Candidate models η_t for the rainfall data

Model 1	$\beta_0 + \beta_1 Y_{t-1}$
Model 2	$\beta_0 + \beta_1 Y_{t-1} + \beta_2 Y_{t-2}$
Model 3	$\beta_0 + \beta_1 Y_{t-1} + \beta_2 Y_{t-2} + \beta_3 Y_{t-3}$
Model 4	$\beta_0 + \beta_1 Y_{t-1} + \beta_3 Y_{t-3}$

Support for model (2.42) is also obtained from the AIC and BIC (see Section 1.6.2) when comparing the models in Table 2.3. From Table 2.4 we see that both model selection criteria are minimized at Model 4 which is the same as (2.42). Recall that WR stands for the sum of squares of the working residuals, MSE is the mean square error from the response residuals,

$$\text{MSE} = \frac{1}{107} \sum_{t=1}^{107} (Y_t - \pi_t(\hat{\beta}))^2,$$

and $\mathcal{X}^2$ is the sum of squares of the Pearson residuals. As before, the AIC and BIC are computed up to an additive constant from $D + 2p$, $D + p \log N$, respectively.

In successful binary prediction $\pi_t(\hat{\beta})$ should be not too far from Y_t, and hence a relatively small MSE is expected. Here, the best model gives an MSE of 0.21. This is relatively large pointing to the difficulty of predicting rainfall occurrence from past occurrences only. The relatively high WR is due to division by the $\hat{\pi}_t(1 - \hat{\pi}_t)$.

2.5.2 Modeling Successive Eruptions

Duration of eruptions of the Old Faithful geyser in Yellowstone National Park, Wyoming, for the period between August 1 to August 15, has been reported in [24]. The data consist of $N = 299$ successive durations recorded either numerically in minutes, or as “L” (long duration), as “M” (medium duration; there were only two M’s) or as “S" (short duration). We have clipped the duration time series to obtain a 0–1 series where 1 stands for durations greater than 3 minutes, or for durations classified as L or M, and using 0 to denote short durations. The binary series, denoted

Table 2.4 Comparison of the models from Table 2.3 using the logistic link for the rainfall data

Model	p	WR	MSE	$\mathcal{X}^2$	D	df	AIC	BIC
1	2	492.01	0.22	107.03	134.85	105	138.85	144.20
2	3	491.43	0.22	106.89	134.70	104	140.70	148.72
3	4	540.39	0.21	106.72	128.54	103	136.54	147.23
4	3	550.06	0.21	107.94	129.67	104	135.67	143.69

Table 2.5 Precipitation at Mt. Washington, NH, 6.1.1971–9.15.1971, in inches and hundredths. *Source*: `http://www.nws.noaa.gov/er/box/dailydata.htm`

t	Y_t	t	Y_t	t	Y_t	t	Y_t	t	Y_t	t	Y_t
1	0.00	19	0.00	37	0.13	55	0.05	73	0.00	91	0.27
2	0.01	20	0.00	38	0.00	56	0.00	74	0.00	92	0.00
3	0.38	21	0.81	39	0.00	57	0.22	75	0.00	93	0.00
4	0.14	22	0.00	40	0.00	58	0.00	76	0.00	94	0.00
5	0.00	23	0.00	41	0.00	59	0.93	77	0.00	95	0.00
6	0.07	24	0.02	42	0.00	60	0.00	78	0.00	96	0.00
7	1.19	25	0.67	43	0.43	61	0.56	79	0.00	97	1.09
8	1.06	26	0.36	44	0.92	62	0.00	80	0.00	98	0.00
9	0.00	27	0.00	45	0.21	63	0.29	81	0.36	99	0.00
10	0.00	28	0.00	46	0.18	64	1.70	82	0.25	100	0.00
11	0.00	29	0.00	47	0.72	65	0.13	83	0.06	101	0.00
12	0.00	30	0.18	48	0.12	66	0.00	84	1.20	102	0.00
13	0.00	31	0.97	49	1.13	67	0.12	85	0.00	103	0.00
14	0.00	32	0.00	50	0.00	68	0.00	86	0.00	104	1.37
15	0.00	33	0.00	51	0.00	69	0.00	87	0.00	105	0.26
16	0.00	34	0.00	52	0.00	70	0.00	88	0.72	106	1.01
17	0.00	35	0.00	53	0.00	71	0.00	89	3.02	107	0.00
18	0.00	36	0.06	54	2.62	72	0.67	90	0.50		

by $\{Y_t\}$, is given in Table 2.6. Interestingly, there are runs of 1's of more than a single symbol but not of 0's, that is, a 0 is always followed by a 1, but a 1 may be followed by either a 0 or a 1 [306].

The binary data were analyzed in [24] and later also in [306] to conclude that a two-state second-order Markov chain fits the data quite well. In fact, out of 7 models considered in [306], the second-order Markov chain had the smallest AIC and BIC values. In light of this finding, it is interesting to see what regression analysis applied to the binary time series brings to the fore.

Table 2.6 Clipped eruption of Old Faithful geyser time series, read across and down. $N = 299$.

101110110101011010110101010111110101010101010101010101011111
010101011010101110111110111010101010101010101010101101010101
011101111111011111011111110101010101011111101010101110101011
010111101010101110101011011011101010101101111111010101111011
01110110101011101010111110111011010111111111101010101010110

We fit regression models to the clipped duration data in Table 2.6 using logistic, probit, and cloglog links with η_t as given in Table 2.7, using only $Y_1, ..., Y_{259}$; the remaining $Y_{260}, ..., Y_{299}$ are used as test data.

Table 2.8 reports the regression results for different η_t using the logistic link. The results using the probit and cloglog links are almost identical to those obtained from logistic regression, except for the working residuals. Corresponding to Models 1,2,3,4, for probit WR=(260.41,271.87,272.36,273.60), and for cloglog

Table 2.7 Candidate models η_t for the Old Faithful data

Model 1	$\beta_0 + \beta_1 Y_{t-1}$
Model 2	$\beta_0 + \beta_1 Y_{t-1} + \beta_2 Y_{t-2}$
Model 3	$\beta_0 + \beta_1 Y_{t-1} + \beta_2 Y_{t-2} + \beta_3 Y_{t-3}$
Model 4	$\beta_0 + \beta_1 Y_{t-1} + \beta_2 Y_{t-2} + \beta_3 Y_{t-3} + \beta_4 Y_{t-4}$

Table 2.8 Comparison of models of Table 2.7 for the Old Faithful data with the logistic link. Similar results are obtained for the probit and cloglog links.

Model	p	WR	MSE	$\mathcal{X}^2$	D	df	AIC	BIC
1	2	665.50	0.16	165.00	227.38	253	231.38	238.46
2	3	717.50	0.15	165.00	215.53	252	221.53	232.15
3	4	719.75	0.15	165.00	215.08	251	223.08	237.24
4	5	725.60	0.15	164.97	213.99	250	223.99	241.69

WR=(374.25,407.71,408.33,411.28). Regardless of the link, both the AIC and BIC are minimized at Model 2, the second-order model, in agreement with [24] and [306].

The estimated logistic regression model is

$$\hat{\pi}_t = \pi_t(\hat{\boldsymbol{\beta}}) = \frac{1}{1 + \exp\left\{-(\hat{\beta}_0 + \hat{\beta}_1 Y_{t-1} + \hat{\beta}_2 Y_{t-2})\right\}}, \tag{2.44}$$

where $\hat{\boldsymbol{\beta}} = (\hat{\beta}_0, \hat{\beta}_1, \hat{\beta}_2)' = (9.10, -9.80, 1.10)'$ with corresponding–quite large–estimated standard errors $(10.51, 10.50, 0.32)$. Testing the hypothesis that $\beta_2 = 0$ using the scaled deviance from Models 1 and 2, gives $D_0 - D_1 = 227.38 - 215.53 = 11.85$ with a single degree of freedom, and hence a p-value of 0.00058. Thus β_2 is significant, supporting the choice of the second-order model.

Similarly, the estimated probit regression model suggested by the AIC and BIC is of second order,

$$\hat{\pi}_t = \pi_t(\hat{\boldsymbol{\beta}}) = \Phi(\hat{\beta}_0 + \hat{\beta}_1 Y_{t-1} + \hat{\beta}_2 Y_{t-2}), \tag{2.45}$$

with $\hat{\boldsymbol{\beta}} = (\hat{\beta}_0, \hat{\beta}_1, \hat{\beta}_2)' = (3.16, -3.59, 0.68)'$, and corresponding estimated standard errors $(2.12, 2.12, 0.20)'$. Also from the AIC and BIC, the estimated complementary log-log model is also of second order,

$$\hat{\pi}_t = \pi_t(\hat{\boldsymbol{\beta}}) = 1 - \exp(-\exp(\hat{\beta}_0 + \hat{\beta}_1 Y_{t-1} + \hat{\beta}_2 Y_{t-2})) \tag{2.46}$$

with $\hat{\boldsymbol{\beta}} = (\hat{\beta}_0, \hat{\beta}_1, \hat{\beta}_2)' = (1.55, -2.45, 0.82)'$, and corresponding estimated standard errors $(1.34, 1.33, 0.24)'$. Again, the deviance values are nearly identical for the three models, meaning that they are practically indistinguishable on the basis of the AIC

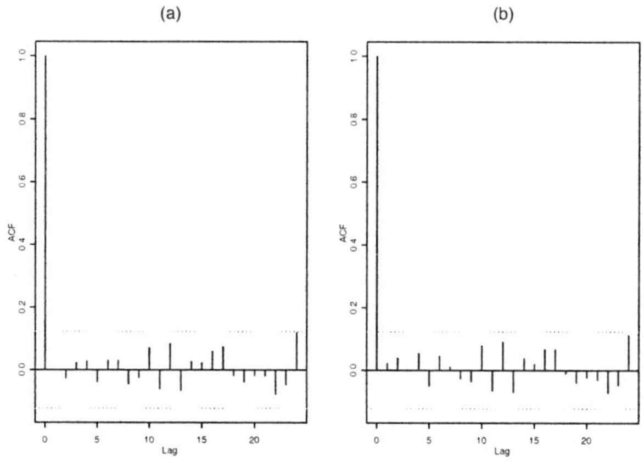

Fig. 2.10 Residual analysis for the logistic model (2.44) obtained from the clipped duration data. (a) Sample autocorrelation function of the raw residuals. (b) Sample autocorrelation function of the deviance residuals.

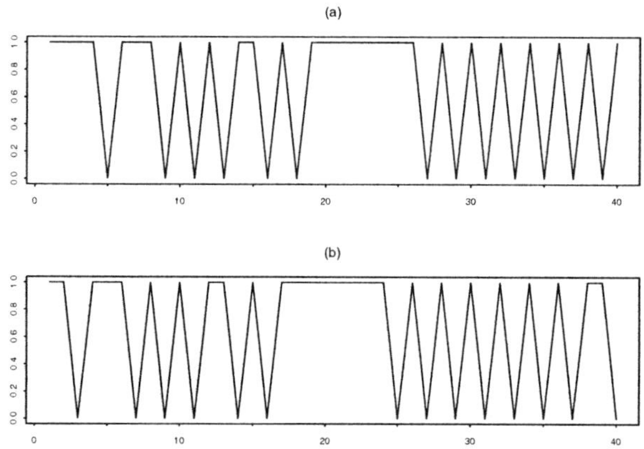

Fig. 2.11 Application of the logistic model (2.44) in prediction of the last 40 observations of the clipped Old Faithful geyser data. (a) Predicted. (b) Observed.

and BIC criteria, and that the above test concerning β_2 also holds for the probit and complementary log-log models.

Further support for a second-order model is derived from the sample autocorrelation function (acf) of the raw and deviance residuals (see Section 1.6.3) depicted in Figure 2.10 and corresponding to (2.44). Both autocorrelation functions display very similar behavior, after the first few lags, one that is consistent with white noise. Similar figures are obtained for the other two links.

Finally, Figure 2.11 shows the predicted and observed values of the test clipped duration data for $t = 260, 261, \ldots, 299$, where prediction at time t from past values is accomplished by logistic regression and the ad hoc rule,

$$\hat{Y}_t = \begin{cases} 1, & \text{if } \hat{\pi}_t \geq 0.5 \\ 0, & \text{if } \hat{\pi}_t < 0.5 \end{cases}. \tag{2.47}$$

Despite of clear resemblance between the upper and lower panels of Figure 2.11, the misclassification rate is approximately 22.5%, mainly due to the problematic prediction at change points where the series shifts from 0 to 1 and vice versa.

2.5.3 Stock Price Prediction

Figure 2.12 shows 100 daily closing prices of Intel, Microsoft, and General Electric stock from 7/2/1999 to 11/22/1999. During this period, for the most part, the three series appear to be fairly stable displaying moderate volatility and therefore seem quite "predictable". Let Z_t represent the Intel series and define a binary time series by clipping at level 79.111,

$$Y_t = \begin{cases} 1, & \text{if } Z_t \geq 79.111 \\ 0, & \text{otherwise.} \end{cases}$$

In running logistic regression of Y_t on past values of stock price, the log–partial likelihood ratio test reveals that past values (lag 1) of Microsoft and General Electric stock are not significant. Similarly, when the model is

$$\text{logit}(\pi_t(\boldsymbol{\beta})) = \beta_0 + \beta_1 Z_{t-1} + \beta_2 Z_{t-2} \tag{2.48}$$

the test

$$\text{H}_0 : \beta_2 = 0 \ \text{ against } \ \text{H}_1 : \beta_2 \neq 0$$

gives a p-value of 0.761. This leaves us with the first-order model

$$\text{logit}(\pi_t(\boldsymbol{\beta})) = \beta_0 + \beta_1 Z_{t-1} \tag{2.49}$$

in agreement with the observation that stock prices tend to behave as a random walk: the best forecast of tomorrow's price is today's [61, p. 169].

For model (2.49), we obtain that $\hat{\beta}_0 = -61.75$, and $\hat{\beta}_1 = 0.78$ with corresponding standard errors 13.8 and 0.17, respectively. The lower (L) and upper (U) 95% prediction limits (2.29) for the last 25 days corresponding to the period 10/19/1999–11/22/1999 are given in Table 2.9. Except for a single case marked with "(*)", the prediction intervals are consistent with the Y–observations. That is, except for day 87, when $Y_t = 0$ the corresponding prediction interval contains values smaller than 0.5, and when $Y_t = 1$ the corresponding prediction interval contains values greater than 0.5. The prediction is decisive in 18 cases out of 25 when the upper limit is smaller than 0.5 for $Y_t = 0$ and when the lower limit is above 0.5 for $Y_t = 1$. The

Table 2.9 Prediction intervals for the probability that Intel stock exceeds 79.111 during 25 days in 10/19/1999–11/22/1999. Limits below 0 and above 1 were rounded to 0 and 1, respectively.

Day	L	U	Y
76	0.000	0.002	0
77	0.000	0.000	0
78	0.000	0.003	0
79	0.000	0.010	0
80	0.000	0.032	0
81	0.000	0.007	0
82	0.000	0.008	0
83	0.000	0.002	0
84	0.000	0.014	0
85	0.045	0.343	0
86	0.000	0.160	0
87	0.050	0.353	(*)1
88	0.388	0.797	1
89	0.707	1.000	1
90	0.810	1.000	1
91	0.781	1.000	1
92	0.447	0.853	0
93	0.250	0.645	1
94	0.329	0.736	0
95	0.000	0.179	0
96	0.000	0.049	0
97	0.000	0.220	0
98	0.000	0.075	0
99	0.169	0.540	1
100	0.411	0.820	1

same information is given in Figure 2.13. Interestingly, from the table and the figure, the prediction becomes decisive after some "learning period" for relatively long stretches of 0's or 1's. At the beginning of a run of like symbols the prediction is poor but it improves as we get further into the run.

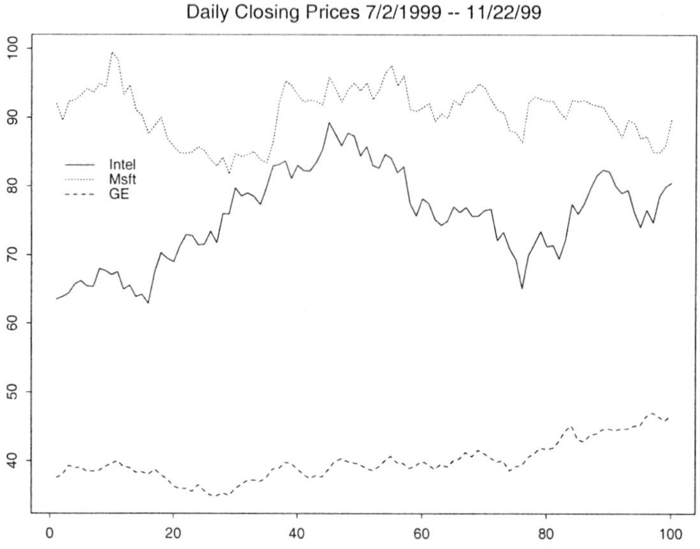

Fig. 2.12 Daily closing prices for Intel, Microsoft, and General Electric stock, 7/2/1999 to 11/22/1999. $N = 100$.

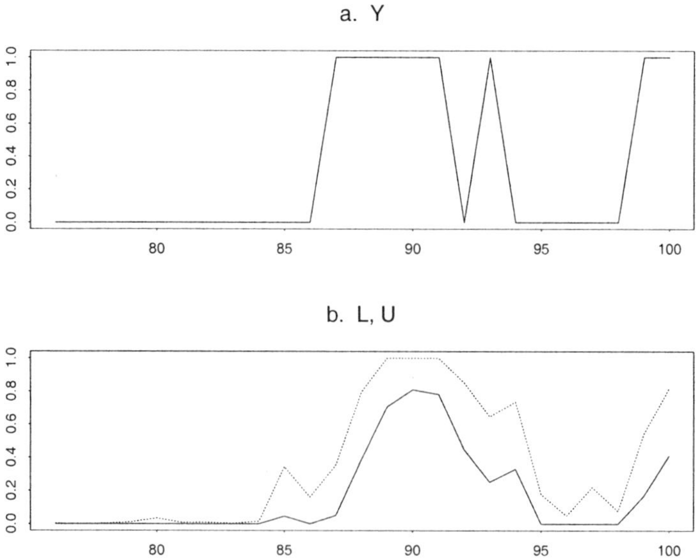

Fig. 2.13 a. Clipped Intel daily closing price, $Y_t \equiv I_{[Z_t \geq 79.111]}$, 10/19/1999 to 11/22/1999. b. Corresponding lower and upper prediction limits.

2.5.4 Modeling Sleep Data

The last data example in this section illustrates the regression methodology for binary time series in the prediction of a future sleep state based on covariate information such as heart rate and temperature. The data consist of clipped measurements of sleep state of a newborn infant recorded as,

$$Y_t = \begin{cases} 1, & \text{if awake} \\ 0, & \text{otherwise} \end{cases}$$

together with heart rate R_t and temperature T_t. The total number of observations equals 1024 and we remove the first two observations in what follows due to lagging.

Table 2.10 Comparison of models for the sleep data with the cloglog link. $N = 700$

Model	η_t	p	D	AIC	BIC
1	$1+Y_{t-1}$	2	85.24	89.24	98.34
2	$1+Y_{t-1} + \log R_t$	3	82.90	88.90	102.56
3	$1+Y_{t-1} + \log R_t + T_t$	4	81.84	89.84	108.05
4	$1+Y_{t-1} + T_t$	3	85.23	91.23	104.89
5	$1+Y_{t-1} + Y_{t-2} + \log R_t$	4	82.72	90.72	108.92
6	$1+Y_{t-1} + \log R_{t-1}$	3	85.23	91.23	104.88
7	$1+\log R_t$	2	654.01	658.01	667.11

Table 2.10 reports the results of fitting several regression models to the binary data using the complementary log–log link function (recall 2.16) employing the first 700 observations, whereas the remaining 322 observations are used in testing the model. The last row of Table 2.10 compared with the second row shows clearly that the lagged value of the response Y_{t-1} is a significant predictor of the sleep state. Notice that the AIC is minimized for Model 2 while the BIC ranks this model as second best among the considered models in Table 2.10. A formal application of the likelihood ratio test when comparing Model 1 to Model 2 shows that $\lambda_N = 85.24 - 82.90 = 2.34$ with p-value equals to 0.1260, which raises uncertainty as to whether or not $\log R_t$ should be included in the model. Aiming for a more prudent model we choose Model 2 as the best possible candidate and use it in the prediction of sleep state. For this particular model we have that

$$\pi_t(\beta) = 1 - \exp\left(-\exp(\beta_0 + \beta_1 Y_{t-1} + \beta_2 \log R_t)\right)$$

with $\hat{\beta}_0 = -13.94$, $\hat{\beta}_1 = 6.12$, $\hat{\beta}_2 = 1.87$ and corresponding standard errors of 6.25, 0.51 and 1.28, respectively. Further support in favor of Model 2 is gained from Figures 2.14 and 2.15. Figure 2.14 displays cumulative periodogram plots ([357]) for both raw and Pearson residuals together with 95% confidence limits. In both cases the residuals are clearly white. Figure 2.15 shows time series plots of both observed and predicted values for the testing data set. The predicted values are obtained by the ad hoc rule (2.47) which gives a misclassification rate of 0.01.

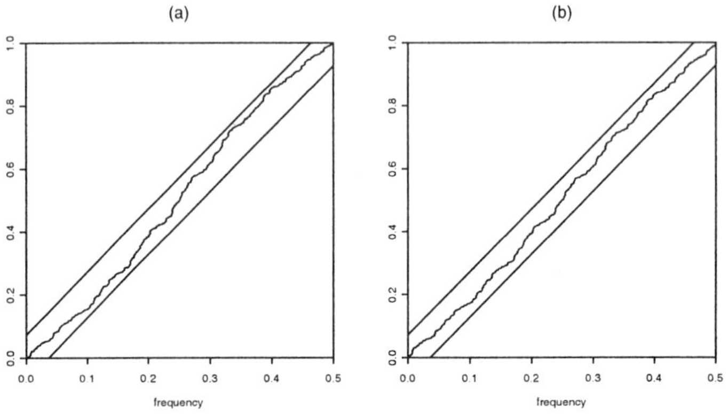

Fig. 2.14 Cumulative periodogram plots of (a) raw and (b) Pearson residuals for Model 2 of Table 2.10 applied to the sleep data with the complementary log–log link.

In Section 3.5.3 we show that Model 2 can be also applied successfully in categorical prediction using an alternative link function, and also provide a more complete description of the sleep data.

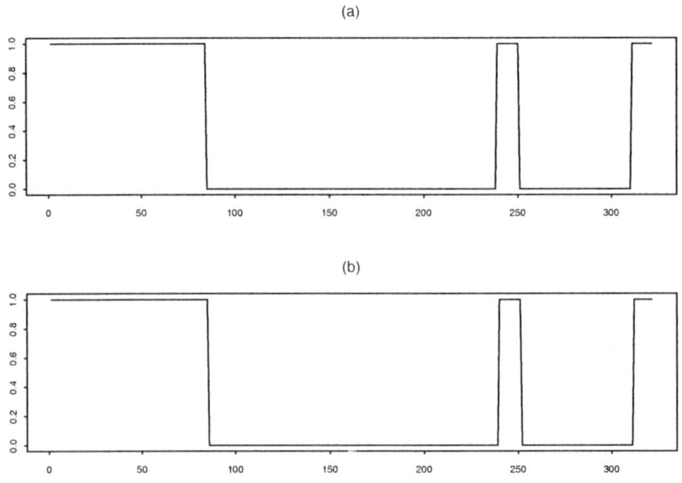

Fig. 2.15 (a) Observed versus (b) predicted sleep state for Model 2 of Table 2.10 applied to the testing sleep data set with the complementary log–log link . $N = 322$.

2.6 PROBLEMS AND COMPLEMENTS

1. *Connection between the logistic link and the exponential family* [248]. Let Y be a binary 0-1 random variable, and $\mathbf{Z}$ a vector of covariates. Assume that the log ratio of the conditional densities $f(\mathbf{z}|Y=1)$ and $f(\mathbf{z}|Y=0)$ is linear,

$$\log\left\{\frac{f(\mathbf{z}|Y=1)}{f(\mathbf{z}|Y=0)}\right\} = \alpha_0 + \boldsymbol{\beta}'\mathbf{h}(\mathbf{z}). \tag{2.50}$$

Here $\mathbf{h}$ is a vector valued function of $\mathbf{z}$. For example $\mathbf{h}(\mathbf{z}) = \mathbf{z}$.

(a) Show that the logistic model holds for some β_0,

$$\log\left\{\frac{f(Y=1|\mathbf{z})}{f(Y=0|\mathbf{z})}\right\} = \beta_0 + \boldsymbol{\beta}'\mathbf{h}(\mathbf{z}).$$

(b) Suppose $\mathbf{Z} = Z$ is a univariate random variable distributed as $\mathrm{N}(\mu_j, \sigma_j^2)$ when $Y = j$, $j = 0, 1$. For $\sigma_0^2 \neq \sigma_1^2$, verify (2.50) and show that $\mathbf{h}(z) = (z, z^2)$. Show that when $\sigma_0^2 = \sigma_1^2$, then $\mathbf{h}(z) = z$.

(c) In relation to (b), show that in the corresponding lognormal case, $\sigma_0^2 \neq \sigma_1^2$ implies $\mathbf{h}(z) = (\log z, \log^2 z)$, while $\sigma_0^2 = \sigma_1^2$ gives $\mathbf{h}(z) = \log z$.

(d) More generally, show that (2.50) holds when Z given Y has a distribution that is a member of the k-parameter exponential family with density,

$$f(z;\boldsymbol{\theta}) = B(\boldsymbol{\theta})h(z)\exp\left\{\sum_{i=1}^{k} R_i(z)Q_i(\boldsymbol{\theta})\right\}, \quad \boldsymbol{\theta} = (\theta_1, ..., \theta_k)',$$

where $\mathbf{h}$ is defined in terms of the $R_i(z)$.

2. *Characterization of the logistic model* [239], [313, p. 113], [355]. Let Y be a binary 0-1 random variable, $\mathbf{x}$ the corresponding random covariate vector, and assume

$$P(Y=1|\mathbf{x}) = \frac{\exp(\alpha + \boldsymbol{\beta}'\mathbf{x})}{1 + \exp(\alpha + \boldsymbol{\beta}'\mathbf{x})}.$$

In *prospective sampling* $\mathbf{x}$ is controlled; we fix $\mathbf{x}$ and then observe Y. The prospective likelihood–assuming independence–is then written in terms of $P(Y_j = 1|\mathbf{x}_j)$. *Retrospective sampling* is used when controlling $\mathbf{x}$ is not feasible or impractical. By this scheme, a pair $(Y, \mathbf{x})$ is chosen from a list of similar pairs (such as a hospital record) where the selection depends on Y only but not on $\mathbf{x}$. More precisely, let Z be equal to 1 when $(Y, \mathbf{x})$ is selected and 0 otherwise. Define

$$P(Z=1|Y=i, \mathbf{x}) = \pi_i, \quad i = 0, 1,$$

where π_0, π_1 do not depend on $\mathbf{x}$. The retrospective likelihood–assuming independence–is then expressed in terms of $P(Y_j = 1|Z_j = 1, \mathbf{x}_j)$.

(a) Use Bayes theorem to show that there exists an α^* such that

$$P(Y = 1|Z = 1, \mathbf{x}) = \frac{\exp(\alpha^* + \boldsymbol{\beta}'\mathbf{x})}{1 + \exp(\alpha^* + \boldsymbol{\beta}'\mathbf{x})}.$$

(b) Conclude that despite retrospective sampling, the logistic model still holds with the same slope $\boldsymbol{\beta}$ but a different intercept α^*, and provide an interpretation for this fact regarding the estimation of the slope and intercept.

(c) Argue that this property is not shared by the probit and complementary log-log models.

3. Consider two instruments I_0 and I_1 measuring the same quantity Z. Let Y be a binary random variable taking the values 0 or 1. When $Y = 0$, Z is an I_0-measurement, and when $Y = 1$, Z is an I_1-measurement. Let $g_0(z)$ be the pdf of Z when $Y = 0$, and $g_1(z)$ the pdf of Z when $Y = 1$. Assuming the logistic model,

$$P(Y = 1|z) = F_l(\alpha^* + \beta h(z))$$

for some real valued function $h(z)$, show that there is an α such that the following exponential tilt holds [155], [156],

$$g_1(z) = \exp(\alpha + \beta h(z)) g_0(z).$$

4. For a general (inverse) link,

$$\pi_t(\boldsymbol{\beta}) \equiv P_{\boldsymbol{\beta}}(Y_t = 1|\mathcal{F}_{t-1}) = F(\boldsymbol{\beta}'\mathbf{Z}_{t-1})$$

show the following facts. Notice that E really means $\mathrm{E}_{\boldsymbol{\beta}}$, an expectation when the parameter value is $\boldsymbol{\beta}$.

(a) Prediction:

$$\pi_t(\boldsymbol{\beta}) = \mathrm{E}[Y_t|\mathcal{F}_{t-1}].$$

(b) Martingale difference property:

$$\mathrm{E}[\mathbf{Z}_{t-1}(Y_t - \pi_t(\boldsymbol{\beta}))|\mathcal{F}_{t-1}] = \mathbf{0}.$$

Hence, $\mathrm{E}[\mathbf{Z}_{t-1}(Y_t - \pi_t(\boldsymbol{\beta}))] = \mathbf{0}$.

(c) Orthogonality: For $s < t$,

$$\mathrm{E}[\mathbf{Z}_{s-1}\mathbf{Z}'_{t-1}(Y_s - \pi_s(\boldsymbol{\beta}))(Y_t - \pi_t(\boldsymbol{\beta}))|\mathcal{F}_{t-1}] = \mathbf{0}.$$

Hence, for $s \neq t$,

$$\mathrm{E}[\mathbf{Z}_{s-1}\mathbf{Z}'_{t-1}(Y_s - \pi_s(\boldsymbol{\beta}))(Y_t - \pi_t(\boldsymbol{\beta}))] = \mathbf{0}.$$

(d) Getting the factor $\pi_s(\beta)(1-\pi_s(\beta))$ under the integral:

$$\mathrm{E}[\mathbf{Z}_{s-1}\mathbf{Z}'_{s-1}(Y_s-\pi_s(\beta))^2] = \mathrm{E}[\mathbf{Z}_{s-1}\mathbf{Z}'_{s-1}\pi_s(\beta)(1-\pi_s(\beta))].$$

(e) Under the logistic regression model (2.12),

$$\nabla\pi_t(\beta) = \mathbf{Z}_{t-1}\pi_t(\beta)(1-\pi_t(\beta)).$$

5. *Binary autoregressive models* [220]. Define a stationary Binary AR(p) by the two-state, 0–1, pth order Markov chain $\{Y_t\}$, $t = 0, \pm 1, \pm 2, ...$, assuming the identity link,

$$P(Y_t = 1|Y_{t-1}, Y_{t-2}, ...) = \lambda + \phi_1 Y_{t-1} + \cdots + \phi_p Y_{t-p}.$$

Discuss restrictions on the parameters that ensure the transition probabilities are between 0 and 1 (e.g. $\lambda \in (0,1)$), and then show the following.

(a) $\mu \equiv P(Y_t = 1) = \lambda/(1-\phi_1-\cdots-\phi_p)$

(b) $\mathrm{E}(Y_t|Y_{t-1}, Y_{t-2}, ...) = \lambda + \phi_1 Y_{t-1} + \cdots + \phi_p Y_{t-p}$

(c) $\gamma(k) = \phi_1\gamma(k-1) + \cdots + \phi_p\gamma(k-p)$, $k = 1, 2, ...$, where $\gamma(k) = \mathrm{Cov}(Y_t, Y_{t-k})$. *Hint*: Use (b) and properties of conditional expectation.

(d) $\gamma(0) = \mu(1-\mu)$

(e) For $p = 1$, $\mathrm{Corr}(Y_t, Y_{t-k}) = \phi_1^k$, $k = 0, 1, 2, ...$

6. *Binary AR(1) with a random coefficient* [316]. Referring to Problem 5 with $p = 1$, consider an alternative formulation leading to a stationary two-state, 0–1, Markov chain $\{Y_n\}$ constructed as follows. Let $\{A_n\}$, $\{B_n\}$, be independent sequences of independent and identically distributed Bernoulli random variables such that $P(A_n = 1) = \alpha$, $P(B_n = 1) = \beta \equiv (1-\alpha)\theta/(1-\theta)$. Define

$$Y_n = (A_n - B_n)Y_{n-1} + B_n, \quad n = 0, \pm 1, ...$$

Verify the following.

(a) $P(Y_n = 1) = \theta = 1 - P(Y_n = 0)$, $P(Y_n = 1|Y_{n-1} = 0) = \beta$, $P(Y_n = 1|Y_{n-1} = 1) = \alpha$.

(b) The autocorrelation function of $\{Y_n\}$ is

$$\rho(k) = \left(\frac{\alpha-\theta}{1-\theta}\right)^k, \quad k = 0, 1, 2, ...$$

7. *Additional binary AR(1) processes with random coefficients.* Let $\{Z_n\}$, $\{U_n\}$, $\{V_n\}$ be independent sequences of i.i.d. Bernoulli random variables such that $P(Z_n = 1) = p$, $P(U_n = 1) = \theta(1-p)/(1-2p\theta)$, and $P(V_n = 1) = \theta$.

   (a) Let $\oplus$ denote addition mod 2. Define a binary AR(1) time series by [244], [316],
   $$Y_n = Z_n(Y_{n-1} \oplus U_n) + (1 - Z_n)U_n.$$
   Show that the autocorrelation function of $\{Y_n\}$ is $\rho(k) = [(1-2\theta)p]^k$, $k = 0, 1, 2, ...$

   (b) A binary discrete mixed autoregressive moving average (DARMA) time series can be defined by [221], [316],
   $$Y_n = Z_n Y_{n-1} + (1 - Z_n)V_n.$$
   Show that the autocorrelation function of $\{Y_n\}$ is $\rho(k) = p^k$, $k = 0, 1, 2, ...$

8. Consider logistic regression. By fixing the values of the covariates, show that $\log \text{PL}(\beta)$ is strictly concave in β (a $d \times 1$ vector). Conclude that if the MPLE $\hat{\beta}$ exists it is unique. (*Hint*: A function $f : \mathcal{R}^d \to \mathcal{R}$ with continuous second-order partial derivatives is strictly concave if $\nabla\nabla' f$ is negative definite.)

9. *Independent binary time series [395].* Suppose there are multiple *independent* binary realizations Y_t^j and corresponding vectors of covariates $\mathbf{Z}_{t-1}^j$, $j = 1, \cdots, m$. This occurs in *longitudinal* studies when there is a short time series from each of many individuals. Define
   $$\pi_t^i(\beta) \equiv P_\beta(Y_t^i = 1|\mathcal{F}_{t-1}) = \frac{1}{1 + \exp[-\beta' \mathbf{Z}_{t-1}^i]}$$
   where $\mathcal{F}_{t-1}$ is generated by all the Y_s^i, $\mathbf{Z}_s^i$, $0 \leq s \leq t-1$.

   (a) Show that the partial likelihood is given by
   $$\text{PL}(\beta) = \prod_{t=1}^{N} \prod_{i=1}^{m} [\pi_t^i(\beta)]^{y_t^i} [1 - \pi_t^i(\beta)]^{1-y_t^i}.$$

   (b) Show that the score vector process given by
   $$\mathbf{S}_t(\beta) = \sum_{s=1}^{t} \sum_{i=1}^{m} \mathbf{Z}_{s-1}^i (Y_s^i - \pi_s^i(\beta)),$$
   $t = 1, 2, \cdots, N$, is a martingale, and that
   $$\mathbf{S}_N(\beta) = \nabla \log \text{PL}(\beta).$$

(c) Show that

$$\begin{aligned} \mathbf{G}_N(\boldsymbol{\beta}) &\equiv \nabla\nabla'(-\log \mathrm{PL}(\boldsymbol{\beta})) \\ &= \sum_{s=1}^{N} \mathrm{Var}_{\boldsymbol{\beta}} \left[\sum_{i=1}^{m} \mathbf{Z}_{s-1}^{i}(Y_s^i - \pi_s^i(\boldsymbol{\beta}))|\mathcal{F}_{s-1} \right]. \end{aligned}$$

(d) Argue that under appropriate conditions the development in the present chapter (for $m = 1$) can be extended straightforwardly to the case of multiple independent binary time series.

10. Supply a statistical proof to the following fact [100]. Let $\mathbf{A}, \mathbf{B}$ be real symmetric positive definite matrices, and write $\mathbf{A} \geq \mathbf{B}$ if $\mathbf{A} - \mathbf{B}$ is positive semidefinite. Then,

$$\mathbf{A} \geq \mathbf{B} \Rightarrow \mathbf{A}^{-1} \leq \mathbf{B}^{-1}.$$

Hint [240]: Create independent random vectors $\mathbf{Y} \sim \mathrm{N}(\boldsymbol{\Theta}, \mathbf{B})$, and noise $\mathbf{Z} \sim \mathrm{N}(\mathbf{0}, \mathbf{A} - \mathbf{B})$. Then $\mathbf{X} \equiv \mathbf{Y} + \mathbf{Z} \sim \mathrm{N}(\boldsymbol{\Theta}, \mathbf{A})$, and writing $\mathbf{I}_{\mathbf{x}}(\boldsymbol{\Theta})$ for Fisher information of $\mathbf{X}$,

$$\mathbf{A}^{-1} = \mathbf{I}_{\mathbf{x}}(\boldsymbol{\Theta}) \leq \mathbf{I}_{\mathbf{y},\mathbf{z}}(\boldsymbol{\Theta}) = \mathbf{I}_{\mathbf{y}}(\boldsymbol{\Theta}) + \mathbf{0} = \mathbf{B}^{-1}.$$

11. *Another proof of the Cramér–Rao inequality* [238]. Suppose that $\mathbf{Y}$ is an n–dimensional random variable such that

$$\mathrm{E}[\mathbf{Y}] = \mathbf{P}\boldsymbol{\theta}$$

and positive definite covariance matrix $\boldsymbol{\Sigma}$ given by

$$\boldsymbol{\Sigma} = \mathrm{E}\left[(\mathbf{Y} - \mathbf{P}\boldsymbol{\theta})(\mathbf{Y} - \mathbf{P}\boldsymbol{\theta})'\right],$$

where $\mathbf{P}$ is an $n \times p$ non–stochastic matrix and $\boldsymbol{\theta}$ a $p \times 1$ vector of parameters. Employing the same notation as in Problem 10 and assuming regularity conditions, show that

$$\mathbf{I}_{\mathbf{Y}} \geq \mathbf{P}'\boldsymbol{\Sigma}^{-1}\mathbf{P}. \tag{2.51}$$

Use (2.51) and the monotonicity of the Fisher information to supply another proof of the Cramér–Rao inequality.

12. *Asymptotic relative efficiency* [395]. Consider the stationary $AR(p)$ process, $X_t = \beta_1 X_{t-1} + \cdots \beta_p X_{t-p} + \epsilon_t$, and let $Y_t = I_{[X_t \geq 0]}$, $\mathbf{Z}_{t-1} = (X_{t-1}, X_{t-2}, \cdots, X_{t-p})'$. Assume that the ϵ_t are i.i.d. standard normal random variables with distribution function Φ. Define the probit model

$$\tilde{p}_t(\boldsymbol{\beta}) \equiv P_{\boldsymbol{\beta}}(Y_t = 1|\mathcal{F}_{t-1}) = \Phi(\boldsymbol{\beta}'\mathbf{Z}_{t-1}).$$

(a) Follow the discussion in Section 2.3.1 to show that

$$\mathbf{G}^L(\boldsymbol{\beta}) = \mathrm{E}_{\boldsymbol{\beta}}[\mathbf{Z}\mathbf{Z}']$$

and

$$\mathbf{G}^{PL}(\beta) = \mathrm{E}_{\beta}\left[\mathbf{Z}\mathbf{Z}'\frac{\phi^2(\beta'\mathbf{Z})}{\Phi(\beta'\mathbf{Z})(1-\Phi(\beta'\mathbf{Z}))}\right].$$

(b) Conclude that the upper bound for asymptotic relative efficiency (ARE) as described in Section 2.3.1 is $2/\pi$.

13. Show that the elements of $\mathbf{A}$ and the jth column of $\mathbf{B}$ are as given in Theorem 2.4.1. *Hint*: Use Assumption A together with dominated convergence and see Problem 4. For $\mathbf{B}$, use (2.22) and the approximation

$$\hat{\beta} - \beta \approx \mathbf{G}^{-1}(\beta)\mathbf{S}_N(\beta)/N.$$

14. *Expected value of quadratic forms.* Recall that the trace, denoted by tr, of a square matrix is the sum of the diagonal elements. Let $\mathbf{Y}$ be an $n \times 1$ random vector with mean vector $\mathbf{m}$ and variance-covariance matrix $\mathbf{\Sigma_y}$, and $\mathbf{Q}$ an $n \times n$ real symmetric matrix. Show that

$$\mathrm{E}[\mathbf{Y}'\mathbf{Q}\mathbf{Y}] = \mathbf{m}'\mathbf{Q}\mathbf{m} + tr(\mathbf{\Sigma_y}\mathbf{Q}).$$

Hint: $\mathbf{Y}'\mathbf{Q}\mathbf{Y}$ is scalar, so that E and tr commute, and use the fact that the trace is invariant under cyclical permutations, $tr(\mathbf{AB}) = tr(\mathbf{BA})$.

15. Use (2.7) to show that the joint distribution of any binary random vector $\mathbf{Y} = (Y_1, ..., Y_N)'$ is given by

$$p(\mathbf{y}) = \exp\{a_0 + \sum_i a_i y_i + \sum_{i<j} a_{ij} y_i y_j + \cdots + a_{12\ldots N} y_1 y_2 \cdots y_N)\}.$$

16. *Sum of Bernoulli trials with dependence* [251], [309]. Suppose $\{Y_t\}_{t=1}^N$ is a stationary Markov 0–1 sequence with parameters

$$P(Y_t = 1) = p, \quad P(Y_t = 1|Y_{t-1} = 1) = \lambda_1.$$

Let $\pi(k, N) = P(S_N = k)$, where

$$S_N = Y_1 + \cdots + Y_N.$$

Put $p_{11} = \lambda_1, p_{01} = 1 - \lambda_1, p_{00} = (1 - 2p + \lambda_1 p)/(1-p), p_{10} = 1 - p_{00}$. Define

$$\alpha(k, N) = P(S_N = k, Y_N = 0)$$
$$\beta(k, N) = P(S_N = k, Y_N = 1).$$

Then, given p, λ_1, and N, $\pi(k, N) = \alpha(k, N) + \beta(k, N)$ can be obtained recursively as follows. First compute

$$\alpha(0,1) = 1 - p, \quad \beta(0,1) = 0, \quad \alpha(1,1) = 0, \quad \beta(1,1) = p.$$

Next, for $N \geq 2$

$$\alpha(0,N) = (1-p)p_{00}^{N-1}, \quad \beta(0,N) = 0$$

and for $k = 1, ..., N-1$

$$\begin{aligned}\alpha(k,N) &= p_{00}\alpha(k,N-1) + p_{01}\beta(k,N-1)\\ \beta(k,N) &= p_{10}\alpha(k-1,N-1) + p_{11}\beta(k-1,N-1).\end{aligned}$$

When $k = N$

$$\alpha(N,N) = 0, \quad \beta(N,N) = pp_{11}^{N-1}.$$

a. Show that $\max(0, (2p-1)/p) \leq \lambda_1 \leq 1$.

b. Show how to extend the algorithm to 2nd and higher order Markov dependence. *Hint*: For second-order, $\pi(k,N)$ is a sum of 2^2 terms.

17. *Probabilities in terms of moments* [251]. Let $Y_1, ..., Y_N$ be any binary time series.

    (a) Show that

    $$P(Y_1 = y_1, ..., Y_n = y_N) = \mathrm{E}\left[\prod_{j=1}^{N}(-1)^{y_j+1}(Y_j + y_j - 1)\right]$$

    (b) Use (a) to show that $P(Y_1 = y_1, ..., Y_n = y_N)$ is a linear combination of probabilities of the form $P(Y_{t_1} = 1, ..., Y_{t_k} = 1)$, $(t_1, ..., t_k) \subset \{1, ..., N\}$, where the coefficients are $+1$ or -1.

18. *Normal orthant probabilities* [251], [338]. Let $X_1, ..., X_n$ be a sequence of normal random variables with mean zero and unit variance. Assume the X_i are equicorrelated: $\mathrm{Corr}(X_i, X_j) = \tau$, $i \neq j$, and $|\tau| < 1$. Define

$$Y_i = \begin{cases} 1 & \text{if} \quad X_i \geq 0 \\ 0 & \text{if} \quad X_i < 0 \end{cases}$$

$i = 1, ..., n$. Then $P(Y_1 = 1, ..., Y_n = 1)$ is given by the formula

$$p_n(\tau) = \frac{1}{2^n}\sum_{k=0}^{[n/2]}\frac{A_n^{2k}}{\pi^k}F_{n,k}(\tau), \tag{2.52}$$

where $[n/2]$ is the integer part of $n/2$, $A_n^{2k} = n!/(n-2k)!$, and

$$F_{n,0}(\tau) = 1, \quad n \geq 1$$

$$F_{n,1}(\tau) = \sin^{-1}\tau, \quad n \geq 2$$

$$F_{n,k}(\tau) = \int_0^\tau F_{n-2,k-1}\left(\frac{x}{1+2x}\right)d\sin^{-1}x, \quad 2 < 2k \leq n.$$

Verify formula (2.52) for $1 \leq n \leq 5$.

3

Regression Models for Categorical Time Series

Figure 3.1 displays the first 300 records of EEG sleep state scores, typical of newborn infants, classified or quantized in four categories as follows.

1 : quiet sleep,
2 : indeterminate sleep,
3 : active sleep,
4 : awake.

Here the sleep state categories or levels are assigned integer values. This is an example of a *categorical time series* $\{Y_t\}$, $t = 1, ..., N$, taking the values $1, ..., 4$.

The EEG plot raises several questions. Is there an apparent "periodic" tendency in the data? What is the best way to predict a future sleep state? Do lagged values of sleep state determine future states? These questions and others show that "ordinary" and categorical time series pose the same basic problems. The present chapter offers some answers to these problems by considering regression models for categorical time series.

The reader should recognize that Markov chains provide a simple but important example of categorical time series where lagged values of the response are instrumental in determining its future states. Recall that a process $\{Y_t\}$, $t = 1, \ldots, N$, defined on $\{1, 2, \ldots, m\}$, is called Markov chain of order p if it satisfies

$$\mathrm{P}\left[Y_t = k \mid Y_{t-1}, Y_{t-2}, \ldots\right] = \mathrm{P}\left[Y_t = k \mid Y_{t-1}, Y_{t-2}, \ldots, Y_{t-p}\right], \quad k = 1, \ldots, m. \tag{3.1}$$

Thus given the past values of $Y_{t-1}, \ldots, Y_{t-p}$, (3.1) provides the conditional probabilities of a future state or category k. Markov modeling in the context of categorical time series can be problematic for two reasons. First, as the order of the Markov chain increases so does the number of free parameters. Indeed, for a Markov chain of order

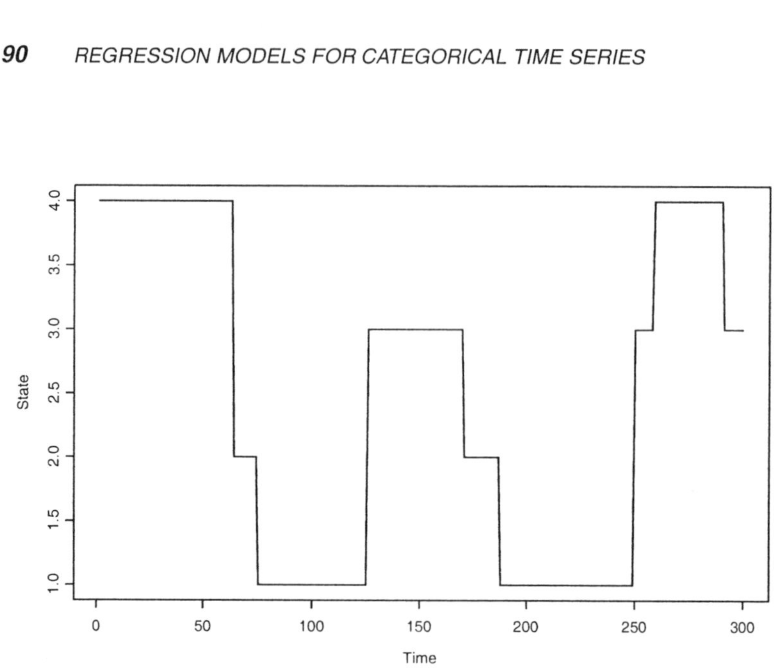

Fig. 3.1 Time Series Plot of the Sleep Data.

p, the total number of parameters needed to be estimated is equal to $m^p(m-1)$. Second, the insistence on using the Markov property requires the specification of the joint dynamics of the response and the covariates. This however may not be always possible. The regression theory for categorical time series resolves both of these issues by *parsimonious* modeling and *incorporation* of random time dependent covariates.

The topic of Markov chains has been studied by many authors (see, e.g., [245], [324]). Problem 2 lists some elementary properties of the first order-Markov chains. The texts [29, Ch. 4], [50], and [188, Ch. 2] present statistical inference theory for Markov chains (see also Problem 4).

3.1 MODELING

Assume that we observe a categorical time series $\{Y_t\}$, $t = 1, \ldots, N$, and let m be the number of categories. In other words, for each t, the possible values of Y_t are $1, 2, \ldots, m-1, m$, where the "first" category is assigned the integer value of 1, the "second" category is assigned the integer value of 2, and so on. In general, the assignment of integer values to the categories is a matter of convenience and hence it is not unique. To clarify this point, suppose we observe the daily choice of transportation for an individual who may travel by either a car, bus, or by train. Then, "car", "bus", and "train" are the observed means of travel, or categories, and a possible assignment of integer values is through the following *arbitrary* correspondence:

$$\begin{array}{lll} \text{car} & \longleftrightarrow & 1 \\ \text{bus} & \longleftrightarrow & 2 \\ \text{train} & \longleftrightarrow & 3. \end{array}$$

However an alternative recording might be

$$\begin{array}{lll} \text{bus} & \longleftrightarrow & 1 \\ \text{train} & \longleftrightarrow & 2 \\ \text{car} & \longleftrightarrow & 3, \end{array}$$

and so on. Apparently we run the risk of multiple interpretations as different assignments may yield different results.

To reduce the amount of arbitrariness incurred by integer assignment to categories, it is helpful to note that the t'th observation of any categorical time series–regardless of the measurement scale–can be expressed by the vector $\mathbf{Y}_t = (Y_{t1}, \ldots, Y_{tq})'$ of length $q = m - 1$, with elements

$$Y_{tj} = \begin{cases} 1, & \text{if the } j\text{th category is observed at time } t \\ 0, & \text{otherwise} \end{cases} \tag{3.2}$$

for $t = 1, \ldots, N$ and $j = 1, \ldots, q$. With this convention, the assignment

$$\begin{array}{lll} \text{bicycle} & \longleftrightarrow & 1 \\ \text{car} & \longleftrightarrow & 2 \\ \text{bus} & \longleftrightarrow & 3 \\ \text{train} & \longleftrightarrow & 4 \end{array}$$

implies that $m = 4$, $q = 3$, and

$$\begin{aligned} \mathbf{Y}_t &= (1,0,0)' \\ \mathbf{Y}_t &= (0,1,0)' \\ \mathbf{Y}_t &= (0,0,1)' \\ \mathbf{Y}_t &= (0,0,0)' \end{aligned}$$

indicate "bicycle", "car", "bus", and (none of the above) "train", respectively. Observe that also here the assignment of the components of $\mathbf{Y}_t$ to the various categories is arbitrary, but this is mitigated by the fact that no matter which component of $\mathbf{Y}_t$ represents, say, "bus", the 0-1 plot of that particular component is always the same.

Denote by $\boldsymbol{\pi}_t = (\pi_{t1}, \ldots \pi_{tq})'$ the vector of conditional probabilities given $\mathcal{F}_{t-1}$ where

$$\pi_{tj} = \mathrm{E}[Y_{tj} \mid \mathcal{F}_{t-1}] = \mathrm{P}(Y_{tj} = 1 \mid \mathcal{F}_{t-1}), \quad j = 1, \ldots, q$$

for every $t = 1, \ldots, N$. At times we refer to the π_{tj} as "transition probabilities". As before, the σ-field $\mathcal{F}_{t-1}$ stands for the whole information up to and including time t. Define

$$Y_{tm} = 1 - \sum_{j=1}^{q} Y_{tj}$$

and

$$\pi_{tm} = 1 - \sum_{j=1}^{q} \pi_{tj}.$$

In addition, put $\{\mathbf{Z}_{t-1}\}$, $t = 1, \ldots, N$, for a $p \times q$ *matrix* that represents a covariate process. In other words, each response Y_{tj} corresponds to a vector of length p of random time dependent covariates which forms the j´th column of $\mathbf{Z}_{t-1}$. The covariate matrix may consist of lagged values of the response process and of any other auxiliary process, as discussed earlier.

Following the theory of generalized linear models, we assume that the vector of transition probabilities–that is the conditional expectation of the response vector–is linked to the covariate process through the equation

$$\boldsymbol{\pi}_t(\boldsymbol{\beta}) = \mathbf{h}(\mathbf{Z}'_{t-1}\boldsymbol{\beta}), \tag{3.3}$$

with $\boldsymbol{\beta}$ a p-dimensional vector of time invariant parameters. Equation (3.3) gives the general form of the *multivariate* generalized linear model for categorical time series,

$$\boldsymbol{\pi}_t(\boldsymbol{\beta}) = \begin{pmatrix} \pi_{t1}(\boldsymbol{\beta}) \\ \pi_{t2}(\boldsymbol{\beta}) \\ \cdots \\ \pi_{tq}(\boldsymbol{\beta}) \end{pmatrix} = \begin{pmatrix} h_1(\mathbf{Z}'_{t-1}\boldsymbol{\beta}) \\ h_2(\mathbf{Z}'_{t-1}\boldsymbol{\beta}) \\ \cdots \\ h_q(\mathbf{Z}'_{t-1}\boldsymbol{\beta}) \end{pmatrix} = \mathbf{h}(\mathbf{Z}'_{t-1}\boldsymbol{\beta}),$$

where the *inverse* link function $\mathbf{h}$ is defined on R^q and takes values in R^q as well. To guarantee that the transition probabilities fall between 0 and 1 we impose the condition that $\mathbf{h}$ maps a subset $H \subseteq R^q$ one-to-one onto $\{(w_1, \ldots, w_q)' : \ w_j > 0, \ j = 1, \ldots, q, \ \sum_{j=1}^{q} w_j < 1\}$. Model (3.3) has been considered by a number of authors including [140], [152], [246], and [359] where the past probability vector $\boldsymbol{\pi}_{t-1}$ is included as a covariate. Some further work can be found in [70], [154]. In addition, recent work by [71] develops a wavelet based method for the analysis of categorical time series.

The important special case of binary time series is obtained when $m = 2$. Then $q = 1$, the covariate matrix reduces to a p-dimensional vector$\{\mathbf{Z}_{t-1}\}$, and equation (3.3) reduces to

$$\pi_t(\boldsymbol{\beta}) = \mathrm{P}[Y_t = 1 | \mathcal{F}_{t-1}] = h(\boldsymbol{\beta}'\mathbf{Z}_{t-1}),$$

with h a monotone function, for example, a cdf from R onto $(0, 1)$. Models for binary time series have been considered in Chapter 2.

3.2 LINK FUNCTIONS FOR CATEGORICAL TIME SERIES

We now turn to the problem of what constitutes a reasonable choice for the inverse link function $\mathbf{h}$ in the context of regression models for categorical time series. We introduce some widely used models for the analysis of categorical time series including the so called *multinomial logit* and *cumulative odds* models. In general, the choice of model depends on one of three measurement scales: *nominal*, *ordinal*, and *interval*. We shall only examine nominal and ordinal time series since interval (e.g. quantized) time series can be handled by methods designed for ordinal data.

3.2.1 Models for Nominal Time Series

By nominal categorical variables we mean variables whose scale of measurement lacks any natural ordering. Daily choice of transportation, as discussed earlier, is an example of a nominal time series. The multinomial logit model defined by the equations

$$\pi_{tj}(\beta) = \frac{\exp(\beta_j' \mathbf{z}_{t-1})}{1 + \sum_{l=1}^{q} \exp(\beta_l' \mathbf{z}_{t-1})}, \qquad j = 1, \ldots, q \tag{3.4}$$

is frequently employed in the analysis of nominal valued time series [2, Ch. 9.2]. Here $\beta_j, j = 1, \ldots, q$ are d-dimensional regression parameters and $\mathbf{z}_{t-1}$ is the corresponding d-dimensional vector of stochastic time dependent covariates independent of j. Obviously

$$\pi_{tm}(\beta) = \frac{1}{1 + \sum_{l=1}^{q} \exp(\beta_l' \mathbf{z}_{t-1})}.$$

Typical examples of $\mathbf{z}_{t-1}$ include

$$\mathbf{z}_{t-1} = (1, W_{t-1}, Y_{t-1}, \log(W_{t-1}))'$$

or when interactions are entertained,

$$\mathbf{z}_{t-1} = (1, W_{t-1}, Y_{t-1}, Y_{t-1} W_{t-1})'$$

and so on, given an auxiliary process $\{W_t\}$.

The multinomial logits model (3.4) is derived either by a straightforward extension of the logistic model or by maximizing a random utility. The first approach defines log–odds ratios relative to π_{tm},

$$\log \frac{\pi_{tj}}{\pi_{tm}} = \beta_j' \mathbf{z}_{t-1}, \quad j = 1, \ldots, q.$$

Then (3.4) follows from the fact that $\sum_{j=1}^{m} \pi_{tj} = 1$. The second line of argument uses the maximization of a random utility function as described in [315] and outlined in Problem 5.

An essential observation is that (3.4) implies

$$\log \frac{\pi_{tj}}{\pi_{ti}} = (\beta_j' - \beta_i') \mathbf{z}_{t-1}.$$

Thus the ratio π_{tj}/π_{ti} for the j´th and i´th categories is the same regardless of the total number of categories m. This property is referred to as *independence of irrelevant alternatives* [305].

Model (3.4) is a special case of (3.3). Indeed, define β to be the qd-vector

$$\beta = (\beta_1', \ldots, \beta_q')',$$

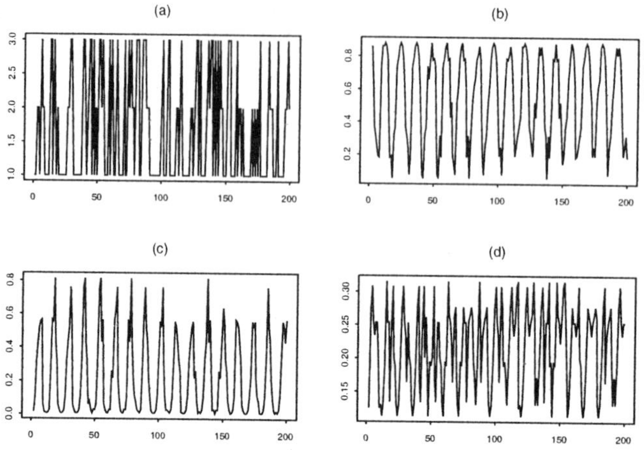

Fig. 3.2 Typical realization of the multinomial logit model (3.4) with 3 categories. Here $\boldsymbol{\beta}_1 = (0.30, 1.25, 0.50, 1.00)'$, $\boldsymbol{\beta}_2 = (-0.20, -2.00, -.75, -1.00)'$, $\mathbf{z}_{t-1} = (1, \cos(2\pi t/12), \mathbf{Y}_{t-1})$ and $N = 200$. (a) Y_t. (b) π_{t1}. (c) π_{t2}. (d) π_{t3}.

and $\mathbf{Z}_{t-1}$ the $qd \times q$ matrix

$$\mathbf{Z}_{t-1} = \begin{bmatrix} \mathbf{z}_{t-1} & \mathbf{0} & \cdots & \mathbf{0} \\ \mathbf{0} & \mathbf{z}_{t-1} & \cdots & \mathbf{0} \\ \vdots & \vdots & \ddots & \vdots \\ \mathbf{0} & \mathbf{0} & \cdots & \mathbf{z}_{t-1} \end{bmatrix}.$$

Let $\mathbf{h}$ stand for the vector valued function whose components h_j, $j = 1, \ldots, q$, are given by

$$\pi_{tj}(\boldsymbol{\beta}) = h_j(\boldsymbol{\eta}_t) = \frac{\exp(\eta_{tj})}{1 + \sum_{l=1}^{q} \exp(\eta_{tl})}, \quad j = 1, \ldots, q$$

with

$$\boldsymbol{\eta}_t = (\eta_{t1}, \ldots, \eta_{tq})' = \mathbf{Z}'_{t-1}\boldsymbol{\beta}.$$

With this notation, equation (3.3) reduces to (3.4) when $p = qd$.

3.2.1.1 Example: Multinomial Logit Model with a Periodic Component

It is instructive to examine closely a simple example by means of simulated data. Figure 3.2 (a) displays a typical realization of a categorical time series with $m = 3$ categories and length $N = 200$. Since $m = 3$, $\mathbf{Y}_t$ has $q = 2$ components: $\mathbf{Y}_t = (Y_{t1}, Y_{t2})'$. The data were generated according to the model

$$\begin{aligned} \log\left(\frac{\pi_{t1}}{\pi_{t3}}\right) &= \boldsymbol{\beta}'_1 \mathbf{z}_{t-1} \\ &= \beta_{10} + \beta_{11} \cos(2\pi t/12) \end{aligned}$$

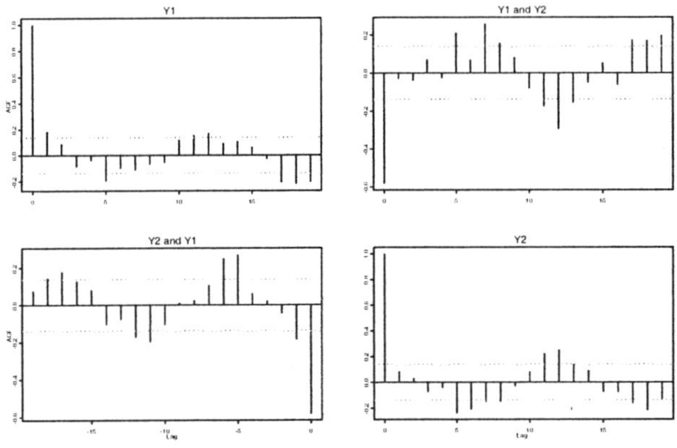

Fig. 3.3 Sample autocorrelation and cross–correlation functions of the simulated data from Figure 3.2.

$$
\begin{aligned}
& + \quad \beta_{12} Y_{(t-1)1} + \beta_{13} Y_{(t-1)2}, \\
\log\left(\frac{\pi_{t2}}{\pi_{t3}}\right) &= \boldsymbol{\beta}_2' \mathbf{z}_{t-1} \\
&= \beta_{20} + \beta_{21}\cos(2\pi t/12) \\
& + \quad \beta_{22} Y_{(t-1)1} + \beta_{23} Y_{(t-1)2} \qquad (3.5)
\end{aligned}
$$

with $\boldsymbol{\beta}_1 = (0.30, 1.25, 0.50, 1.00)'$, $\boldsymbol{\beta}_2 = (-0.20, -2.00, -.75, -1.00)'$ while $\mathbf{z}_{t-1} = (1, \cos(2\pi t/12), \mathbf{Y}_{t-1})$. In other words, the simulated model incorporates a sinusoidal component and a lagged value of order 1. One starts with arbitrary values for the Y_{tj} to get the π_{tj}, and then use the π_{tj} to generate the Y_{tj}, and so on. Figures 3.2 b,c,d display the transition probabilities of each of the categories, respectively. Figure 3.3 displays the sample auto- and cross-correlation functions of the simulated data. The upper left and lower right panels display plots of the sample autocorrelation functions of Y_{t1} and Y_{t2}, respectively. The other plots depict the sample cross-correlation function between Y_{t1} and Y_{t2} for positive (upper right) and negative (lower left) lags. In all these plots the sinusoidal component is apparent. Similar information is depicted in Figures 3.4 and 3.5 for $N = 200$ and $\boldsymbol{\beta}_1 = (-0.30, 2.00, -1.00, 0.25)'$ and $\boldsymbol{\beta}_2 = (0.20, 1.75, 1.25, -0.75)'$. Notice that in both Figures 3.2 (a) and 3.4 (a) the assignment of values to the three categories, namely 1,2,3, is arbitrary, but this has no bearing on the final results due to (3.2).

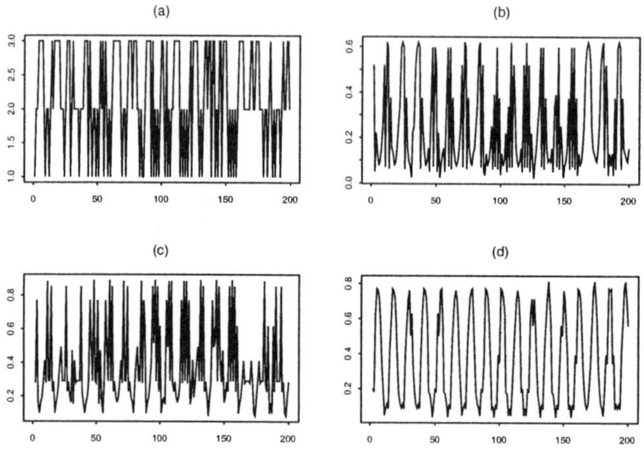

Fig. 3.4 Typical realization of the multinomial logit model (3.4) with 3 categories. Here $\boldsymbol{\beta}_1 = (-0.30, 2.00, -1.00, 0.25)'$, $\boldsymbol{\beta}_2 = (0.20, 1.75, 1.25, -0.75)'$, $\mathbf{z}_{t-1} = (1, \cos(2\pi t/12), \mathbf{Y}_{t-1})$ and $N = 200$. (a) Y_t. (b) π_{t1}. (c) π_{t2}. (d) π_{t3}.

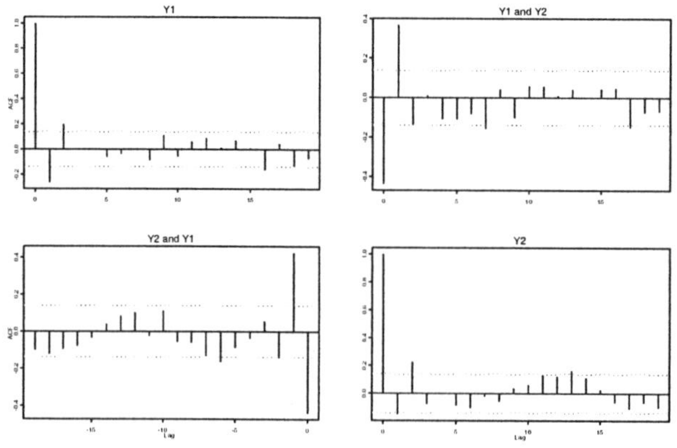

Fig. 3.5 Sample autocorrelation and cross–correlation functions of the simulated data from Figure 3.4.

3.2.2 Models for Ordinal Time Series

Ordinal categorical variables–such as blood pressure classified as low, normal, and high–are measured on a scale endowed with a natural ordering. Thus, the hourly blood pressure of an individual charted as low, normal, and high, constitutes an ordinal time series. The cumulative odds model ([311], [397]) is often used in applications for the analysis of ordinal data. The derivation of this model is better understood by means of a latent or auxiliary variable. That is, we assume the observed data result from the following threshold mechanism. Put

$$X_t = -\gamma' \mathbf{z}_{t-1} + e_t,$$

where e_t is a sequence of i.i.d random variables with continuous cdf F, γ is a d-dimensional vector of parameters, and $\mathbf{z}_{t-1}$ is a covariate vector of the same dimension. The process $\{X_t\}$, referred to as a "latent" process, may or may not be observed, but regardless of whether it is observed or not, the same calculations persist. Define a categorical time series $\{Y_t\}, t = 1, \ldots, N$, from the levels of $\{X_t\}$,

$$Y_t = j \Longleftrightarrow Y_{tj} = 1 \Longleftrightarrow \theta_{j-1} \leq X_t < \theta_j$$

for $j = 1, \ldots, m$, where $\{\theta_0, \theta_1, ..., \theta_m\}$ is a set of *threshold parameters* satisfying

$$-\infty = \theta_0 < \theta_1 < \ldots < \theta_m = \infty.$$

Then

$$\begin{aligned} \pi_{tj} &= \mathrm{P}(\theta_{j-1} \leq X_t < \theta_j \mid \mathcal{F}_{t-1}) \\ &= F(\theta_j + \gamma' \mathbf{z}_{t-1}) - F(\theta_{j-1} + \gamma' \mathbf{z}_{t-1}), \end{aligned} \tag{3.6}$$

for $j = 1, \ldots, m$. In other words,

$$\mathrm{P}(Y_t \leq j \mid \mathcal{F}_{t-1}) = F(\theta_j + \gamma' \mathbf{z}_{t-1}) \quad j = 1, \ldots, m. \tag{3.7}$$

From the estimates of (3.6) we obtain estimates for (3.7), since the set of the cumulative probabilities corresponds one to one to the set of the response probabilities. Many different special cases arise for various choices for F. For example, the cumulative logistic or *proportional odds* model is obtained when F is the logistic distribution function,

$$F_l(x) = \frac{1}{1 + \exp(-x)}.$$

Then we have

$$\log \left\{ \frac{\mathrm{P}[Y_t \leq j | \mathcal{F}_{t-1}]}{\mathrm{P}[Y_t > j | \mathcal{F}_{t-1}} \right\} = \theta_j + \gamma' \mathbf{z}_{t-1} \tag{3.8}$$

for $j = 1, \ldots, q$. Other choices for F include the standard normal cumulative distribution function

$$F \equiv \Phi,$$

the extreme minimal distribution function

$$F \equiv 1 - \exp(-\exp(x)),$$

and the extreme maximal distribution function

$$F \equiv \exp(-\exp(-x)).$$

In principle, any inverse link function appropriate for binary time series can be used when entertaining a cumulative odds model.

To recognize that model (3.7) is a special case of (3.3), let $\boldsymbol{\beta}$ denote the $q + d$ vector

$$\boldsymbol{\beta} = (\theta_1, \ldots, \theta_q, \boldsymbol{\gamma}')'$$

and $\mathbf{Z}_{t-1}$ the $(q + d) \times q$ matrix

$$\mathbf{Z}_{t-1} = \begin{bmatrix} 1 & 0 & \cdots & 0 \\ 0 & 1 & \cdots & 0 \\ \vdots & \vdots & \ddots & \vdots \\ 0 & 0 & \cdots & 1 \\ \mathbf{z}_{t-1} & \mathbf{z}_{t-1} & \cdots & \mathbf{z}_{t-1} \end{bmatrix}.$$

Now set

$$\mathbf{h} = (h_1, \ldots, h_q)',$$

with

$$\begin{aligned} \pi_{t1}(\boldsymbol{\beta}) = h_1(\boldsymbol{\eta}_t) &= F(\eta_{t1}), \\ \pi_{tj}(\boldsymbol{\beta}) = h_j(\boldsymbol{\eta}_t) &= F(\eta_{tj}) - F(\eta_{t(j-1)}), \quad j = 2, \ldots, q, \end{aligned}$$

where

$$\boldsymbol{\eta}_t = (\eta_{t1}, \ldots, \eta_{tq})' = \mathbf{Z}_{t-1}'\boldsymbol{\beta}.$$

It is clear that (3.3) is satisfied with this notation and $p = q + d$.

Some other models worth mentioning for the analysis of ordinal responses include the *continuation ratio* model specified by

$$F^{-1}\left(\frac{\pi_{tj}(\boldsymbol{\beta})}{\pi_{t(j+1)}(\boldsymbol{\beta}) + \ldots + \pi_{tm}(\boldsymbol{\beta})}\right) = \boldsymbol{\beta}'\mathbf{z}_{t-1}, \tag{3.9}$$

and the *adjacent categories logits* model given by

$$\mathrm{P}(Y_t = j | Y_t \in \{r, r+1\}, \mathcal{F}_{t-1}) = F(\boldsymbol{\beta}'\mathbf{z}_{t-1}), \tag{3.10}$$

where F stands for a continuous cdf. Various authors have considered the so called *two step* and *mean response* models–the latter for the analysis of interval response variables. The reader is referred to [2], [144, Ch. 3], [232] for further details on modeling aspects of ordinal and interval data.

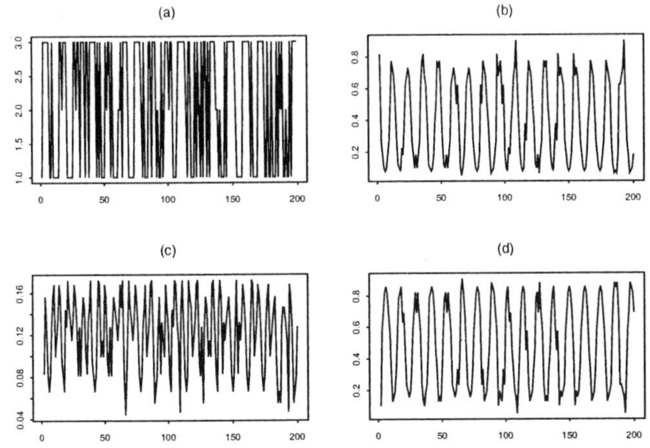

Fig. 3.6 Typical realization of the proportional odds model (3.8) with three categories. Here $\theta_1 = -0.50$, $\theta_2 = 0.20$, $\gamma = (2.00, -0.50, 1.00)'$, $\mathbf{z}_{t-1} = (\cos(2\pi t/12), \mathbf{Y}_{t-1})$, and $N = 200$. (a) Y_t. (b) π_{t1}. (c) π_{t2}. (d) π_{t3}.

3.2.2.1 Example: Proportional Odds Model with a Periodic Component

Figure 3.6 (a) shows a typical realization of a categorical time series of length $N = 200$ with $m = 3$ categories generated by the following proportional odds model (3.8):

$$\begin{aligned}\log\left\{\frac{\mathrm{P}[Y_t \le 1|\mathcal{F}_{t-1}]}{\mathrm{P}[Y_t > 1|\mathcal{F}_{t-1}}\right\} &= \theta_1 + \gamma'\mathbf{z}_{t-1} \\ &= \theta_1 + \gamma_1\cos(2\pi t/12) + \gamma_2 Y_{(t-1)1} + \gamma_3 Y_{(t-1)2},\end{aligned}$$

and

$$\begin{aligned}\log\left\{\frac{\mathrm{P}[Y_t \le 2|\mathcal{F}_{t-1}]}{\mathrm{P}[Y_t > 2|\mathcal{F}_{t-1}}\right\} &= \theta_2 + \gamma'\mathbf{z}_{t-1} \\ &= \theta_2 + \gamma_1\cos(2\pi t/12) + \gamma_2 Y_{(t-1)1} + \gamma_3 Y_{(t-1)2}.\end{aligned}$$

The model parameters are $\theta_1 = -0.50$, $\theta_2 = 0.20$, $\gamma = (2.00, -0.50, 1)'$, and the covariate vector $\mathbf{z}_{t-1} = (\cos(2\pi t/12), \mathbf{Y}_{t-1})$ consists of a sinusoidal component and a lagged value of order 1. Figures 3.6 b,c,d display the corresponding transition probabilities of each of the categories, respectively. Figure 3.7 displays the sample autocorrelation function of the simulated data. The sinusoidal component is manifested clearly especially in the upper left plot of Figure 3.7. Figures 3.8 and 3.9 display similar information but for different parameters. Here, we use $\theta_1 = -0.20$, $\theta_2 = 1.00$, $\gamma = (2.00, 1, 0.50)'$.

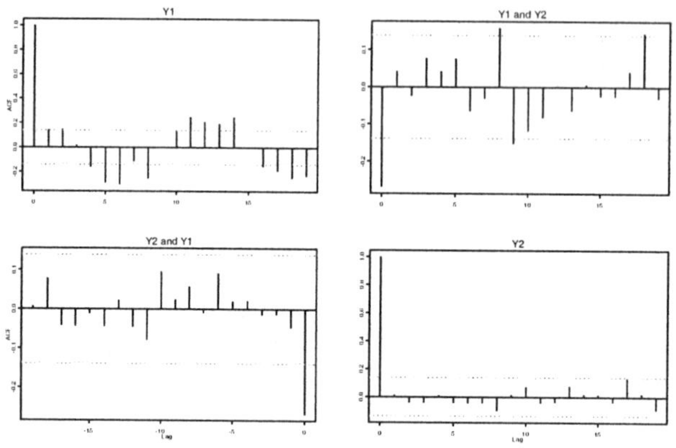

Fig. 3.7 Sample autocorrelation and cross–correlation functions of the simulated data from Figure 3.6.

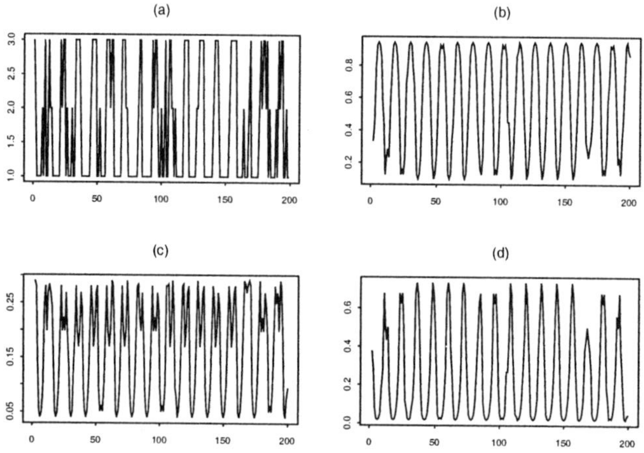

Fig. 3.8 Typical realization of the proportional odds model (3.8) with three categories. Here $\theta_1 = -0.20$, $\theta_2 = 1.00$, $\boldsymbol{\gamma} = (-2.00, 1.00, 0.50)'$, $\mathbf{z}_{t-1} = (\cos(2\pi t/12), \mathbf{Y}_{t-1})$, and $N = 200$. (a) Y_t. (b) π_{t1}. (c) π_{t2}. (d) π_{t3}.

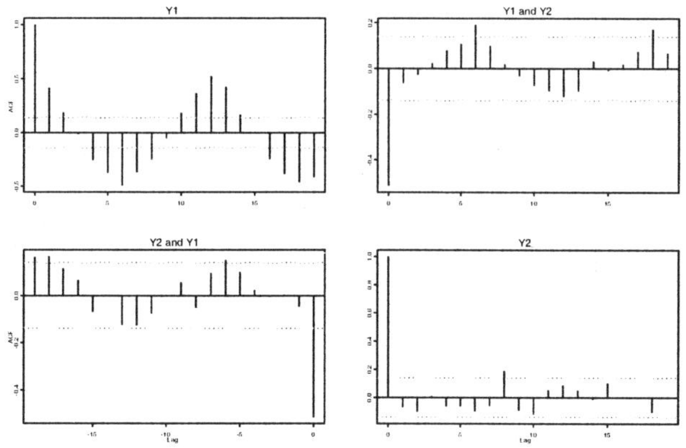

Fig. 3.9 Sample autocorrelation and cross–correlation functions of the simulated data from Figure 3.8.

3.3 PARTIAL LIKELIHOOD ESTIMATION

The estimation of the parameter vector $\boldsymbol{\beta}$ in the general regression model (3.3) follows in principle the partial likelihood methodology described in the previous chapters. The theoretical development resembles that of binary time series except that there are considerable technical complications due to the multivariate nature of the problem. Hence it is instructive to consider first the special case $m = 3$ where the number of response categories is equal to 3.

3.3.1 Inference for m=3

When $m = 3$, $\mathbf{Y}_t$ reduces to a two-dimensional vector $(Y_{t1}, Y_{t2})'$ with

$$Y_{tj} = \begin{cases} 1, & \text{if the } j\text{th category is observed at time } t \\ 0, & \text{otherwise} \end{cases}$$

and $\boldsymbol{\pi}_t = (\pi_{t1}, \pi_{t2})'$. It follows that $Y_{t3} = 1 - (Y_{t1} + Y_{t2})$ and $\pi_{t3} = 1 - (\pi_{t1} + \pi_{t2})$. Furthermore, we obtain

$$\begin{aligned} \text{Var}\left[\mathbf{Y}_t \mid \mathcal{F}_{t-1}\right] &\equiv \boldsymbol{\Sigma}_t(\boldsymbol{\beta}) \\ &= \begin{bmatrix} \pi_{t1}(\boldsymbol{\beta})(1 - \pi_{t1}(\boldsymbol{\beta})) & -\pi_{t1}(\boldsymbol{\beta})\pi_{t2}(\boldsymbol{\beta}) \\ -\pi_{t1}(\boldsymbol{\beta})\pi_{t2}(\boldsymbol{\beta}) & \pi_{t2}(\boldsymbol{\beta})(1 - \pi_{t2}(\boldsymbol{\beta})) \end{bmatrix}. \end{aligned} \tag{3.11}$$

The partial likelihood is a product of the multinomial probabilities

$$\prod_{j=1}^{3} \pi_{tj}^{y_{tj}}(\boldsymbol{\beta}),$$

that is,

$$\mathrm{PL}(\boldsymbol{\beta}) = \prod_{t=1}^{N} \prod_{j=1}^{3} \pi_{tj}^{y_{tj}}(\boldsymbol{\beta}). \tag{3.12}$$

Hence the partial log-likelihood is given by

$$l(\boldsymbol{\beta}) \equiv \log \mathrm{PL}(\boldsymbol{\beta}) = \sum_{t=1}^{N} \sum_{j=1}^{3} y_{tj} \log \pi_{tj}(\boldsymbol{\beta}), \tag{3.13}$$

or

$$\begin{aligned}
l(\boldsymbol{\beta}) &= \sum_{t=1}^{N} \sum_{j=1}^{3} y_{tj} \log \pi_{tj}(\boldsymbol{\beta}) \\
&= \sum_{t=1}^{N} \{ y_{t1} \log \pi_{t1}(\boldsymbol{\beta}) + y_{t2} \log \pi_{t2}(\boldsymbol{\beta}) + y_{t3} \log \pi_{t3}(\boldsymbol{\beta}) \} \\
&= \sum_{t=1}^{N} \Big\{ y_{t1} \log \pi_{t1}(\boldsymbol{\beta}) + y_{t2} \log \pi_{t2}(\boldsymbol{\beta}) \\
&\quad + (1 - y_{t1} - y_{t2}) \log \left(1 - \pi_{t1}(\boldsymbol{\beta}) - \pi_{t2}(\boldsymbol{\beta})\right) \Big\} \\
&= \sum_{t=1}^{N} \left\{ y_{t1} \log \left(\frac{\pi_{t1}(\boldsymbol{\beta})}{1 - \pi_{t1}(\boldsymbol{\beta}) - \pi_{t2}(\boldsymbol{\beta})} \right) + y_{t2} \log \left(\frac{\pi_{t2}(\boldsymbol{\beta})}{1 - \pi_{t1}(\boldsymbol{\beta}) - \pi_{t2}(\boldsymbol{\beta})} \right) \right. \\
&\quad \left. + \log \left(1 - \pi_{t1}(\boldsymbol{\beta}) - \pi_{t2}(\boldsymbol{\beta})\right) \right\}.
\end{aligned} \tag{3.14}$$

Substituting

$$\begin{aligned}
\boldsymbol{\theta}_t(\boldsymbol{\beta}) &= (\theta_{t1}(\boldsymbol{\beta}), \theta_{t2}(\boldsymbol{\beta}))' \\
&= \left(\log \left(\frac{\pi_{t1}(\boldsymbol{\beta})}{1 - \pi_{t1}(\boldsymbol{\beta}) - \pi_{t2}(\boldsymbol{\beta})} \right), \log \left(\frac{\pi_{t2}(\boldsymbol{\beta})}{1 - \pi_{t1}(\boldsymbol{\beta}) - \pi_{t2}(\boldsymbol{\beta})} \right) \right)',
\end{aligned}$$

equation (3.14) is rewritten as

$$\begin{aligned}
l(\boldsymbol{\beta}) &= \sum_{t=1}^{N} \{ y_{t1}\theta_{t1}(\boldsymbol{\beta}) + y_{t2}\theta_{t2}(\boldsymbol{\beta}) - \log\left[1 + \exp\left(\theta_{t1}(\boldsymbol{\beta})\right) + \exp\left(\theta_{t2}(\boldsymbol{\beta})\right)\right] \} \\
&= \sum_{t=1}^{N} l_t(\boldsymbol{\beta}).
\end{aligned} \tag{3.15}$$

To calculate the partial score, recall the chain rule for *multivariate functions* to obtain

$$\frac{\partial l_t}{\partial \boldsymbol{\beta}'} = \frac{\partial l_t}{\partial \boldsymbol{\theta}_t'} \frac{\partial \boldsymbol{\theta}_t}{\partial \boldsymbol{\pi}_t'} \frac{\partial \boldsymbol{\pi}_t}{\partial \boldsymbol{\eta}_t'} \frac{\partial \boldsymbol{\eta}_t}{\partial \boldsymbol{\beta}'}, \tag{3.16}$$

where $\eta_t = \mathbf{Z}'_{t-1}\beta$. A straightforward calculation shows

$$\begin{aligned}\frac{\partial l_t}{\partial \boldsymbol{\theta}'_t} &= \left(\frac{\partial l_t}{\partial \theta_{t1}}, \frac{\partial l_t}{\partial \theta_{t2}}\right) \\ &= (Y_{t1} - \pi_{t1}, Y_{t2} - \pi_{t2}) = (\mathbf{Y}_t - \boldsymbol{\pi}_t)' \end{aligned} \tag{3.17}$$

and

$$\begin{aligned}\frac{\partial \boldsymbol{\theta}_t}{\partial \boldsymbol{\pi}'_t} &= \begin{bmatrix} \dfrac{\partial \theta_{t1}}{\partial \pi_{t1}} & \dfrac{\partial \theta_{t1}}{\partial \pi_{t2}} \\ \dfrac{\partial \theta_{t2}}{\partial \pi_{t1}} & \dfrac{\partial \theta_{t2}}{\partial \pi_{t2}} \end{bmatrix} \\ &= \begin{bmatrix} \dfrac{1-\pi_{t2}}{\pi_{t1}(1-\pi_{t1}-\pi_{t2})} & \dfrac{1}{1-\pi_{t1}-\pi_{t2}} \\ \dfrac{1}{1-\pi_{t1}-\pi_{t2}} & \dfrac{1-\pi_{t1}}{\pi_{t2}(1-\pi_{t1}-\pi_{t2})} \end{bmatrix}. \end{aligned} \tag{3.18}$$

Hence, from (3.11)

$$\boldsymbol{\Sigma}_t^{-1} = \frac{\partial \boldsymbol{\theta}_t}{\partial \boldsymbol{\pi}'_t}. \tag{3.19}$$

Since $\eta_t = \mathbf{Z}'_{t-1}\beta$ is a 2×1 vector, we obtain

$$\mathbf{D}'_t \equiv \frac{\partial \boldsymbol{\pi}_t}{\partial \boldsymbol{\eta}'_t} = \begin{bmatrix} \dfrac{\partial \pi_{t1}}{\partial \eta_{t1}} & \dfrac{\partial \pi_{t1}}{\partial \eta_{t2}} \\ \dfrac{\partial \pi_{t2}}{\partial \eta_{t1}} & \dfrac{\partial \pi_{t2}}{\partial \eta_{t2}} \end{bmatrix} = \frac{\partial \mathbf{h}(\boldsymbol{\eta}_t)}{\partial \boldsymbol{\eta}'_t}. \tag{3.20}$$

The last equality is a consequence of (3.3). In addition,

$$\frac{\partial \boldsymbol{\eta}_t}{\partial \beta'} = \mathbf{Z}'_{t-1}. \tag{3.21}$$

Substitution of equations (3.17), (3.19), (3.20) and (3.21) into (3.16) shows that

$$\frac{\partial l_t}{\partial \beta'} = (\mathbf{Y}_t - \boldsymbol{\pi}_t)' \, \boldsymbol{\Sigma}_t^{-1} \mathbf{D}'_t \mathbf{Z}'_{t-1}, \tag{3.22}$$

and therefore the partial score as an explicit function of β is given by

$$\begin{aligned}\mathbf{S}_N(\beta) &= \nabla l(\beta) = \left(\frac{\partial l(\beta)}{\partial \beta_1}, \ldots, \frac{\partial l(\beta)}{\partial \beta_p}\right)' \\ &= \sum_{t=1}^{N} \mathbf{Z}_{t-1}\mathbf{D}_t(\beta)\boldsymbol{\Sigma}_t^{-1}(\beta)\left(\mathbf{Y}_t - \boldsymbol{\pi}_t(\beta)\right). \end{aligned} \tag{3.23}$$

An alternative form of (3.23) is

$$\mathbf{S}_N(\beta) = \sum_{t=1}^{N} \mathbf{Z}_{t-1}\mathbf{U}_t(\beta)\left(\mathbf{Y}_t - \boldsymbol{\pi}_t(\beta)\right) \tag{3.24}$$

with

$$\begin{aligned} \mathbf{U}_t(\boldsymbol{\beta}) &= \mathbf{D}_t(\boldsymbol{\beta})\boldsymbol{\Sigma}_t^{-1}(\boldsymbol{\beta}) \\ &= \frac{\partial \mathbf{u}(\boldsymbol{\eta}_t)}{\partial \boldsymbol{\eta}_t}, \end{aligned} \tag{3.25}$$

where $\mathbf{u}$ is the composition of $\boldsymbol{\theta}_t$ and $\mathbf{h}$. In other words,

$$\begin{aligned} \mathbf{u} &= (u_1, u_2)' = (\theta_{t1}(\mathbf{h}), \theta_{t2}(\mathbf{h}))' \\ &= \left(\log\left(\frac{h_1(\boldsymbol{\eta}_t)}{1 - h_1(\boldsymbol{\eta}_t) - h_2(\boldsymbol{\eta}_t)}\right), \log\left(\frac{h_2(\boldsymbol{\eta}_t)}{1 - h_1(\boldsymbol{\eta}_t) - h_2(\boldsymbol{\eta}_t)}\right)\right)' \end{aligned} \tag{3.26}$$

The reader should verify (see Problem 8) that the observed information matrix is given by

$$\mathbf{H}_N(\boldsymbol{\beta}) = -\nabla\nabla' l(\boldsymbol{\beta}) = \mathbf{G}_N(\boldsymbol{\beta}) - \mathbf{R}_N(\boldsymbol{\beta}), \tag{3.27}$$

where

$$\mathbf{G}_N(\boldsymbol{\beta}) = \sum_{t=1}^{N} \mathbf{Z}_{t-1}\mathbf{U}_t(\boldsymbol{\beta})\boldsymbol{\Sigma}_t(\boldsymbol{\beta})\mathbf{U}_t'(\boldsymbol{\beta})\mathbf{Z}_{t-1}',$$

and

$$\mathbf{R}_N(\boldsymbol{\beta}) = \sum_{t=1}^{N}\sum_{r=1}^{2} \mathbf{Z}_{t-1}\mathbf{W}_{tr}(\boldsymbol{\beta})\mathbf{Z}_{t-1}' \left(Y_{tr} - \pi_{tr}(\boldsymbol{\beta})\right)$$

with

$$\mathbf{W}_{tr}(\boldsymbol{\beta}) = \frac{\partial^2 u_r(\boldsymbol{\eta}_t)}{\partial \boldsymbol{\eta}_t \partial \boldsymbol{\eta}_t'},$$

for $r = 1, 2$, following the same arguments of the proof of (1.33) and (1.34).

3.3.2 Inference for m>3

The above results generalize to the case $m > 3$ following the same steps. Let $\{Y_t\}$ be a categorical time series with $m = q+1$ categories, and recall that $\mathbf{Y}_t = (Y_{t1}, \ldots, Y_{tq})'$ where

$$Y_{tj} = \begin{cases} 1, & \text{if the } j\text{th category is observed at time } t \\ 0, & \text{otherwise.} \end{cases}$$

Set $\boldsymbol{\pi}_t = (\pi_{t1}, \ldots, \pi_{tq})'$ for the vector of conditional probabilities where $\pi_{tj} = \mathrm{P}(Y_{tj} = 1 \mid \mathcal{F}_{t-1}),\ \ j = 1, \ldots, q$. Introduce the multinomial probability

$$f(\mathbf{y}_t; \boldsymbol{\beta} \mid \mathcal{F}_{t-1}) = \prod_{j=1}^{m} \pi_{tj}(\boldsymbol{\beta})^{y_{tj}}.$$

The partial likelihood is a product of the multinomial probabilities,

$$\begin{aligned} \mathrm{PL}(\beta) &= \prod_{t=1}^{N} f(\mathbf{y}_t;\beta|\mathcal{F}_{t-1}) \\ &= \prod_{t=1}^{N}\prod_{j=1}^{m} \pi_{tj}^{y_{tj}}(\beta), \end{aligned} \tag{3.28}$$

so that the partial log-likelihood is given by

$$l(\beta) \equiv \log \mathrm{PL}(\beta) = \sum_{t=1}^{N}\sum_{j=1}^{m} y_{tj}\log \pi_{tj}(\beta). \tag{3.29}$$

It is useful to introduce the *logit* function by

$$\mathbf{logit}(\mathbf{x}) = \left(\log(\frac{x_1}{1-\sum_{j=1}^{q} x_j}), \ldots, \log(\frac{x_q}{1-\sum_{j=1}^{q} x_j})\right)', \tag{3.30}$$

for a q–dimensional vector $\mathbf{x}$ which belongs in the set $\{(x_1,\ldots,x_q)' : x_j > 0, j = 1,\ldots,q, \sum_{j=1}^{q} x_j < 1\}$. It can be shown (see Problem 7) that the logit function corresponds to the inverse canonical link of the multinomial distribution.

Computation of the maximum partial likelihood estimator (MPLE) $\hat{\beta}$ is carried out by maximizing the partial log–likelihood (3.29). This, in turn, implies that if the MPLE $\hat{\beta}$ exists then it is given as solution of the partial score equations

$$\nabla l(\beta) = \nabla \log \mathrm{PL}(\beta) = \mathbf{0} \tag{3.31}$$

assuming differentiability. The solution of the partial score equations (3.31) is obtained by Fisher scoring, as described in Section 1.3.2.

Following the derivation of (3.23), we obtain the *partial score* by differentiating (3.29)

$$\begin{aligned} \mathbf{S}_N(\beta) = \nabla l(\beta) &= \left(\frac{\partial l(\beta)}{\partial \beta_1}, \ldots, \frac{\partial l(\beta)}{\partial \beta_p}\right)' \\ &= \sum_{t=1}^{N} \mathbf{Z}_{t-1}\mathbf{D}_t(\beta)\boldsymbol{\Sigma}_t^{-1}(\beta)\left(\mathbf{Y}_t - \boldsymbol{\pi}_t(\beta)\right). \end{aligned} \tag{3.32}$$

Recall (3.25) and set

$$\mathbf{U}_t(\beta) = \mathbf{D}_t(\beta)\boldsymbol{\Sigma}_t^{-1}(\beta),$$

where

$$\mathbf{D}_t(\beta) = \frac{\partial \mathbf{h}(\boldsymbol{\eta}_t)}{\partial \boldsymbol{\eta}_t}$$

and $\boldsymbol{\Sigma}_t(\beta)$ is the conditional covariance matrix of $\mathbf{Y}_t$ with generic elements

$$\sigma_t^{(ij)}(\beta) = \begin{cases} -\pi_{ti}(\beta)\pi_{tj}(\beta) & \text{if } i \neq j \\ \pi_{ti}(\beta)(1-\pi_{ti}(\beta)) & \text{if } i = j \end{cases}$$

for $i, j = 1 \ldots, q$. It follows that the partial score (3.32) can also be expressed by the equation

$$\mathbf{S}_N(\boldsymbol{\beta}) = \sum_{t=1}^{N} \mathbf{Z}_{t-1}\mathbf{U}_t(\boldsymbol{\beta})\left(\mathbf{Y}_t - \boldsymbol{\pi}_t(\boldsymbol{\beta})\right), \tag{3.33}$$

where

$$\mathbf{U}_t(\boldsymbol{\beta}) = \frac{\partial \mathbf{u}(\boldsymbol{\eta}_t)}{\partial \boldsymbol{\eta}_t}$$

is now a $q \times q$ matrix and $\boldsymbol{\eta}_t = \mathbf{Z}'_{t-1}\boldsymbol{\beta}$, as before. The function $\mathbf{u} = (u_1, \ldots, u_q)'$ is the composition of the functions **logit** in (3.30) and $\mathbf{h}$ in (3.3) (recall (3.26)). In other words

$$\mathbf{u} = (u_1, \ldots u_q)' = \left(\log\left(\frac{h_1(\eta_t)}{1 - \sum_{j=1}^{q} h_j(\eta_t)}\right), \ldots, \log\left(\frac{h_q(\eta_t)}{1 - \sum_{j=1}^{q} h_j(\eta_t)}\right)\right)'.$$

In what follows we shall use (3.33) rather than (3.32).

As before, to carry on the large sample theory, we introduce the following basic quantities. The *partial score process* is defined through the partial sum

$$\mathbf{S}_t(\boldsymbol{\beta}) = \sum_{s=1}^{t} \mathbf{Z}_{s-1}\mathbf{U}_s(\boldsymbol{\beta})\left(\mathbf{Y}_s - \boldsymbol{\pi}_s(\boldsymbol{\beta})\right), \tag{3.34}$$

and the conditional information matrix is

$$\begin{aligned}\mathbf{G}_N(\boldsymbol{\beta}) &= \sum_{t=1}^{N} \operatorname{Cov}\left[\mathbf{Z}_{t-1}\mathbf{U}_t(\boldsymbol{\beta})\left(\mathbf{Y}_t - \boldsymbol{\pi}_t(\boldsymbol{\beta})\right) | \mathcal{F}_{t-1}\right] \\ &= \sum_{t=1}^{N} \mathbf{Z}_{t-1}\mathbf{U}_t(\boldsymbol{\beta})\boldsymbol{\Sigma}_t(\boldsymbol{\beta})\mathbf{U}'_t(\boldsymbol{\beta})\mathbf{Z}'_{t-1}. \end{aligned} \tag{3.35}$$

The unconditional information matrix is given by

$$\mathbf{F}_N(\boldsymbol{\beta}) = \mathrm{E}\left[\mathbf{G}_N(\boldsymbol{\beta})\right] \tag{3.36}$$

and the second derivative of the partial log likelihood multiplied by -1 is

$$\mathbf{H}_N(\boldsymbol{\beta}) = -\nabla\nabla' l(\boldsymbol{\beta}) = \mathbf{G}_N(\boldsymbol{\beta}) - \mathbf{R}_N(\boldsymbol{\beta}), \tag{3.37}$$

where

$$\mathbf{R}_N(\boldsymbol{\beta}) = \sum_{t=1}^{N}\sum_{r=1}^{q} \mathbf{Z}_{t-1}\mathbf{W}_{tr}(\boldsymbol{\beta})\mathbf{Z}'_{t-1}\left(Y_{tr} - \pi_{tr}(\boldsymbol{\beta})\right)$$

with

$$\mathbf{W}_{tr}(\boldsymbol{\beta}) = \frac{\partial^2 u_r(\boldsymbol{\eta}_t)}{\partial \boldsymbol{\eta}_t \partial \boldsymbol{\eta}'_t},$$

for $r = 1, \ldots, q$.

3.3.3 Large Sample Theory

The large sample properties of the MPLE $\hat{\beta}$ are studied along the lines put forth in earlier chapters. In this regard we prove in the appendix to this chapter the following three basic facts needed in order to establish consistency and asymptotic normality for the MPLE. First, the score process (3.34) is a zero mean square integrable martingale that satisfies the conditions of the martingale central limit theorem (Lemma A.0.1). Second, as is the case in the theory of linear models, the minimum eigenvalue of $\mathbf{F}_N(\beta)$ diverges (Lemma A.0.2), and third, Lemma A.0.3 demonstrates that for values of β in a small neighborhood around its true value we have

$$\frac{\mathbf{R}_N(\beta)}{N} \to \mathbf{0}$$

in probability, as $N \to \infty$ (see Section 1.4.2).

The covariate matrix process $\{\mathbf{Z}_{t-1}\}$ belongs now to $R^{p\times q}$ and this requires a slight modification of (A.2) and (A.5) in Assumption **A** as follows:

(A.2) The covariate matrix $\mathbf{Z}_{t-1}$ almost surely lies in a nonrandom compact subset Γ of $R^{p\times q}$ such that $P[\sum_{t=1}^{N} \mathbf{Z}_{t-1}\mathbf{Z}'_{t-1} > \mathbf{0}] = 1$. Furthermore we assume that $\mathbf{Z}'_{t-1}\beta$ lies almost surely in the domain H of $\mathbf{h}$ for all $\mathbf{Z}_{t-1} \in \Gamma$ and $\beta \in B$.

(A.5) There is a probability measure ν on $R^{p\times q}$ such that $\int_{R^{p\times q}} \mathbf{Z}\mathbf{Z}'\nu(d\mathbf{Z})$ is positive definite, and such that under (3.3), for Borel sets $A \subset R^{p\times q}$ we have

$$\frac{1}{N}\sum_{t=1}^{N} I_{[\mathbf{Z}_{t-1}\in A]} \to \nu(A)$$

in probability, as $N \to \infty$.

The modified assumption is still referred to as Assumption A throughout this chapter. Recalling that integration with respect to a matrix means integration with respect to each element of the matrix, A.5 shows that the conditional information matrix $\mathbf{G}_N(\beta)$ has a non-random limit. Namely,

$$\frac{\mathbf{G}_N(\beta)}{N} \to \int_{R^{p\times q}} \mathbf{Z}\mathbf{U}(\beta)\boldsymbol{\Sigma}(\beta)\mathbf{U}'(\beta)\mathbf{Z}'\nu(d\mathbf{Z}) = \mathbf{G}(\beta) \tag{3.38}$$

in probability, as $N \to \infty$, where

$$\mathbf{U}(\beta) = \frac{\partial \mathbf{u}(\boldsymbol{\eta})}{\partial \boldsymbol{\eta}}$$

with $\boldsymbol{\eta} = \mathbf{Z}'\beta$, and $\boldsymbol{\Sigma}(\beta)$ has generic elements

$$\sigma^{(ij)}(\beta) = \begin{cases} -h_i(\mathbf{Z}'\beta)h_j(\mathbf{Z}'\beta) & \text{if } i \neq j \\ h_i(\mathbf{Z}'\beta)(1 - h_i(\mathbf{Z}'\beta)) & \text{if } i = j \end{cases}$$

for $i, j = 1 \ldots, q$. Furthermore, (A.5) implies that $\mathbf{G}(\boldsymbol{\beta})$ is a positive definite matrix at the true parameter value and therefore its inverse exists.

As was noted in [152], this approach is quite general and does not call for any Markov assumption. Previous related work on *conditional likelihood* estimation in categorical time series can be found in [140] and [246]. The latter reference provides a rigorous treatment of consistency, asymptotic normality, and efficiency of the maximum conditional likelihood estimator.

We now state the main theorem the proof of which is given in the appendix to this chapter.

Theorem 3.3.1 Under Assumption A, the probability that a locally unique maximum partial likelihood estimator exists converges to one. Moreover there exists a sequence of maximum partial likelihood estimators $\hat{\boldsymbol{\beta}}$ which is consistent and asymptotically normal:

$$\sqrt{N}(\hat{\boldsymbol{\beta}} - \boldsymbol{\beta}) \rightarrow \mathcal{N}_p \left(\mathbf{0}, \mathbf{G}^{-1}(\boldsymbol{\beta})\right)$$

in distribution, as $N \to \infty$.

Theorem 3.3.1 implies that if each component of the link function is log-concave, that is, $\log h_j$ is concave for every $j = 1, \ldots, m$ with $h_m = 1 - \sum_{j=1}^{q} h_j$ and the parameter space B is R^p, then we obtain the following:

Corollary 3.3.1 Suppose Assumption A holds. Assume further that $\log h_j$ is concave for $j = 1, \ldots, m$. Then the probability that a unique maximum partial likelihood estimator exists converges to one. Any such sequence is consistent and asymptotically normal as in Theorem 3.3.1.

Another straightforward application of Theorem 3.3.1 leads to the construction of prediction intervals for $\boldsymbol{\pi}_t(\boldsymbol{\beta})$ given the data. An application of the delta method (see [366, p.338]) shows that:

Theorem 3.3.2 Under Assumption A we have

$$\sqrt{N}\left(\boldsymbol{\pi}_t(\hat{\boldsymbol{\beta}}) - \boldsymbol{\pi}_t(\boldsymbol{\beta})\right) \rightarrow \mathcal{N}_q \left(\mathbf{0}, \mathbf{Z}_{t-1}\mathbf{D}_t(\boldsymbol{\beta})\mathbf{G}^{-1}(\boldsymbol{\beta})\mathbf{D}_t'(\boldsymbol{\beta})\mathbf{Z}_{t-1}'\right)$$

in distribution, as $N \to \infty$, where

$$\mathbf{D}_t(\boldsymbol{\beta}) = \frac{\partial \mathbf{h}(\boldsymbol{\eta}_t)}{\partial \boldsymbol{\eta}_t}.$$

3.3.4 Inference for the Multinomial Logit Model

The aforementioned results can be illustrated quite well in terms of the multinomial logit model (3.4). For this important special case, $\mathbf{u}$ is the identity function since it is the composition of the logit and inverse logit functions. Thus $\mathbf{U}_t(\boldsymbol{\beta}) = \mathbf{I}$. Alternatively, a straightforward differentiation shows that

$$\mathbf{D}_t(\boldsymbol{\beta}) = \boldsymbol{\Sigma}_t(\boldsymbol{\beta}).$$

Thus the score equation (3.33) reduces to

$$\mathbf{S}_N(\boldsymbol{\beta}) = \sum_{t=1}^{N} \mathbf{Z}_{t-1}\left(\mathbf{Y}_t - \boldsymbol{\pi}_t(\boldsymbol{\beta})\right) \tag{3.39}$$

and the conditional information matrix (3.35) becomes

$$\mathbf{G}_N(\boldsymbol{\beta}) = \sum_{t=1}^{N} \mathbf{Z}_{t-1}\boldsymbol{\Sigma}_t(\boldsymbol{\beta})\mathbf{Z}'_{t-1}. \tag{3.40}$$

In addition $\mathbf{R}_N(\boldsymbol{\beta}) = \mathbf{0}$ and therefore the equality $\mathbf{H}_N(\boldsymbol{\beta}) = \mathbf{G}_N(\boldsymbol{\beta})$. The sample information matrix per single observation $\mathbf{G}_N(\boldsymbol{\beta})/N$ converges in probability to a special case of the limit (3.38)

$$\frac{\mathbf{G}_N(\boldsymbol{\beta})}{N} \to \mathbf{G}(\boldsymbol{\beta}) = \int_{R^{p\times q}} \mathbf{Z}\boldsymbol{\Sigma}(\boldsymbol{\beta})\mathbf{Z}'\nu(d\mathbf{Z}). \tag{3.41}$$

Table 3.1 summarizes simulation results from 1000 runs of model (3.5) for different sample sizes and parameter values as indicated in Section 3.2.1.1. The first column lists the true values of $\boldsymbol{\beta} = (\boldsymbol{\beta}'_1, \boldsymbol{\beta}'_2)'$ and the remaining columns give the estimated parameter values formed by averaging the 1000 estimates, and the corresponding standard errors. The standard errors were computed in two ways, by inverting the conditional information matrix $\mathbf{G}_N$ (indicated by the heading $\mathbf{G}_N^{-1}$), and from the sample variance of 1000 estimates (indicated by the heading $\hat{\text{SE}}$).

Table 3.1 Theoretical versus rounded experimental results for multinomial logits. The first part of the $\boldsymbol{\beta}$ column gives $\boldsymbol{\beta}_1$, the second $\boldsymbol{\beta}_2$.

	$N = 200$			$N = 300$			$N = 500$		
$\boldsymbol{\beta}$	$\hat{\boldsymbol{\beta}}$	$\mathbf{G}_N^{-1}$	$\hat{\text{SE}}$	$\hat{\boldsymbol{\beta}}$	$\mathbf{G}_N^{-1}$	$\hat{\text{SE}}$	$\hat{\boldsymbol{\beta}}$	$\mathbf{G}_N^{-1}$	$\hat{\text{SE}}$
0.30	0.37	0.46	0.49	0.34	0.40	0.37	0.30	0.28	0.27
1.25	1.30	0.34	0.36	1.29	0.27	0.30	1.29	0.21	0.22
0.50	0.42	0.53	0.56	0.45	0.44	0.42	0.48	0.32	0.31
1.00	0.99	0.62	0.66	1.01	0.54	0.53	1.03	0.39	0.40
−0.20	−0.17	0.49	0.57	−0.20	0.44	0.42	−0.20	0.32	0.32
−2.00	−2.13	0.52	0.60	−2.07	0.42	0.46	−2.04	0.33	0.35
−0.75	−0.86	0.57	0.62	−0.80	0.49	0.49	−0.77	0.38	0.36
−1.00	−1.15	0.60	0.70	−1.05	0.54	0.53	−1.03	0.41	0.39

3.3.5 Testing Hypotheses

In applications it is often necessary to test the general linear hypotheses

$$\text{H}_0 : \mathbf{C}\boldsymbol{\beta} = \boldsymbol{\beta}_0 \text{ against } \text{H}_1 : \mathbf{C}\boldsymbol{\beta} \neq \boldsymbol{\beta}_0, \tag{3.42}$$

where $\mathbf{C}$ is an appropriate known matrix with full rank, say $r \le p$. To this end, it is convenient to denote by $\tilde{\beta}$ the restricted partial maximum likelihood estimator under the hypothesis (3.42) while $\hat{\beta}$ is the unrestricted MPLE. Then the most commonly used test statistics for testing (3.42) are,

- The partial likelihood ratio statistic

$$\lambda_N = 2\left\{l(\hat{\beta}) - l(\tilde{\beta})\right\}. \tag{3.43}$$

- The Wald statistic

$$w_N = \{\mathbf{C}\hat{\beta} - \beta_0\}' \left\{\mathbf{C}\mathbf{G}^{-1}(\hat{\beta})\mathbf{C}'\right\}^{-1} \{\mathbf{C}\hat{\beta} - \beta_0\}. \tag{3.44}$$

- The partial score statistic

$$c_N = \frac{1}{N}\mathbf{S}'_N(\tilde{\beta})\mathbf{G}^{-1}(\tilde{\beta})\mathbf{S}_N(\tilde{\beta}). \tag{3.45}$$

The following theorem states the asymptotic distribution of these statistics (see [135], and Theorem 1.5.1)

Theorem 3.3.3 Under Assumption A the test statistics λ_N, w_N and c_N are asymptotically equivalent. Furthermore, under the null hypotheses in (3.42), their asymptotic distribution is chi–square with r degrees of freedom.

The behavior of all three test statistics under a sequence of alternatives is examined in [140] where Theorem 3.3.3 is applied in testing homogeneity and order of a Markov chain, structural change and independence of two parallel time series.

3.4 GOODNESS OF FIT

As we have seen previously, a central issue that arises after fitting a regression model is that of goodness of fit. For most generalized linear models goodness of fit is examined either by the Pearson goodness of fit statistic or by the scaled deviance (see Section 1.6.1). For categorical time series, the scaled deviance takes the following form

$$D = -2\sum_{t=1}^{N}\sum_{j=1}^{m} Y_{tj} \log \pi_{tj}(\hat{\beta}), \tag{3.46}$$

whereas the Pearson goodness of fit statistic becomes (see Problem 13)

$$\begin{aligned} \mathcal{X}^2 &= \sum_{t=1}^{N}\left(\mathbf{Y}_t - \boldsymbol{\pi}_t(\hat{\beta})\right)' \boldsymbol{\Sigma}_t^{-1}(\hat{\beta})\left(\mathbf{Y}_t - \boldsymbol{\pi}_t(\hat{\beta})\right) \\ &= \sum_{t=1}^{N}\sum_{j=1}^{m}\frac{\left(Y_{tj} - \pi_{tj}(\hat{\beta})\right)^2}{\pi_{tj}(\hat{\beta})}. \end{aligned} \tag{3.47}$$

It can be shown that under suitable regularity conditions, the asymptotic distribution of both (3.46) and (3.47) approaches the chi–square distribution with $Nq - p$ degrees of freedom. However the latter approximation may fail for a number of reasons echoing the discussion of Section 2.4.1. Hence, some other means and techniques are needed for goodness of fit diagnostics. In this section we first generalize the goodness of fit test proposed in Section 2.4.2 for binary time series, and then introduce the *power divergence family* of goodness of fit statistics.

3.4.1 Goodness of Fit Based on Response Classification

Recall the goodness of fit test for binary time series models based on response classification defined in terms of the covariates (see Section 2.4.2). We implement a similar method for categorical time series by classifying the responses $\{\mathbf{Y}_t\}$ according to mutually exclusive events expressed in terms of the covariate matrices $\{\mathbf{Z_{t-1}}\}$ for $t = 1, \ldots, N$ ([152], [378], [395]).

Suppose that $C_1, \ldots, C_k$ is a partition of $R^{p\times q}$ and for $l = 1, \ldots, k$ define the q-dimensional vectors

$$\mathbf{M}_l = \sum_{t=1}^{N} I_{[\mathbf{Z}_{t-1}\in C_l]}\mathbf{Y}_t$$

and

$$\mathbf{E}_l(\boldsymbol{\beta}) = \sum_{t=1}^{N} I_{[\mathbf{Z}_{t-1}\in C_l]}\boldsymbol{\pi}_t(\boldsymbol{\beta}),$$

where $I_{[\mathbf{Z}_{t-1}\in C_l]}$ is the indicator function of the set $\{\mathbf{Z}_{t-1} \in C_l\}$, for $l = 1, \ldots, k$. Furthermore, define the kq-dimensional vectors

$$\mathbf{M} = \left(\mathbf{M}_1', \ldots, \mathbf{M}_k'\right)'$$

and

$$\mathbf{E}(\boldsymbol{\beta}) = \left(\mathbf{E}_1'(\boldsymbol{\beta}), \ldots, \mathbf{E}_k'(\boldsymbol{\beta})\right)'.$$

If we let $\mathbf{I}_{t-1} = \left(I_{[\mathbf{Z}_{t-1}\in C_1]}, \ldots, I_{[\mathbf{Z}_{t-1}\in C_k]}\right)'$, then

$$\mathbf{d}(\boldsymbol{\beta}) \equiv \mathbf{M}_N - \mathbf{E}_N(\boldsymbol{\beta}) = \sum_{t=1}^{N} \mathbf{I}_{t-1} \otimes (\mathbf{Y}_t - \boldsymbol{\pi}_t(\boldsymbol{\beta})),$$

where $\otimes$ denotes Kronecker product. The following theorem generalizes Theorem 2.4.1 for binary time series. Its proof is given in the appendix.

Theorem 3.4.1 Consider the general regression model (3.3). Suppose that Assumption A holds and let $C_1, \ldots, C_k$ be a partition of $R^{p\times q}$. Then we have.

1.

$$\sqrt{N}\left(\frac{\mathbf{d}'(\boldsymbol{\beta})}{N}, (\hat{\boldsymbol{\beta}} - \boldsymbol{\beta})'\right)' \to \mathcal{N}_{p+kq}(\mathbf{0}, \boldsymbol{\Gamma}(\boldsymbol{\beta}))$$

in distribution, as $N \to \infty$, with $\mathbf{\Gamma}(\boldsymbol{\beta})$ a square matrix of dimension $p + kq$

$$\mathbf{\Gamma}(\boldsymbol{\beta}) = \begin{bmatrix} \mathbf{C}(\boldsymbol{\beta}) & \mathbf{B}'(\boldsymbol{\beta}) \\ \mathbf{B}(\boldsymbol{\beta}) & \mathbf{G}^{-1}(\boldsymbol{\beta}) \end{bmatrix}.$$

Here $\mathbf{C}(\boldsymbol{\beta}) = \oplus_{l=1}^{k} \mathbf{C}_l(\boldsymbol{\beta})$ is the direct sum of k matrices with $\mathbf{C}_l(\boldsymbol{\beta})$ a $q \times q$ symmetric matrix given by

$$\mathbf{C}_l(\boldsymbol{\beta}) = \begin{bmatrix} \int_{C_l} h_1(\boldsymbol{\beta})(1 - h_1(\boldsymbol{\beta}))\nu(d\mathbf{Z}) & \cdots & -\int_{C_l} h_1(\boldsymbol{\beta})h_q(\boldsymbol{\beta})\nu(d\mathbf{Z}) \\ \vdots & \ddots & \vdots \\ -\int_{C_l} h_1(\boldsymbol{\beta})h_q(\boldsymbol{\beta})\nu(d\mathbf{Z}) & \cdots & \int_{C_l} h_q(\boldsymbol{\beta})(1 - h_q(\boldsymbol{\beta}))\nu(d\mathbf{Z}) \end{bmatrix},$$

$\mathbf{G}(\boldsymbol{\beta})$ is the limiting $p \times p$ information matrix, and the l'th column of $\mathbf{B}(\boldsymbol{\beta})$ is given by the matrix

$$\mathbf{G}^{-1}(\boldsymbol{\beta}) \int_{C_l} \mathbf{Z}\mathbf{U}(\boldsymbol{\beta})\mathbf{\Sigma}(\boldsymbol{\beta})\nu(d\mathbf{Z}).$$

2. As $N \to \infty$,

$$\frac{\mathbf{E}(\hat{\boldsymbol{\beta}}) - \mathbf{E}(\boldsymbol{\beta})}{\sqrt{N}} - \sqrt{N}\mathbf{B}'(\boldsymbol{\beta})\mathbf{G}(\boldsymbol{\beta})(\hat{\boldsymbol{\beta}} - \boldsymbol{\beta}) \to \mathbf{0}$$

in probability, as $N \to \infty$.

3. As $N \to \infty$, the asymptotic distribution of the statistic

$$\chi^2(\boldsymbol{\beta}) = \frac{1}{N} \sum_{l=1}^{k} \mathbf{d}_l'(\boldsymbol{\beta})\mathbf{C}_l^{-1}(\boldsymbol{\beta})\mathbf{d}_l(\boldsymbol{\beta}) \tag{3.48}$$

is chi-square with kq degrees of freedom. Here $\mathbf{d}_l(\boldsymbol{\beta}) = \mathbf{M}_l - \mathbf{E}_l(\boldsymbol{\beta})$, and $\mathbf{C}_l(\boldsymbol{\beta})$ is as defined in 1, for $l = 1, \ldots, k$.

The test statistic (3.48) is a diagnostic tool that can be used for checking model adequacy. However, since the choice of the partition is quite flexible, and since the value of (3.48) depends on the partition of the covariate space, there is no escape from a certain degree of arbitrariness in the goodness of fit results. It seems therefore that the use of additional alternatives is sensible.

3.4.2 Power Divergence Family of Goodness of Fit Tests

The power divergence family of goodness of fit tests has been introduced, originally for checking model adequacy in independent data, in [110] as a generalization of the well-known Pearson and likelihood ratio test statistics. Specifically, let α_λ denote the deviation–or power divergence–between observed and expected counts defined by the distance measure,

$$\alpha_\lambda(\text{observed}, \text{expected}) = \frac{2}{\lambda(\lambda+1)}\text{observed}\left[\left(\frac{\text{observed}}{\text{expected}}\right)^\lambda - 1\right].$$

Then the power divergence family of test statistics indexed by a parameter $\lambda \in R$, $I(\lambda)$, is defined as the sum of deviations over all cells, namely,

$$I(\lambda) = \sum_{\text{cells}} \alpha_\lambda(\text{observed}, \text{expected}).$$

It is straightforward to verify that the statistic $I(\lambda)$ reduces to Pearson chi square when $\lambda = 1$ and to the likelihood ratio when $\lambda \to 0$. Some other interesting cases worth mentioning include $\lambda \to -1$, $\lambda = -1/2$ and $\lambda = -2$. For these particular values we obtain the minimum discrimination information statistic, the Freeman–Tukey statistic and the Neyman–modified chi square statistic, respectively (see Problem 10).

A thorough discussion of the properties, including large sample properties, of the power divergence family of goodness of fit tests is given in [368]. The use of $\lambda = 2/3$ is suggested there for checking model adequacy for independent data. Notice that when $\lambda = 2/3$ the resulting test statistic lies between the Pearson and likelihood ratio test statistics.

Large sample properties of the test statistic $I(\lambda)$ can be studied under either fixed or increasing cells assumptions ([368, Ch.4]). Under the assumption of fixed cells the number of cells remains fixed while all group sizes tend suitably to infinity. Under the assumption of increasing cells–or the so called sparseness assumption–the number of cells tends to infinity. These assumptions lead to different asymptotic results. Under fixed cells the asymptotic distribution of the test statistic is approximated by a chi–square distribution, with the appropriate number of degrees of freedom, given that the model is true, while the increasing cells assumption implies that the asymptotic null distribution of the test statistic is approximately normal with mean and variance depending on the parameter λ (see [340]).

3.4.3 A Family of Goodness of Fit Tests

A natural way to extend the family of power divergence test statistics to categorical time series, where the data are dependent, is to introduce the quantity

$$u_t(\beta) \equiv \frac{2}{\lambda(\lambda+1)} \sum_{j=1}^{m} Y_{tj}\left[\left(\frac{Y_{tj}}{\pi_{tj}(\beta)}\right)^\lambda - 1\right]. \tag{3.49}$$

Notice that $u_t(\beta)$ also depends on λ, but we have dropped the latter to simplify the notation. Equation (3.49) simply states that at each time instance t we calculate the deviation between the observed components of $\mathbf{Y}_t$ and the corresponding transition probabilities. Clearly, since each Y_{tj}, $j = 1, \ldots, m$, $t = 1, \ldots, N$, takes the values

0 or 1, we see that equation (3.49) is well defined for $\lambda > -1$. Summing up all the deviations over t, the quantity $\sum_{t=1}^{N} u_t(\beta)$ is the analog of the power divergence statistic adapted to dependent categorical data. Thus, if we calculate the conditional expectation of $u_t(.)$, $t = 1, \ldots, N$, under the correct model, the residual process should evolve around 0. That is, if we define

$$\begin{aligned} e_t(\beta) &\equiv \mathrm{E}\left[u_t(\beta) \mid \mathcal{F}_{t-1}\right] \\ &= \frac{2}{\lambda(\lambda+1)} \sum_{j=1}^{m} \pi_{tj}(\beta) \left[\left(\frac{1}{\pi_{tj}(\beta)} \right)^{\lambda} - 1 \right], \end{aligned} \tag{3.50}$$

the conditional expectation of $u_t(\beta)$ given the past, then the difference

$$\sum_{t=1}^{N} u_t(\beta) - \sum_{t=1}^{N} e_t(\beta)$$

is clearly a zero mean martingale by construction. It follows that if the model is correct then the centered process should fluctuate around 0. Therefore a large value of $|\sum_{t=1}^{N} u_t(\beta) - \sum_{t=1}^{N} e_t(\beta)|$ is evidence against the null hypothesis.

In applications we need to replace β by $\hat{\beta}$, the maximum partial likelihood estimator. It turns out that the quantity

$$I_N(.) = \sum_{t=1}^{N} \left[u_t(.) - e_t(.) \right] \tag{3.51}$$

evaluated at the maximum partial likelihood estimator $\hat{\beta}$, is approximated by a zero mean square integrable martingale that satisfies all the conditions for an application of a central limit theorem for martingales. In addition, the asymptotic variance of this martingale can be calculated explicitly by

$$\xi_N(\beta) = \frac{4}{\lambda^2(\lambda+1)^2} \left\{ \sum_{t=1}^{N} v_t'(\beta) \tilde{\Sigma}_t(\beta) v_t(\beta) - c_N'(\beta) \mathbf{G}_N^{-1}(\beta) c_N(\beta) \right\} \tag{3.52}$$

with

$$c_N(\beta) = \sum_{t=1}^{N} \sum_{j=1}^{m} \frac{\partial \pi_{tj}(\beta)}{\partial \beta} \left[\left(\frac{1}{\pi_{tj}(\beta)} \right)^{\lambda} - 1 \right], \tag{3.53}$$

$v_t(\beta)$ an $m \times 1$ vector with components

$$v_{tj}(\beta) = \frac{1}{\pi_{tj}^{\lambda}(\beta)} - 1, \quad j = 1, \ldots, m \tag{3.54}$$

and $\tilde{\Sigma}_t(\beta)$ with generic elements

$$\tilde{\sigma}_t^{(ij)}(\beta) = \begin{cases} -\pi_{ti}(\beta)\pi_{tj}(\beta) & \text{if } i \neq j \\ \pi_{ti}(\beta)(1 - \pi_{ti}(\beta)) & \text{if } i = j \end{cases}$$

for $i, j = 1 \ldots, m$.

The following theorem states that the sum (3.51) normalized by the expression (3.52) converges to the standard normal distribution for each fixed λ [150].

Theorem 3.4.2 Under assumption A and that model (3.3) holds, the following is true for $\lambda > -1$.

$$\mathbf{I}_{\lambda} \equiv \frac{I_N(\hat{\beta})}{\sqrt{\xi_N(\hat{\beta})}} \to \mathcal{N}$$

in distribution, as $N \to \infty$, where $\mathcal{N}$ is the standard normal random variable.

Hence, the null hypothesis is rejected for large values of $|\mathbf{I}_{\lambda}|$ relative to the standard normal distribution.

Although the asymptotic distribution of $\mathbf{I}_{\lambda}$ is independent of λ, to evaluate I_N and ξ_N we still need to choose λ. Values that fall between -1 and 1 are sensible [368]. However, the normal approximation becomes problematic for $\lambda > 1$ when the transition probabilities are small. This is so since values of $x^{-\lambda}$ for x too close to 0 and $\lambda \geq 1$ are large.

Questions regarding the power provided by the test statistic $\mathbf{I}_{\lambda}$ are largely untouched and further work remains to be done. We briefly mention that even in the case of independent data, optimality of the test has been proved in certain cases only including the equiprobable model; see [368, Ch 8.].

3.4.4 Further Diagnostic Tools

Model selection criteria and residual analysis offer some further diagnostic tools in the analysis of regression models for categorical time series.

Recall (see Section 1.6.2) that the AIC model selection criterion is defined up to a constant by $D + 2p$, where D denotes the scaled deviance (3.46) and p is the number of estimated parameters. The BIC criterion is defined by $D + p \log N$, where N is the length of the observed time series.

Residual analysis for categorical regression time series models is based either on the so called *raw* residuals

$$\hat{\mathbf{e}}_t = \begin{pmatrix} \hat{e}_{t1} \\ \hat{e}_{t2} \\ \vdots \\ \hat{e}_{tq} \end{pmatrix} = \mathbf{Y}_t - \hat{\boldsymbol{\pi}}_t = \begin{pmatrix} Y_{t1} - \hat{\pi}_{t1} \\ Y_{t2} - \hat{\pi}_{t2} \\ \vdots \\ Y_{tq} - \hat{\pi}_{tq} \end{pmatrix}, \tag{3.55}$$

or the *squared Pearson* residuals

$$\hat{r}_t = (\mathbf{Y}_t - \hat{\boldsymbol{\pi}}_t)' \hat{\boldsymbol{\Sigma}}_t^{-1} (\mathbf{Y}_t - \hat{\boldsymbol{\pi}}_t), \tag{3.56}$$

where $\hat{\boldsymbol{\Sigma}}_t = \boldsymbol{\Sigma}_t(\hat{\beta})$, for $t = 1, \ldots, N$. Notice that formally the Pearson residuals can be defined by

$$\hat{\mathbf{r}}_t = \hat{\boldsymbol{\Sigma}}_t^{-1/2} (\mathbf{Y}_t - \hat{\boldsymbol{\pi}}_t), \tag{3.57}$$

where $\hat{\mathbf{r}}_t$ is a q-dimensional vector.

3.5 EXAMPLES

The regression methodology of categorical time series can be employed in diverse applications as this section illustrates in terms of DNA, soccer and sleep data. Moreover, models such as (3.4) and (3.8) offer great flexibility and accommodate dependence by inclusion of past values of the response and other covariates when available.

3.5.1 Explanatory Analysis of DNA Sequence Data

Regression models for categorical time series can be used in explanatory analysis of DNA sequence data as shown next by model fitting and testing conditional on past response values.

A DNA sequence consists of four nucleotides differing only in the nitrogenous base, whose order determines the genetic information of each organism. The four nucleotides are given one letter shorthand abbreviation as follows:

- A is for adenine
- G is for guanine
- C is for cytosine
- T is for thymine

Adenine and guanine are purines–the larger of the two types of bases found in DNA–whereas cytosine and thymine are pyrimidines.

Thus a strand of DNA can be represented as a sequence of letters from { A, C, G, T } and can be viewed as a *nominal* categorical time series with the assignment A=1, C=2, G=3 and T=4. For more see [421].

We present an explanatory analysis for DNA sequence data of the gene BNRF1 of the Epstein–Barr Virus (see [388, Sec. 5.9]) considering only the first 1000 observations–the whole data set is 3954 long. Table 3.2 reports the first 100 observations of these data. The idea is to apply the multinomial logits model (3.4) by fitting a series of various order models. For example a first-order model is given by

$$\log\left(\frac{\pi_{ti}(\beta)}{\pi_{t4}(\beta)}\right) = \beta_{i0} + \beta_{i1}Y_{(t-1)1} + \beta_{i2}Y_{(t-1)2} + \beta_{i3}Y_{(t-1)3}, \tag{3.58}$$

for $i = 1, 2, 3$, and is denoted by $1 + \mathbf{Y}_{t-1}$. A second-order model is labeled $1+\mathbf{Y}_{t-1}+\mathbf{Y}_{t-2}$ and consists of (3.58) plus a linear combination in terms of $Y_{(t-2)1}$, $Y_{(t-2)2}$, $Y_{(t-2)3}$, and so on.

Table 3.3 lists the models applied to the DNA sequence data and Table 3.4 reports the inferential results where the last two columns give the values of the likelihood ratio test statistic (3.43) together with its p-values for testing the order of the model. Thus the value 13.48 is not significant and therefore we might not include a lagged value of the response of order 2. In other words, the hypothesis that $\mathbf{Y}_{t-2}$ should not enter the regression equation is accepted. Similarly, the p-value of 0.2121 casts a

Table 3.2 First 100 records from the gene BNRF1 of the Epstein–Barr virus DNA sequence data. Here A=1, C=2, G=3 and T=4. Read across and down. *Source*: European Molecular Biology Laboratory, `http://www.embl-heidelberg.de/`.

1	4	3	3	1	1	3	1	3	1	3	3	3	3	2
1	3	3	3	1	1	1	2	3	2	1	1	1	4	3
2	2	3	3	4	4	3	2	2	2	3	3	4	1	4
3	3	3	3	3	2	2	2	3	4	4	4	1	4	4
1	4	3	3	4	1	1	3	3	2	4	2	4	4	2
3	3	3	2	1	1	3	1	4	3	3	1	3	1	3
3	2	1	1	1	2	1	4	1	2					

Table 3.3 Candidate models for the gene BNRF1 of the Epstein–Barr virus DNA sequence data

Model 1	$1 + \mathbf{Y}_{t-1}$
Model 2	$1 + \mathbf{Y}_{t-1} + \mathbf{Y}_{t-2}$
Model 3	$1 + \mathbf{Y}_{t-1} + \mathbf{Y}_{t-2} + \mathbf{Y}_{t-3}$
Model 4	$1 + \mathbf{Y}_{t-1} + \mathbf{Y}_{t-2} + \mathbf{Y}_{t-3} + \mathbf{Y}_{t-4}$

doubt on the inclusion of $\mathbf{Y}_{t-4}$, while a p-value of 0.0698 indicates that the inclusion of $\mathbf{Y}_{t-3}$ is reasonable. The last column is constructed by appealing to the chi–square distribution with 9 degrees of freedom (why?) Throughout the analysis we use $N = 996$ observations.

Table 3.4 Comparison of different multinomial logits models for the gene BNRF1 of the Epstein–Barr virus DNA sequence data. $N = 996$.

Model	p	D	AIC	BIC	λ_N	p-value
Independence	3	2711.31	2717.31	2732.02		
Model 1	12	2677.75	2701.75	2760.60	33.56	0.0001
Model 2	21	2664.27	2706.27	2809.25	13.48	0.1420
Model 3	30	2648.41	2708.41	2855.52	15.86	0.0698
Model 4	39	2639.39	2714.39	2905.63	12.02	0.2121

The results from Table 3.4 show that the AIC criterion is minimized for Model 1 whereas the BIC criterion is minimized for the independence model. However the likelihood ratio test also indicates that an adequate model for the data at hand consists of an intercept, $\mathbf{Y}_{t-1}$ and $\mathbf{Y}_{t-3}$. For $1 + \mathbf{Y}_{t-1} + \mathbf{Y}_{t-3}$ the number of estimated parameters is $p = 21$, $D = 2663.20$, BIC=2808.17, AIC=2705.20 and its p-value is 0.0968 when compared with Model 3. The estimated parameters for this model are given in Table 3.5 together with their standard errors in parentheses.

Figure 3.10 shows the sample autocorrelation plot of the squared Pearson residuals (3.56), suggesting that the fit is quite reasonable. Further evidence of this fact is manifested in Table 3.6 where the values of the power divergence statistic $\mathbf{I}_\lambda$ (recall Theorem 3.4.2) are tabulated for different λ. It is seen that, all the $\mathbf{I}_\lambda$ values are less that 1.96 in absolute value, which confirms from another point of view that model $1 + \mathbf{Y}_{t-1} + \mathbf{Y}_{t-3}$ is adequate.

Table 3.5 Estimated parameters β_{ij}, $i = 1, 2, 3$, $j = 0, \ldots, 6$ for model $1 + \mathbf{Y}_{t-1} + \mathbf{Y}_{t-3}$ with their standard errors for the gene BNRF1 of the Epstein–Barr virus DNA sequence data.

i	1	$Y_{(t-1)1}$	$Y_{(t-1)2}$	$Y_{(t-1)3}$	$Y_{(t-3)1}$	$Y_{(t-3)2}$	$Y_{(t-3)3}$
1	−0.908 (0.326)	0.541 (0.342)	0.665 (0.303)	1.071 (0.321)	0.534 (0.320)	0.167 (0.301)	0.787 (0.293)
2	−0.438 (0.290)	0.423 (0.301)	0.288 (0.269)	0.904 (0.285)	0.558 (0.301)	0.486 (0.274)	0.784 (0.277)
3	0.165 (0.268)	0.266 (0.283)	−0.412 (0.264)	0.584 (0.271)	0.262 (0.293)	0.320 (0.262)	0.422 (0.270)

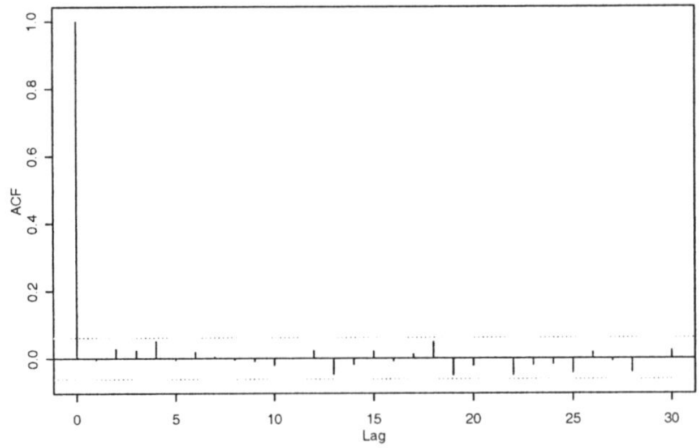

Fig. 3.10 Sample autocorrelation function of the squared Pearson residuals corresponding to model $1 + \mathbf{Y}_{t-1} + \mathbf{Y}_{t-3}$ for the gene BNRF1 of the Epstein–Barr virus DNA sequence data.

The transition probabilities

$$\mathrm{P}(Y_t = i \mid Y_{t-1} = j, Y_{t-3} = l),$$

for $i, j, k = 1, 2, 3, 4$ are estimated by substitution of the maximum partial likelihood estimators into the regression equation of π_{ti} using (3.2). Table 3.7 reports the

Table 3.6 Values of the power divergence statistic $\mathbf{I}_\lambda$ for model $1 + \mathbf{Y}_{t-1} + \mathbf{Y}_{t-3}$ for the gene BNRF1 of the Epstein–Barr virus DNA sequence data.

λ	−0.800	−0.400	−0.100	0.300	0.600	0.800	1.000	1.20(
Value	−1.235	−1.056	−0.938	−0.774	−0.640	−0.556	−0.475	−0.39(

Table 3.7 Estimated transition matrix from model $1 + \mathbf{Y}_{t-1} + \mathbf{Y}_{t-3}$ for the for the gene BNRF1 of the Epstein–Barr virus DNA sequence data.

Y_{t-3}	Y_t	Y_{t-1} A	C	G	T
A	A	0.2004	0.2756	0.2352	0.1583
	C	0.2915	0.3097	0.3257	0.2592
	G	0.3389	0.2089	0.3219	0.3526
	T	0.1692	0.2058	0.1172	0.2299
C	A	0.1479	0.2107	0.1763	0.1149
	C	0.2889	0.3179	0.3279	0.2526
	G	0.3828	0.2443	0.3692	0.3916
	T	0.1804	0.2271	0.1266	0.2409
G	A	0.2167	0.2972	0.2511	0.1738
	C	0.3069	0.3251	0.3384	0.2770
	G	0.3342	0.2053	0.3135	0.3531
	T	0.1422	0.1724	0.0970	0.1961
T	A	0.1643	0.2289	0.2001	0.1249
	C	0.2335	0.2513	0.2715	0.1997
	G	0.3652	0.2279	0.3596	0.3656
	T	0.2370	0.2919	0.1688	0.3098

transition probabilities among the different states where, for example, if $Y_{t-3} = \text{A}$ and $Y_{t-1} = \text{T}$, then the transition probability to $Y_t = \text{C}$ is equal to 0.2592.

3.5.2 Soccer Forecasting

A popular weekly game in Greece is that of forecasting soccer games outcomes. Each week a list of 13 soccer games is published by the Greek Organization of Forecasting Soccer Games in the form "Team A vs. Team B", where Team A plays at home. The 13 pairs vary every week. The published list usually consists of games played by the Greek First National League but occasionally some other games, either from the Greek Second National League or from a foreign league, enter the list. A potential bettor is challenged to forecast either 13, 12 or 11 correct outcomes by using the symbols "1" (Team A wins), "X" (a tie) or "2" (Team B" wins).

The data consist of the true outcomes of the games for the first four positions on the list starting from 3/5/1995 and ending on 10/29/2000, a total of 289 sequential observations[1]. Somewhat oddly, we record "X" as 3, and point out that there are some weeks when the gambling game didn't run on schedule. However, for data analysis purposes, we view these data as a regular *ordinal* time series with ordered categories "1", "X" and "2".

Table 3.8 reports the frequencies of the categories "1", "X", "2". Thus in the first position, among all the soccer games corresponding to the first four positions on the list, 166 games ended up a win for the home team, 63 games were a tie and 60 games were a loss. Figure 3.11 depicts time series plots of the first 150 outcomes for these data for the different positions. For each position, the weakly occurrence of the ordinal values "1", "X"(="3"), "2" defines a categorical time series.

Table 3.8 Frequencies for the Soccer Forecasting Data

	"1"	"X"	"2"
Position 1	166	63	60
Position 2	150	71	68
Position 3	156	57	76
Position 4	155	63	71

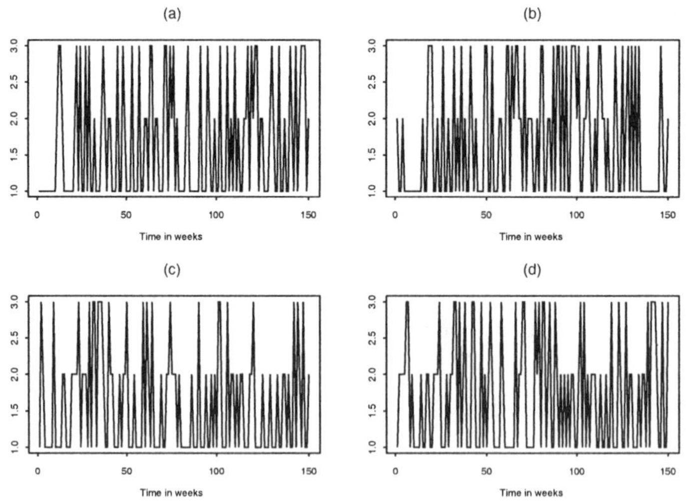

Fig. 3.11 Time series plot of the first 150 observations for the soccer forecasting data. (a) Outcomes of first position. (b) Outcomes of second position. (c) Outcomes of third position. (d) Outcomes of fourth position.

[1]Many thanks to the Greek Organization for Forecasting Soccer Games for providing the data.

Table 3.9 Comparison of different proportional odds models for the soccer forecasting data. $N = 287$.

Time Series	Model	p	D	AIC	BIC
Position 1	Independence	2	562.43	566.43	573.74
	$1+\mathbf{Y}_{t-1}$	4	562.15	570.15	584.79
	$1+\mathbf{Y}_{t-1}+\mathbf{Y}_{t-2}$	6	561.73	573.73	595.69
Position 2	Independence	2	588.63	592.63	599.95
	$1+\mathbf{Y}_{t-1}$	4	586.84	594.84	609.47
	$1+\mathbf{Y}_{t-1}+\mathbf{Y}_{t-2}$	6	578.09	590.09	612.05
Position 3	Independence	2	575.97	579.97	587.28
	$1+\mathbf{Y}_{t-1}$	4	574.13	582.13	596.77
	$1+\mathbf{Y}_{t-1}+\mathbf{Y}_{t-2}$	6	569.18	581.18	603.13
Position 4	Independence	2	580.33	584.33	591.65
	$1+\mathbf{Y}_{t-1}$	4	580.12	588.12	602.75
	$1+\mathbf{Y}_{t-1}+\mathbf{Y}_{t-2}$	6	579.93	591.93	613.88

We investigate whether there is dependence among the games by analyzing these categorical time series. For each time series we fit a proportional odds model (3.8) with lagged values of the response up to order 2 as covariates. The results with $N = 287$ are summarized in Table 3.9. Let us consider the first-order model fitted for position 2. According to (3.8) and with suggestive notation,

$$\log\left[\frac{\mathrm{P}(Y_t \leq \text{"1"} \mid \mathcal{F}_{t-1})}{\mathrm{P}(Y_t > \text{"1"} \mid \mathcal{F}_{t-1})}\right] = \theta_1 + \gamma_1 Y_{(t-1)1} + \gamma_2 Y_{(t-1)2},$$

$$\log\left[\frac{\mathrm{P}(Y_t \leq \text{"X"} \mid \mathcal{F}_{t-1})}{\mathrm{P}(Y_t > \text{"X"} \mid \mathcal{F}_{t-1})}\right] = \theta_2 + \gamma_1 Y_{(t-1)1} + \gamma_2 Y_{(t-1)2}.$$

The corresponding estimators are $\hat{\theta}_1 = 0.251$, $\hat{\theta}_2 = 1.368$, $\hat{\gamma}_1 = -0.131$ and $\hat{\gamma}_2 = -0.415$ and their standard errors are $0.244, 0.257, 0.287$ and 0.327, respectively.

From Table 3.9, the BIC criterion is minimized for the independence model for all positions, while the AIC criterion is minimized for the independence model with the single exception of position 2. Similarly, the deviance is not reduced significantly when entering the lagged regressors into the model equation. It seems therefore reasonable to conclude that the independence model is quite adequate for the soccer forecasting data and that the betting game is fair. As expected from home games, "1" is more frequent than "X" and "2": "1" appears roughly 50% of the times, while the relative frequency of "X" and "2" is about 25% each.

3.5.3 Sleep Data Revisited

The sleep data which have been discussed briefly–see Section 2.5.4 and Figure 3.1–consist of sleep state measurements of a newborn infant together with his heart rate

(R_t) and temperature (T_t) sampled every 30 seconds[2]. Recall that the sleep states are classified as

$$\begin{array}{rl} 1: & \text{quiet sleep,} \\ 2: & \text{indeterminate sleep,} \\ 3: & \text{active sleep,} \\ 4: & \text{awake.} \end{array}$$

The total number of observations is equal to 1024 and a plot of the data is displayed in Figure 3.12. The objective is to predict–or classify–the sleep state based on covariate information. In this respect, Figure 3.12 shows that sleep state depends on heart rate–higher values of heart rate tend to correspond to state 4.

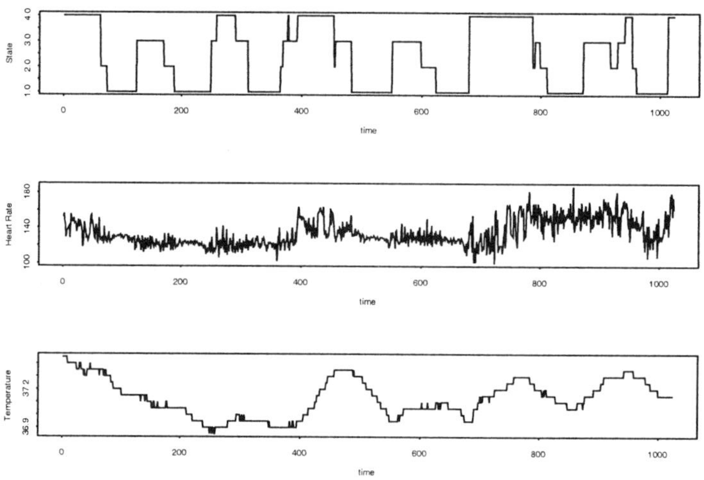

Fig. 3.12 Time series plot for the sleep data. $N = 1024$.

To begin analyzing these data, notice that the response–the sleep state, say Y_t–is an ordered time series in the sense that "4" < "1" < "2" < "3"; that is, the response increases from awake to active sleep. By removing the first two observations we fit several proportional odds models (3.8) to the data using only the subsequent 700 observations. The remaining 322 data records are used as a testing data set. The results of the analysis are summarized in Table 3.10.

Our analysis indicates that a sensible model for the sleep data includes $\mathbf{Y}_{t-1}$ and the logarithm of heart rate ($\log R_t$). Comparing Model 1 with Model 7, we notice that $\mathbf{Y}_{t-1}$ is clearly a significant predictor. In addition the deviance difference between Model 2 and Model 1 is 2.05 (p-value=0.1522) suggesting that the logarithm of heart rate may be included in the model. Models 3,4,5, and 6 do not substantially enhance the fitted model leading to the conclusion that temperature, and higher order lagged

[2]Many thanks to F. Sapatinas for providing the sleep data.

Table 3.10 Comparison of different proportional odds models for the sleep data. $N = 700$.

Model	Covariates	p	D	AIC	BIC
1	$1+\mathbf{Y}_{t-1}$	6	389.56	401.56	428.86
2	$1+\mathbf{Y}_{t-1}+\log R_t$	7	387.51	401.51	433.37
3	$1+\mathbf{Y}_{t-1}+\log R_t+T_t$	8	387.32	403.32	439.73
4	$1+\mathbf{Y}_{t-1}+T_t$	7	389.52	403.52	435.38
5	$1+\mathbf{Y}_{t-1}+\mathbf{Y}_{t-2}+\log R_t$	10	387.28	407.28	452.79
6	$1+\mathbf{Y}_{t-1}+\log R_{t-1}$	7	389.40	403.40	435.26
7	$1+\log R_t$	4	1684.31	1692.31	1710.51

values of the response, are not significant predictors. These considerations lead to the following model

$$\log\left[\frac{\mathrm{P}(Y_t \leq \text{"4"} \mid \mathcal{F}_{t-1})}{\mathrm{P}(Y_t > \text{"4"} \mid \mathcal{F}_{t-1})}\right] = \theta_1 + \gamma_1 Y_{(t-1)1} + \gamma_2 Y_{(t-1)2} + \gamma_3 Y_{(t-1)3} + \gamma_4 \log R_t,$$

$$\log\left[\frac{\mathrm{P}(Y_t \leq \text{"1"} \mid \mathcal{F}_{t-1})}{\mathrm{P}(Y_t > \text{"1"} \mid \mathcal{F}_{t-1})}\right] = \theta_2 + \gamma_1 Y_{(t-1)1} + \gamma_2 Y_{(t-1)2} + \gamma_3 Y_{(t-1)3} + \gamma_4 \log R_t,$$

$$\log\left[\frac{\mathrm{P}(Y_t \leq \text{"2"} \mid \mathcal{F}_{t-1})}{\mathrm{P}(Y_t > \text{"2"} \mid \mathcal{F}_{t-1})}\right] = \theta_3 + \gamma_1 Y_{(t-1)1} + \gamma_2 Y_{(t-1)2} + \gamma_3 Y_{(t-1)3} + \gamma_4 \log R_t,$$

with $\hat{\theta}_1 = -30.352$, $\hat{\theta}_2 = -23.493$, $\hat{\theta}_3 = -20.349$, $\hat{\gamma}_1 = 16.718$, $\hat{\gamma}_2 = 9.533$, $\hat{\gamma}_3 = 4.755$ and $\hat{\gamma}_4 = 3.556$. The corresponding standard errors are 12.051, 12.012, 11.985, 0.872, 0.630, 0.501 and 2.470.

Model 2 is applied to the testing data set which consists of 322 measurements. Figure 3.13 displays time series plots of the observed and predicted sleep states for the testing data set. The predicted responses are obtained by the following simple rule

$$Y_t = j \leftrightarrow \max_k \hat{\pi}_{tk} = \hat{\pi}_{tj},$$

that is, category j is chosen if and only if its estimated transition probability is the maximum among the estimated transition probabilities. The misclassification rate of Model 2 is 0.034.

Table 3.11 Values of the power divergence statistic $\mathbf{I}_\lambda$ for Model 2 of Table 3.10 applied to sleep data.

λ	-0.500	5.000	5.500	6.000	6.500	7.000
Value	8.813	4.581	2.767	1.667	1.002	0.601

Some further diagnostics for Model 2 are given in Figure 3.14 and Table 3.11. Figure 3.14 displays cumulative periodogram plots ([357]) for the Pearson residuals defined by (3.57) together with 95% confidence bands. Notice that for the sleep data $m = 4$, $q = 3$, so Figure 3.14 (a) corresponds to the cumulative periodogram of

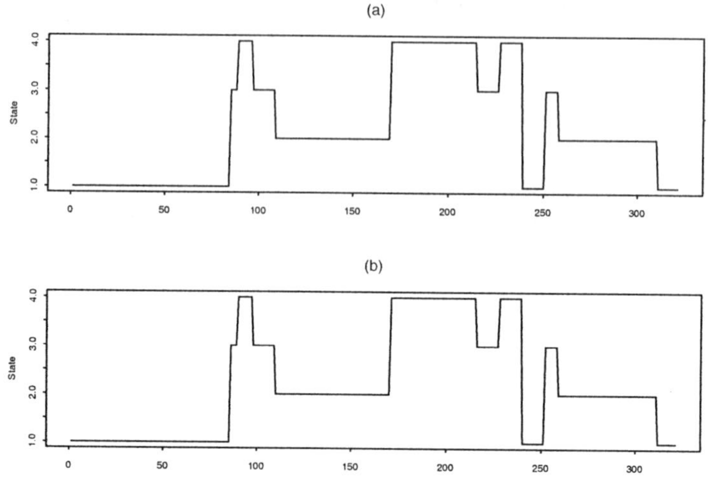

Fig. 3.13 (a) Observed versus (b) predicted sleep states for Model 2 of Table 3.10 applied to the testing data set. $N = 322$.

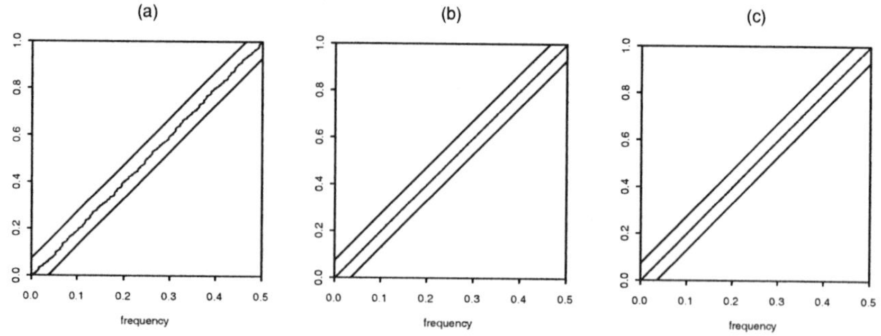

Fig. 3.14 Cumulative Periodogram plots for the Pearson residuals from Model 2 of Table 3.10 applied to sleep data.

the Pearson residuals for the first category and so on. In all these cases we observe that the residual processes correspond to white noise. Table 3.11 lists the values of the power divergence statistic $\mathbf{I}_\lambda$ for different λ. We notice that for some λ the test is reassuring but this conclusion is not uniform. For an alternative approach to the problem of sleep state prediction using wavelet methods see [335].

3.6 ADDITIONAL TOPICS

3.6.1 Alternative Modeling

Models for discrete valued time series provide alternative approaches to categorical time series modeling. Important examples include higher order Markov chains [25], [365],[113], and discrete autoregressive moving average (DARMA) [221], [222]. Another useful class is that of variable length Markov chains (VLMC) defined on a finite state space, where the Markov property is retained with a variable order [76].

Another source of models are various transformations of an underlying process. Notable examples are categorical time series generated by "clipping" or "hard limiting" of a Gaussian process [251], [252]. For an interesting extension of this to "discrete images" obtained by quantizing a Gaussian random field see [277] and [255]. Interestingly, under stationarity, parameters in the original series/field can be estimated quite effectively from the quantized data using very few (e.g. 3) quantization levels. We mention [258] as another example whereby a binary time series is generated according to an underlying strictly stationary but unobserved process. The connection between Hidden Markov models and categorical time series has been explored in [306].

3.6.2 Spectral Analysis

Spectral analysis, a topic indigenous to time series, deserves serious consideration especially when the goal is to discover periodic components in the data ([357]). Thus we are led to consider the spectrum of a categorical time series. However, due to the qualitative nature of nominal data, the notion of spectrum is problematic. Recent work in this area in [401] and [402] attacks the problem by introducing the notions of scaling, assigning numerical values to the categories, and that of spectral envelope for selecting scales.

3.6.3 Longitudinal Data

Various authors have considered the analysis of longitudinal categorical data. For a survey of results in this area see the early article by [23] and the recent works [3], [4], [327], and [346]. Inference for longitudinal multinomial data is mostly based on generalized estimating equation approach (see Section 1.7.1). Some key references include [403] where the authors suggest a method for comparing ordered categorical responses in two groups of subjects observed repeatedly allowing for time dependent covariates and missing observations, and more recently [101], [143], [206], [325], [408], and [433]. The authors in [207] propose the *lorelogram*, which can be used as data analysis tool for exploring dependence in longitudinal categorical responses, and in [275] the authors develop a Markov chain model for repeated ordinal data in continuous time.

3.7 PROBLEMS AND COMPLEMENTS

1. *Multinomial logistic regression and semiparametric estimation* [156].

   Consider a categorical random variable y such that $P(y = j) = \pi_j$, $j = 1, ..., m$, and $\sum_{j=1}^{m} \pi_j = 1$. Let x be a continuous random variable, and define the conditional densities $f(x|y = j) = g_j(x)$, $j = 1, ..., m$. Assume that

$$P(y = j|x) = \frac{\exp(\alpha_j^* + \beta_j h(x))}{1 + \sum_{k=1}^{q} \exp(\alpha_k^* + \beta_k h(x))}, \quad j = 1, \ldots, m,$$

   where $h(x)$ is an arbitrary but known function of x.

   (a) Use Bayes theorem to show that

$$\frac{g_j(x)}{g_m(x)} = \exp(\alpha_j + \beta_j h(x)),\ j = 1, ..., q$$

   with $\alpha_j = \alpha_j^* + \log[\pi_m/\pi_j]$, $j = 1, ..., q$, $q = m - 1$.

   (b) Let $g_m(x) \equiv g(x)$, and let $G(x)$ be the corresponding cumulative distribution function (cdf). Suppose $\mathbf{x}_j = (x_{j1}, ..., x_{jn_j})'$ is a sample of size n_j from $g_j(x)$, $j = 1, ..., m$. Denote the combined data from the m samples by $\mathbf{t}$, $\mathbf{t} = (t_1, ..., t_n)' = (\mathbf{x}_1', ..., \mathbf{x}_q', \mathbf{x}_m')'$ where $n = n_1 + \cdots + n_q + n_m$. Use the combined data $\mathbf{t}$ to obtain a nonparametric estimator for $G(x)$. That is, argue that the maximum likelihood estimator of $G(x)$ can be obtained from the combined data $\mathbf{t} = (t_1, ..., t_n)'$ by maximizing the likelihood over the class of step cdf's with jumps at the observed values $t_1, ..., t_n$. Accordingly, if $p_i = dG(t_i)$, $i = 1, .., n$, the likelihood becomes,

$$\mathcal{L}(\boldsymbol{\alpha}, \boldsymbol{\beta}, G) = \prod_{i=1}^{n} p_i \prod_{j=1}^{n_1} \exp(\alpha_1 + \beta_1 h(x_{1j})) \cdots \prod_{j=1}^{n_q} \exp(\alpha_q + \beta_q h(x_{qj}))$$

   and the problem is to estimate the p_i subject to the $q + 1 = m$ constraints

$$\sum_{i=1}^{n} p_i = 1,\ \sum_{i=1}^{n} p_i[w_1(t_i) - 1] = 0, ..., \sum_{i=1}^{n} p_i[w_q(t_i) - 1] = 0$$

   where $w_j(t) = \exp(\alpha_j + \beta_j h(t))$, $j = 1, ..., q$, while holding $\boldsymbol{\alpha} = (\alpha_1, ..., \alpha_q)'$, $\boldsymbol{\beta} = (\beta_1, ..., \beta_q)'$ fixed.

   (c) Show that

$$\hat{p}_i = \frac{1}{n_m} \cdot \frac{1}{1 + \rho_1 \exp(\hat{\alpha}_1 + \hat{\beta}_1 h(t_i)) + \cdots + \rho_q \exp(\hat{\alpha}_q + \hat{\beta}_q h(t_i))}$$

   and therefore,

$$\hat{G}(t) = \frac{1}{n_m} \cdot \sum_{i=1}^{n} \frac{I(t_i \leq t)}{1 + \rho_1 \exp(\hat{\alpha}_1 + \hat{\beta}_1 h(t_i)) + \cdots + \rho_q \exp(\hat{\alpha}_q + \hat{\beta}_q h(t_i))},$$

where $\hat{\alpha}, \hat{\beta}$, are obtained from the profile log-likelihood (up to a constant),

$$l = - \sum_{i=1}^{n} \log[1 + \rho_1 w_1(t_i) + \cdots + \rho_q w_q(t_i)] + \sum_{j=1}^{n_1}[\alpha_1 + \beta_1 h(x_{1j})] + \cdots + \sum_{j=1}^{n_q}[\alpha_q + \beta_q h(x_{qj})]$$

and $\rho_j = n_j/n_m, j = 1, ..., q$.

(d) Argue that the maximum likelihood estimators $\hat{\alpha}, \hat{\beta}$, are asymptotically normal,

$$\sqrt{n}\begin{pmatrix} \hat{\alpha} - \alpha \\ \hat{\beta} - \beta \end{pmatrix} \rightarrow \mathcal{N}_{2q}(\mathbf{0}, \mathbf{\Sigma})$$

in distribution as $n \rightarrow \infty$, where $\mathbf{\Sigma} = \mathbf{S}^{-1}\mathbf{V}\mathbf{S}^{-1}$, for some matrices $\mathbf{S}$ and $\mathbf{V}$.

(e) Obtain a test for the hypothesis $\mathrm{H}_0 : \beta_1 = \cdots = \beta_q = 0$, and interpret the results.

2. *Some properties of Markov Chains* [306]. Suppose that $\{Y_t\}$ is an irreducible homogeneous Markov chain on $\{1, 2, \ldots, m\}$ with transition probability matrix $\mathbf{Q}$. In other words, $\mathbf{Q} = (q_{ij})$, with

$$q_{ij} = \mathrm{P}\left[Y_t = j \mid Y_{t-1} = i\right].$$

Since $\{Y_t\}$ is irreducible, there exists unique, strictly positive stationary distribution which we denote by $\boldsymbol{\delta} = (\delta_1, \delta_2 \ldots, \delta_m)'$. Assume further that $\{Y_t\}$ is stationary so that $\boldsymbol{\delta}$ is the distribution of $\{Y_t\}$ for all t. If $\mathbf{u} = (1, 2, \ldots, m)'$ and $\mathbf{U} = \mathrm{diag}(\mathbf{u})$, show that

(a)

$$\mathrm{E}\left[Y_t\right] = \boldsymbol{\delta}'\mathbf{u}$$

and

$$\mathrm{E}\left[Y_t Y_{t+k}\right] = \boldsymbol{\delta}'\mathbf{U}\mathbf{Q}^k\mathbf{u}.$$

(b) Suppose further that $\mathbf{Q}$ can be written as

$$\mathbf{Q} = \mathbf{V}\mathbf{\Lambda}\mathbf{V}^{-1}$$

with $\mathbf{\Lambda} = \mathrm{diag}(1, \lambda_2, \ldots, \lambda_m)$, where λ_i, $i = 2, \ldots, m$ are the eigenvalues of $\mathbf{Q}$ other than 1, and the columns (respectively rows) of $\mathbf{V}$ (respectively $\mathbf{V}^{-1}$) are the right (respectively left) eigenvectors of $\mathbf{Q}$. Show that

$$\mathrm{Cov}\left(Y_t, Y_{t+k}\right) = \sum_{i=2}^{m} a_i b_i \lambda_i^k,$$

for suitable a_i, b_i. What is the form of the correlation function in this case?

3. Suppose that $\{Y_t\}$ denotes a Markov chain of order p and m states. Show that the total number of free parameters needed to be estimated is equal to $m^p(m-1)$.

4. *Inference for Markov chains* [306]. Let $\{Y_t\}$ denote a first-order Markov chain as in Problem 2. Estimation of the unknown transition probabilities is accomplished by conditional likelihood. Letting n_{ij} to be the number of transitions from state i to state j, argue that the conditional likelihood of the observed data given the first observation is

$$\prod_{i=1}^{m}\prod_{j=1}^{m} q_{ij}^{n_{ij}}.$$

Conclude that the conditional maximum likelihood estimators of q_{ij} are

$$\frac{n_{ij}}{\sum_{k=1}^{m} n_{ik}}.$$

5. *Multinomial logits derived by utility* [315]. Let $V_1, V_2, \ldots, V_m$ be random variable such that

$$V_i = v_i + \epsilon_i$$

with v_i being a fixed value and ϵ_i, $i = 1, \ldots, m$ independent and identically distributed random variables with common cumulative distribution function F and density f. The *principle of maximum random utility* states that, if V_i, $i = 1, \ldots, m$ are unobservable, then an observable categorical random variable Y takes on its values according to

$$Y = i \Longleftrightarrow V_i = \max_{l=1,\ldots,m} V_l. \tag{3.59}$$

(a) Show that (3.59) implies

$$P(Y = i) = \int_{-\infty}^{\infty} \prod_{s \neq i} F(v_i - v_s + \epsilon) f(\epsilon) d\epsilon. \tag{3.60}$$

(b) Put

$$F(x) = \exp\left(-\exp(-x)\right)$$

in (3.60) to prove that

$$P(Y = i) = \frac{\exp(v_i)}{\sum_{l=1}^{m} \exp(v_l)}.$$

In other words, the principle of the maximum utility leads to the multinomial logit model.

6. Show that the continuation ratio model (3.9) is a special case of (3.3).

7. Prove that the logit function (3.30) defines the canonical link for a generalized linear model when the random component is the multinomial distribution.

8. Following the calculations in Section 3.3.1, prove that the observed information matrix is given by

$$\mathbf{H}_N(\boldsymbol{\beta}) = -\nabla\nabla' l(\boldsymbol{\beta}) = \mathbf{G}_N(\boldsymbol{\beta}) - \mathbf{R}_N(\boldsymbol{\beta}),$$

where

$$\mathbf{G}_N(\boldsymbol{\beta}) = \sum_{t=1}^{N} \mathbf{Z}_{t-1}\mathbf{U}_t(\boldsymbol{\beta})\boldsymbol{\Sigma}_t(\boldsymbol{\beta})\mathbf{U}'_t(\boldsymbol{\beta})\mathbf{Z}'_{t-1},$$

and

$$\mathbf{R}_N(\boldsymbol{\beta}) = \sum_{t=1}^{N}\sum_{r=1}^{2} \mathbf{Z}_{t-1}\mathbf{W}_{tr}(\boldsymbol{\beta})\mathbf{Z}'_{t-1}\left(Y_{tr} - \pi_{tr}(\boldsymbol{\beta})\right)$$

with

$$\mathbf{W}_{tr}(\boldsymbol{\beta}) = \left[\frac{\partial^2 u_r(\boldsymbol{\eta}_t)}{\partial\boldsymbol{\eta}_t\partial\boldsymbol{\eta}'_t}\right],$$

for $r = 1, 2$.

9. Suppose that $\{X_N\}_{N=1}^{\infty}$ and $\{Y_N\}_{N=1}^{\infty}$ are both sequences of positive random variables. In addition, assume that $Y_N \to c$ in probability, as $N \to \infty$, with c denoting a constant. Show that

$$|\mathrm{P}(X_N \le Y_N) - P(X_N \le c)| \to 0$$

as $N \to \infty$.

10. *Power divergence family of tests* [110]. Recall that the power divergence family of test statistics indexed by a parameter $\lambda \in R$, $I(\lambda)$, is defined as the sum of deviations over all cells, namely,

$$I(\lambda) = \sum_{\text{cells}} \alpha_\lambda(\text{observed}, \text{expected}),$$

where

$$\alpha_\lambda(\text{observed}, \text{expected}) = \frac{2}{\lambda(\lambda+1)}\text{observed}\left[\left(\frac{\text{observed}}{\text{expected}}\right)^{\lambda} - 1\right].$$

Verify that $I(\lambda)$ reduces to Pearson X^2 when $\lambda = 1$ and to the likelihood ratio when $\lambda \to 0$. In addition show that as $\lambda \to -1$, $\lambda = -1/2$ and $\lambda = -2$ we obtain the minimum discrimination information statistic, the Freeman–Tukey statistic and the Neyman–modified X^2 statistic, respectively.

11. Show that Lindeberg's condition (A.1) is satisfied following the proof of Lemma A.0.1.

12. Prove assertion A.16.

13. Prove equations 3.46 and 3.47.

Appendix: Asymptotic Theory

This appendix describes in detail the essential steps for proving Theorem 3.3.1. We follow [152] where the asymptotic normality of the maximum partial likelihood estimator has been recently established extending the results in [246]. We emphasize that the method of proof can be used quite generally for any multivariate generalized linear model provided that Assumption A is satisfied.

The notation $\mathbf{B}^{1/2}$ stands for the right Cholesky square root of a positive definite matrix $\mathbf{B}$. That is, $\mathbf{B}^{1/2}$ is the unique upper triangular matrix with positive elements such that $\mathbf{B} = (\mathbf{B}^{1/2})'\mathbf{B}^{1/2}$. We put $\mathbf{B}^{t/2} = (\mathbf{B}^{1/2})'$.

We first show that the score process (3.34) is a zero–mean square integrable martingale that converges in distribution, after appropriate normalization, to a normal random variable. Recall that a stochastic process $\{X_t\}$, $t = 1, 2, \ldots$, with a finite expectation is a martingale if it has the property, $\mathrm{E}\left[X_t|X_1, \ldots, X_{t-1}\right] = X_{t-1}$ almost surely (e.g. [51]) . Square integrability means that given a martingale $\{X_t\}$, $t = 1, 2, \ldots$, then $\mathrm{E}\left[||X_t||^2\right] < \infty$.

The following lemma asserts that the score process satisfies the conditions of the central limit theorem for martingales.

Lemma A.0.1 Consider the model (3.3) and suppose that Assumption A holds. Then the partial score process $\{\mathbf{S}_t(\beta), \mathcal{F}_t\}$ $t = 1, 2, \ldots$, is a zero mean square integrable martingale such that

$$\mathbf{F}_N(\beta)^{-1/2}\mathbf{S}_N(\beta) \rightarrow \mathcal{N}_p$$

in distribution as $N \to \infty$, with $\mathcal{N}_p$ denoting a standard p–dimensional normal random vector.

Proof: The fact that the partial score process is a zero mean square integrable martingale follows from (3.33) and assumption A.2. To show it actually converges in distribution, we consider $\phi_N(\beta) = \lambda'\mathbf{S}_N(\beta)$, with $\lambda \in R^p$, having in mind the Cramér-Wold device (e.g. [51]). Then $\phi_N(\beta)$ is a univariate zero-mean martingale. Its conditional and unconditional covariance matrices are $\lambda'\mathbf{G}_N(\beta)\lambda$ and $\lambda'\mathbf{F}_N(\beta)\lambda$ respectively. Thus

$$\frac{\lambda'\mathbf{G}_N(\beta)\lambda}{\lambda'\mathbf{F}_N(\beta)\lambda} = \frac{\lambda'\mathbf{G}_N(\beta)\lambda/N}{\lambda'\mathbf{F}_N(\beta)\lambda/N} \rightarrow \frac{\lambda'\mathbf{G}(\beta)\lambda}{\lambda'\mathbf{G}(\beta)\lambda} = 1$$

in probability, upon invoking A.2 and A.5. Furthermore, by letting $I_{Nt}(\epsilon)$ be the indicator of the set $\{|\lambda' a_t(\beta)| \geq (\lambda'\mathbf{F}_N(\beta)\lambda)^{\frac{1}{2}}\epsilon\}$ with $a_t(\beta) = \mathbf{S}_t(\beta) - \mathbf{S}_{t-1}(\beta)$, we get (see Problem 11)

$$\frac{1}{\lambda' \mathbf{F}_N(\beta)\lambda} \sum_{t=1}^{N} \mathrm{E}[|\lambda' a_t(\beta)|^2 I_{Nt}(\epsilon)|\mathcal{F}_{t-1}] \to 0. \tag{A.1}$$

Therefore Lindeberg's condition holds.

The conclusion of the lemma follows from the central limit theorem for martingales ([191, Corollary 3.1]).

The next lemma, a consequence of the Lindeberg's condition, parallels the well-known result from linear models ([283]). That is, the minimum eigenvalue of the unconditional information matrix grows to infinity.

Lemma A.0.2 Under Assumption A we have that

$$\lambda_{\min}(\mathbf{F}_N(\beta)) \to \infty$$

as $N \to \infty$, where $\lambda_{\min}$ is the minimum eigenvalue of the unconditional information matrix $\mathbf{F}_N(\beta)$.

Proof: Recall that if $\mathbf{A}$ and $\mathbf{B}$ are positive definite matrices,

$$|\lambda_{\min}(\mathbf{A}) - \lambda_{\min}(\mathbf{B})| \leq c||\mathbf{A} - \mathbf{B}|| \tag{A.2}$$

where the positive constant depends only on the norm of the matrix. Then, by the proof of Lemma A.0.1, we get

$$|\lambda_{\min}(\frac{\mathbf{F}_N(\beta)}{N}) - \lambda_{\min}(\mathbf{G}(\beta))| \to 0.$$

It follows that $\lambda_{\min}(\mathbf{F}_N(\beta)) = O(N)$ and the claim is proved.

The next step is to verify the so called continuity condition. Namely, we show that the matrix $\mathbf{H}_N(\beta)$ in (3.37) approaches the conditional information matrix $\mathbf{G}_N(\beta)$ (see 3.35) in a small neighborhood of the true value. This implies the fact that

$$\frac{\mathbf{R}_N(\beta)}{N} \to \mathbf{0},$$

in probability, as $N \to \infty$. Thus, the remainder term converges to $\mathbf{0}$, a fact used in previous chapters.

Lemma A.0.3 Under Assumption A the following continuity condition holds,

$$\sup_{\tilde{\beta} \in O_N(\delta)} ||\mathbf{F}_N(\beta)^{-\frac{1}{2}}(\mathbf{H}_N(\tilde{\beta}) - \mathbf{G}_N(\beta))\mathbf{F}_N(\beta)^{-\frac{t}{2}}|| \to 0,$$

in probability as $N \longrightarrow \infty$. Here $O_N(\delta) = \{\tilde{\beta} : ||\mathbf{F}_N(\beta)^{\frac{t}{2}}(\tilde{\beta} - \beta)|| \leq \delta\}$, for any $\delta > 0$ and any matrix norm.

Proof: Let $\lambda \in R^p$, with $\lambda \neq 0$ and assume without loss of generality that $||\lambda|| = 1$. We will show the equivalent condition, for any $\delta > 0$

$$\sup_{\tilde{\beta} \in O_N(\delta)} \lambda' \mathbf{F}_N(\beta)^{-\frac{1}{2}} \left(\mathbf{H}_N(\tilde{\beta}) - \mathbf{G}_N(\beta) \right) \mathbf{F}_N(\beta)^{-\frac{t}{2}} \lambda \rightarrow 0 \tag{A.3}$$

in probability using the Cramér-Wold device. By decomposing

$$\mathbf{H}_N(\tilde{\beta}) = \mathbf{G}_N(\tilde{\beta}) - \mathbf{R}_N(\tilde{\beta})$$

we really need to show that

$$g_N(\beta) = \sup_{\tilde{\beta} \in O_N(\delta)} \lambda' \mathbf{F}_N(\beta)^{-\frac{1}{2}} \left(\mathbf{G}_N(\tilde{\beta}) - \mathbf{G}_N(\beta) \right) \mathbf{F}_N(\beta)^{-\frac{t}{2}} \lambda \rightarrow 0, \tag{A.4}$$

in probability, as $N \to \infty$, and

$$\sup_{\tilde{\beta} \in O_N(\delta)} \lambda' \mathbf{F}_N(\beta)^{-\frac{1}{2}} \mathbf{R}_N(\tilde{\beta}) \mathbf{F}_N(\beta)^{-\frac{t}{2}} \lambda \rightarrow 0 \tag{A.5}$$

in probability, as $N \to \infty$, hold simultaneously. Let us define the vectors $w'_{tN}(\beta) = \lambda' \mathbf{F}_N(\beta)^{-\frac{1}{2}} \mathbf{Z}_{t-1}$ for $1 \leq t \leq N$, and $w_N(\beta) = \sum_{t=1}^N w'_{tN}(\beta) w_{tN}(\beta)$. Then we have

$$g_N(\beta) = \sup_{\tilde{\beta} \in O_N(\delta)} \sum_{t=1}^N w'_{tN}(\beta) \left(\mathbf{L}_{t-1}(\tilde{\beta}) - \mathbf{L}_{t-1}(\beta) \right) w_{tN}(\beta),$$

where $\mathbf{L}_{t-1}(\beta) = \mathbf{U}_t(\beta) \mathbf{\Sigma}_t(\beta) \mathbf{U}_t(\beta)'$ for $t = 1, \ldots, N$. It follows that

$$g_N(\beta) \leq w_N(\beta) \sup_{\tilde{\beta} \in O_N(\delta), t} ||\mathbf{L}_{t-1}(\tilde{\beta}) - \mathbf{L}_{t-1}(\beta)||.$$

Using A.2 , $\sup_t ||\mathbf{L}_{t-1}(\tilde{\beta}) - \mathbf{L}_{t-1}(\beta)||$ can be estimated from above by a continuous function of $\tilde{\beta}$ with a zero at $\tilde{\beta} = \beta$. Notice that $\{O_N(\delta)\}$ shrinks to β. Hence

$$\sup_{\tilde{\beta} \in O_N(\delta), t} ||\mathbf{L}_{t-1}(\tilde{\beta}) - \mathbf{L}_{t-1}(\beta)|| \to 0.$$

By applying Markov's inequality we obtain that

$$\begin{aligned}
\mathrm{P}\left[|g_N(\beta)| \geq \epsilon\right] &\leq \frac{\mathrm{E}\left[|g_N(\beta)|\right]}{\epsilon} \\
&\leq \mathrm{E}\left[|w_N(\beta)|\right] \sup_{\tilde{\beta} \in O_N(\delta), t} ||\mathbf{L}_{t-1}(\tilde{\beta}) - \mathbf{L}_{t-1}(\beta)|| \\
&\leq M \sup_{\tilde{\beta} \in O_N(\delta), t} ||\mathbf{L}_{t-1}(\tilde{\beta}) - \mathbf{L}_{t-1}(\beta)|| \to 0,
\end{aligned}$$

where M is a bound for $w_N(\beta)$. Such a bound exists because of assumption A.2 and the convergence of $\mathbf{F}_N(\beta)$. By further decomposition we obtain

$$\sup_{\tilde{\beta}\in O_N(\delta)} \sum_{t=1}^{N} w'_{tN}(\beta)\left(\mathbf{W}_{tj}(\tilde{\beta}) - \mathbf{W}_{tj}(\beta)\right) w_{tN}(\beta)\left(Y_{tj} - \pi_{tj}(\beta)\right) \to 0, \quad \text{(A.6)}$$

$$\sup_{\tilde{\beta}\in O_N(\delta)} \sum_{t=1}^{N} w'_{tN}(\beta)\mathbf{W}_{tj}(\tilde{\beta})w_{tN}(\beta)\left(\pi_{tj}(\beta) - \pi_{tj}(\tilde{\beta})\right) \to 0, \quad \text{(A.7)}$$

$$\sum_{t=1}^{N} w'_{tN}(\beta)\mathbf{W}_{tj}w_{tN}(\beta)\left(Y_{tj} - \pi_{tj}(\beta)\right) \to 0 \quad \text{(A.8)}$$

in probability, for any $j = 1, \ldots .q$, are jointly sufficient for (A.5). The proofs of (A.6) and (A.7) are the same as that of (A.4). To prove (A.8), consider the increments

$$u_{tN}(\beta) = w'_{tN}\mathbf{W}_{tj}(\beta)w_{tN}\left(Y_{tj} - \pi_{tj}(\beta)\right).$$

Then we see that

$$\mathrm{E}\left[u_{tN}(\beta)|\mathcal{F}_{t-1}\right] = 0$$

and

$$\begin{aligned}\mathrm{Var}\left[u_{tN}(\beta)|\mathcal{F}_{t-1}\right] &= w'_{tN}(\beta)\mathbf{W}_{tj}(\beta)w_{tN}(\beta)\\ &\quad \mathrm{Var}\left[Y_{tj} - \pi_{tj}(\beta)|\mathcal{F}_{t-1}\right] w'_{tN}(\beta)\mathbf{W}'_{tj}(\beta)w_{tN}(\beta)\\ &\le K(w'_{tN}(\beta)w_{tN}(\beta))^2,\end{aligned}$$

where K is a bound on $||\mathbf{W}_{tj}(\beta)||^2\mathrm{Var}\left[Y_{tj} - \pi_{tj}(\beta)|\mathcal{F}_{t-1}\right]$. The last two relations show that $\{u_{tN}(\beta),\ t = 1, \ldots, N\}$ are the orthogonal increments of a square integrable zero mean martingale. It follows that

$$\mathrm{E}\left(\sum_{t=1}^{N} u_{tN}(\beta)\right) = 0$$

and

$$\begin{aligned}\mathrm{Var}\left[\sum_{t=1}^{N} u_{(t-1)N}(\beta)\right] &\le K\sum_{t=1}^{N}\mathrm{E}\left[(w'_{tN}(\beta)w_{tN}(\beta))^2\right]\\ &\le K\sup_t \mathrm{E}\left[w'_{tN}(\beta)w_{tN}(\beta)\right]\mathrm{E}[w_N(\beta)].\end{aligned}$$

However,

$$\begin{aligned}\sup_t \mathrm{E}\left(w'_{tN}(\beta)w_{tN}(\beta)\right) &= \sup_t \lambda'\mathbf{F}_N(\beta)^{-\frac{1}{2}}\mathrm{E}\left(\mathbf{Z}_{t-1}\mathbf{Z}'_{t-1}\right)\mathbf{F}_N(\beta)^{-\frac{t}{2}}\lambda\\ &\le \lambda'\mathbf{F}_N(\beta)^{-1}\lambda\sup_t ||\mathrm{E}(\mathbf{Z}_{t-1})||^2\\ &\le \frac{\sup_t ||\mathrm{E}(\mathbf{Z}_{t-1})||^2}{\lambda_{\min}(\mathbf{F}_N(\beta))} \to 0.\end{aligned}$$

Since $\mathrm{E}[w_N]$ is bounded, from its convergence, relation (A.8) holds and therefore the continuity condition was established.

We now prove Theorem 3.3.1, the main result of this chapter. The approach is classical in the sense that we first show there exists a consistent solution of the score equations, and then we obtain its asymptotic normality.

Proof of Theorem (3.3.1): From Lemma A.0.1,

$$\left(\mathbf{F}_N(\boldsymbol{\beta})^{-\frac{1}{2}}\mathbf{S}_N(\boldsymbol{\beta}),\ \ \mathbf{F}_N(\boldsymbol{\beta})^{-\frac{1}{2}}\mathbf{G}_N(\boldsymbol{\beta})\mathbf{F}_N(\boldsymbol{\beta})^{-\frac{t}{2}}\right) \rightarrow (\mathcal{N}_p, \mathbf{I})$$

in distribution. Choosing $\mathbf{G}_N(\boldsymbol{\beta})^{\frac{1}{2}}$ such that $\mathbf{F}_N(\boldsymbol{\beta})^{-\frac{1}{2}}\mathbf{G}_N(\boldsymbol{\beta})^{\frac{1}{2}}$ is the Cholesky square root of $\mathbf{F}_N(\boldsymbol{\beta})^{-\frac{1}{2}}\mathbf{G}_N(\boldsymbol{\beta})\mathbf{F}_N(\boldsymbol{\beta})^{-\frac{t}{2}}$, we have from the continuity of the square root that

$$(\mathbf{F}_N(\boldsymbol{\beta})^{-\frac{1}{2}}\mathbf{S}_N(\boldsymbol{\beta}),\ \ \mathbf{F}_N(\boldsymbol{\beta})^{-\frac{1}{2}}\mathbf{G}_N(\boldsymbol{\beta})^{\frac{1}{2}}) \rightarrow (\mathcal{N}, \mathbf{I})$$

in distribution.

We first prove asymptotic existence and consistency. By Taylor expansion,

$$l_N(\tilde{\boldsymbol{\beta}}) = l_N(\boldsymbol{\beta}) + (\tilde{\boldsymbol{\beta}} - \boldsymbol{\beta})'\mathbf{S}_N(\boldsymbol{\beta}) - \frac{1}{2}(\tilde{\boldsymbol{\beta}} - \boldsymbol{\beta})'\mathbf{H}_N(\tilde{\tilde{\boldsymbol{\beta}}})(\tilde{\boldsymbol{\beta}} - \boldsymbol{\beta}),$$

where $\tilde{\tilde{\boldsymbol{\beta}}}$ lies between $\tilde{\boldsymbol{\beta}}$ and $\boldsymbol{\beta}$. Equivalently,

$$\begin{aligned} l_N(\tilde{\boldsymbol{\beta}}) - l_N(\boldsymbol{\beta}) &= (\tilde{\boldsymbol{\beta}} - \boldsymbol{\beta})'\mathbf{S}_N(\boldsymbol{\beta}) \\ &- \frac{1}{2}(\tilde{\boldsymbol{\beta}} - \boldsymbol{\beta})'\mathbf{H}_N(\tilde{\tilde{\boldsymbol{\beta}}})(\tilde{\boldsymbol{\beta}} - \boldsymbol{\beta}). \end{aligned} \tag{A.9}$$

Let now $\tilde{\boldsymbol{\lambda}} = \mathbf{F}_N(\boldsymbol{\beta})^{\frac{t}{2}}(\tilde{\boldsymbol{\beta}} - \boldsymbol{\beta})/\delta$. Then it follows that $(\tilde{\boldsymbol{\beta}} - \boldsymbol{\beta})' = \tilde{\boldsymbol{\lambda}}'\mathbf{F}_N(\boldsymbol{\beta})^{-\frac{1}{2}}\delta$ and $\tilde{\boldsymbol{\lambda}}'\tilde{\boldsymbol{\lambda}} = 1$. With this notation (A.9) becomes

$$\begin{aligned} l_N(\tilde{\boldsymbol{\beta}}) - l_N(\boldsymbol{\beta}) &= \delta\tilde{\boldsymbol{\lambda}}'\mathbf{F}_N(\boldsymbol{\beta})^{-\frac{1}{2}}\mathbf{S}_N(\boldsymbol{\beta}) \\ &- \frac{\delta^2}{2}\tilde{\boldsymbol{\lambda}}'\mathbf{F}_N(\boldsymbol{\beta})^{-\frac{1}{2}}\mathbf{H}_N(\tilde{\tilde{\boldsymbol{\beta}}})\mathbf{F}_N(\boldsymbol{\beta})^{-\frac{t}{2}}\tilde{\boldsymbol{\lambda}}. \end{aligned} \tag{A.10}$$

We are going to prove that for every $\eta > 0$ there exist N and δ such that

$$\mathrm{P}\left[l_N(\tilde{\boldsymbol{\beta}}) - l_N(\boldsymbol{\beta}) < 0 \ \ \forall\tilde{\boldsymbol{\beta}} \in \partial O_N(\delta)\right] \geq 1 - \eta. \tag{A.11}$$

This shows that, with probability tending to one, there exists a local maximum inside $O_N(\delta)$. From (A.10) we recognize that it is sufficient to show

$$\mathrm{P}\left[||\mathbf{F}_N(\boldsymbol{\beta})^{-\frac{1}{2}}\mathbf{S}_N(\boldsymbol{\beta})||^2 \leq \delta^2\frac{\lambda^2_{\min}(\mathbf{F}_N(\boldsymbol{\beta})^{-\frac{1}{2}}\mathbf{H}_N(\tilde{\tilde{\boldsymbol{\beta}}})\mathbf{F}_N(\boldsymbol{\beta})^{-\frac{t}{2}})}{4}\right] \geq 1 - \eta. \tag{A.12}$$

This is so because of the inequality

$$\begin{aligned}\tilde{\lambda}\mathbf{F}_N(\beta)^{-\frac{1}{2}}\mathbf{S}_N(\beta) &- \frac{\delta}{2}\tilde{\lambda}\mathbf{F}_N(\beta)^{-\frac{1}{2}}\mathbf{H}_N(\tilde{\tilde{\beta}})\mathbf{F}_N(\beta)^{-\frac{t}{2}}\tilde{\lambda}\\ &\leq ||\mathbf{F}_N(\beta)^{-\frac{1}{2}}\mathbf{S}_N(\beta)||^2\\ &- \frac{\delta}{2}\lambda_{\min}(\mathbf{F}_N(\beta)^{-\frac{1}{2}}\mathbf{H}_N(\tilde{\tilde{\beta}})\mathbf{F}_N(\beta)^{-\frac{t}{2}}).\end{aligned}$$

Taking into account (A.2) and lemma A.0.3, we have that for sufficiently large N,

$$\mathrm{P}\left[||\mathbf{F}_N(\beta)^{-\frac{1}{2}}\mathbf{S}_N(\beta)||^2 \leq \delta^2\frac{\lambda^2_{\min}(\mathbf{F}_N(\beta)^{-\frac{1}{2}}\mathbf{H}_N(\tilde{\tilde{\beta}})\mathbf{F}_N(\beta)^{-\frac{t}{2}})}{4}\right]$$

$$\geq \mathrm{P}\left[||\mathbf{F}_N(\beta)^{-\frac{1}{2}}\mathbf{S}_N(\beta)||^2 \leq \frac{\delta^2}{4}\right] - \frac{\eta}{2} \geq 1 - \frac{4\mathrm{E}[||\mathbf{F}_N(\beta)^{-\frac{1}{2}}\mathbf{S}_N(\beta)||^2]}{\delta^2} - \frac{\eta}{2}$$

by using Problem 9. Since $\mathrm{E}[||\mathbf{F}_N(\beta)^{-\frac{1}{2}}\mathbf{S}_N(\beta)||^2] = p$, the above expression can become arbitrarily small. Asymptotic existence therefore was established. More specifically, there exists a sequence of MPLE's $\hat{\beta} = \hat{\beta}_N$ such that for any $\eta > 0$, there is are δ and N_1 such that

$$\mathrm{P}\left[\hat{\beta} \in O_N(\delta)\right] \geq 1 - \eta \;\; \forall N \geq N_1. \tag{A.13}$$

From Lemmas A.0.1 and A.0.3 we obtain that $\mathbf{H}_N(\beta)$ is positive definite throughout $O_N(\delta)$ with probability converging to 1. Therefore the MPLE $\hat{\beta}_N$ is also locally unique. Consistency was established as well, upon noting

$$\begin{aligned}1-\eta &\leq \mathrm{P}\left[||\mathbf{F}_N(\beta)^{\frac{t}{2}}(\hat{\beta}-\beta)|| \leq \delta\right]\\ &\leq \mathrm{P}\left[||\hat{\beta}-\beta|| \leq \frac{\delta}{\lambda_{\min}(\mathbf{F}_N(\beta))}\right].\end{aligned}$$

We prove now asymptotic normality. By Taylor expansion around $\hat{\beta}$, and using the mean value theorem for multivariate function we obtain

$$\mathbf{S}_N(\beta) = \tilde{\mathbf{H}}_N(\hat{\beta}-\beta), \tag{A.14}$$

where

$$\tilde{\mathbf{H}}_N = \int_0^1 \mathbf{H}_N\left(\beta + s(\hat{\beta}-\beta)\right) ds$$

and the integration is taken elementwise. We need to show that

$$\mathbf{F}_N(\beta)^{-\frac{1}{2}}\tilde{\mathbf{H}}_N\mathbf{F}_N(\beta)^{-\frac{t}{2}} \rightarrow \mathbf{I}, \tag{A.15}$$

in probability, as $N \to \infty$. But

$$\begin{aligned}\mathbf{F}_N(\beta)^{-\frac{1}{2}}\tilde{\mathbf{H}}_N\mathbf{F}_N(\beta)^{-\frac{t}{2}} &= \mathbf{F}_N(\beta)^{-\frac{1}{2}}(\tilde{\mathbf{H}}_N - \mathbf{G}_N(\beta))\mathbf{F}_N(\beta)^{-\frac{t}{2}}\\ &+ \mathbf{F}_N(\beta)^{-\frac{1}{2}}\mathbf{G}_N(\beta)\mathbf{F}_N(\beta)^{-\frac{t}{2}}\\ &\rightarrow \mathbf{0} + \mathbf{I} = \mathbf{I},\end{aligned}$$

in probability, as $N \to \infty$. Indeed, for large N the difference between $\tilde{\mathbf{H}}_N - \mathbf{G}_N(\beta)$ and $\mathbf{H}_N(\beta) - \mathbf{G}_N(\beta)$ approaches $\mathbf{0}$ due to the consistency of the MPLE. Invoking Lemmas A.0.1 and A.0.3 it follows that (A.15) holds. Therefore, expression (A.14) becomes

$$\mathbf{F}_N(\beta)^{-\frac{1}{2}}\mathbf{S}_N(\beta) = \left(\mathbf{F}_N(\beta)^{-\frac{1}{2}}\tilde{\mathbf{H}}_N\mathbf{F}_N(\beta)^{-\frac{t}{2}}\right)\left(\mathbf{F}_N(\beta)^{\frac{t}{2}}(\hat{\beta} - \beta)\right).$$

Thus

$$\mathbf{F}_N(\beta)^{\frac{t}{2}}\left(\hat{\beta} - \beta\right) \to \mathcal{N}_p.$$

Therefore also

$$\mathbf{G}_N(\beta)^{\frac{t}{2}}\left(\hat{\beta} - \beta\right) = \mathbf{G}_N(\beta)^{\frac{t}{2}}\mathbf{F}_N(\beta)^{-\frac{t}{2}}\mathbf{F}_N(\beta)^{\frac{t}{2}}\left(\hat{\beta} - \beta\right) \to \mathcal{N}_p$$

since $\mathbf{G}_N(\beta)^{\frac{t}{2}}\mathbf{F}_N(\beta)^{-\frac{t}{2}} \to \mathbf{I}$, in probability as $N \to \infty$. From the continuity of the square root

$$\frac{\mathbf{G}_N(\beta)^{\frac{t}{2}}}{\sqrt{N}} \to \mathbf{G}^{\frac{t}{2}}(\beta),$$

in probability, as $N \to \infty$. An application of Slutsky's theorem yields the conclusion of the theorem.

Proof of Theorem (3.4.1): Recall that $\mathbf{d}(\beta) = \mathbf{M} - \mathbf{E}(\beta)$. It is easy to check that under Assumption A the process $\{\mathbf{d}_t(\beta), \mathcal{F}_t\}$, $t = 1, \ldots, N$, is a zero mean square integrable martingale such that

$$\frac{\mathbf{d}(\beta)}{\sqrt{N}} \to \mathcal{N}_{kq}(0, \mathbf{C}) \tag{A.16}$$

in distribution, as $N \to \infty$. Here $\mathbf{C}(\beta) = \oplus_{l=1}^{k}\mathbf{C}_l(\beta)$, is the direct sum of k matrices with $\mathbf{C}_l(\beta)$ a $q \times q$ symmetric matrix given by

$$\mathbf{C}_l(\beta) = \begin{bmatrix} \int_{C_l} h_1(\beta)(1 - h_1(\beta))\nu(d\mathbf{Z}) & \cdots & -\int_{C_l} h_1(\beta)h_q(\beta)\nu(d\mathbf{Z}) \\ \vdots & \ddots & \vdots \\ -\int_{C_l} h_1(\beta)h_q(\beta)\nu(d\mathbf{Z}) & \cdots & \int_{C_l} h_q(\beta)(1 - h_q(\beta))\nu(d\mathbf{Z}) \end{bmatrix},$$

a direct consequence of the fact that $\mathbf{M}_l - \mathbf{E}_l(\beta)$ is a zero–mean square integrable martingale with asymptotic variance $N\mathbf{C}_l(\beta)$ and that for $l \neq l'$, $\mathbf{M}_l - \mathbf{E}_l(\beta)$ and $\mathbf{M}_{l'} - \mathbf{E}_{l'}(\beta)$ are orthogonal.

Theorem 3.3.1 shows that for some integer N greater than N_0,

$$\frac{1}{\sqrt{N}}\left(\mathbf{d}'(\beta), (\hat{\beta} - \beta)'\right)' \approx \frac{1}{\sqrt{N}}\left(\mathbf{d}'(\beta), \mathbf{G}^{-1}(\beta)\mathbf{S}_N(\beta)\right)'. \tag{A.17}$$

Now since both processes $\mathbf{d}(\beta)$ and $\mathbf{S}_N(\beta)$ are martingales which satisfy the conditions of the central limit theorem for martingales, it follows that jointly (using again the Cramer–Wold device) the vector on the right-hand side of (A.17) converges to a normal random variable as $N \to \infty$. We only need to compute the asymptotic covariance matrix of its components. We have

$$
\begin{aligned}
&\frac{1}{N}\mathbf{G}^{-1}(\beta)\mathbf{S}_N(\beta)\sum_{t=1}^{N} I_{[\mathbf{Z}_{t-1}\in A_l]}(\mathbf{Y}_t - \boldsymbol{\pi}_t(\beta)) \\
&\quad = \frac{1}{N}\mathbf{G}^{-1}(\beta)\sum_{s=1}^{N}\mathbf{Z}_{s-1}\mathbf{U}_s(\beta)(\mathbf{Y}_s - \boldsymbol{\pi}_s(\beta))\sum_{t=1}^{N} I_{[\mathbf{Z}_{t-1}\in A_l]}\otimes(\mathbf{Y}_t - \boldsymbol{\pi}_t(\beta)).
\end{aligned}
$$

But for $s < t$

$$
\begin{aligned}
&\mathrm{E}\left[\mathbf{Z}_{s-1}\mathbf{U}_s(\beta)(\mathbf{Y}_s - \boldsymbol{\pi}_s(\beta))I_{[\mathbf{Z}_{s-1}\in C_l]}\otimes(\mathbf{Y}_t - \boldsymbol{\pi}_t(\beta))\right] \\
&\quad = \mathrm{E}\left\{\mathbf{Z}_{s-1}\mathbf{U}_s(\beta)(\mathbf{Y}_s - \boldsymbol{\pi}_s(\beta))I_{[\mathbf{Z}_{s-1}\in C_l]}\otimes\mathrm{E}\left[(\mathbf{Y}_t - \boldsymbol{\pi}_t(\beta))|\mathcal{F}_{t-1}\right]\right\} = \mathbf{0}.
\end{aligned}
$$

Therefore, we have from assumption A.5 that

$$
\begin{aligned}
&\mathrm{E}\left[\frac{1}{N}\mathbf{G}^{-1}(\beta)\mathbf{S}_N(\beta)\sum_{t=1}^{N} I_{[\mathbf{Z}_{t-1}\in A_l]}\otimes(\mathbf{Y}_t - \boldsymbol{\pi}_t(\beta))\right] \\
&= \mathrm{E}\left[\frac{1}{N}\mathbf{G}^{-1}(\beta)\sum_{s=1}^{N}\mathbf{Z}_{s-1}\mathbf{U}_s(\beta)(\mathbf{Y}_s - \boldsymbol{\pi}_s(\beta))\sum_{t=1}^{N} I_{[\mathbf{Z}_{t-1}\in A_l]}\otimes(\mathbf{Y}_t - \boldsymbol{\pi}_t(\beta))\right] \\
&= \frac{1}{N}\mathbf{G}^{-1}(\beta)\sum_{t=1}^{N} I_{[\mathbf{Z}_{t-1}\in A_l]}\mathbf{Z}_{t-1}\mathbf{U}_s(\beta)\boldsymbol{\Sigma}_t(\beta) \to \mathbf{G}^{-1}(\beta)\int_{C_l}\mathbf{Z}\mathbf{U}(\beta)\boldsymbol{\Sigma}(\beta)\nu d(\mathbf{Z}).
\end{aligned}
$$

Thus the first part of the theorem follows.

To prove the second part, we have by Taylor's expansion

$$
\begin{aligned}
\mathbf{E}_l(\hat{\beta}) &\approx \mathbf{E}_l(\beta) + \left[\frac{\partial \mathbf{E}_l(\beta)}{\partial \beta}\right]_{\beta}(\hat{\beta} - \beta) + o_p(||\hat{\beta} - \beta||) \\
&= \mathbf{E}_l(\beta) + \left[\sum_{t=1}^{N} I_{[\mathbf{Z}_{t-1}\in A_l]}\frac{\partial \boldsymbol{\pi}_t}{\partial \beta}\right](\hat{\beta} - \beta) + o_p(||\hat{\beta} - \beta||) \\
&= \mathbf{E}_l(\beta) + \left[\sum_{t=1}^{N} I_{[\mathbf{Z}_{t-1}\in A_l]}\mathbf{Z}_{t-1}\mathbf{U}_t\boldsymbol{\Sigma}_t\right](\hat{\beta} - \beta) + o_p(||\hat{\beta} - \beta||).
\end{aligned}
$$

So the desired result follows.

The final part of the theorem is a straightforward corollary of (A.16).

4

Regression Models for Count Time Series

The monthly number of tourist arrivals in Cyprus from January 1979 to December 2000[1], 264 observations in thousands of arrivals, is displayed in Figure 4.1. This is an example of a non–negative integer–valued time series or *count time series*. Similar examples of count data are abundant in daily life. Thus, the hourly number of visits to a web site, the monthly number of claims reported by an insurance company, the daily number of clients that connect to a computer server, the monthly number responses to a marketing strategy, and the annual number of Israeli couples who marry in Cyprus, are all examples of count time series. Quite often these count time series are observed together with covariate data such as weather data in the case of insurance claims and customer income and age in the marketing example, and it is of interest to be able to put together all this information in a regression model for the purpose of prediction and hypothesis testing. The goal of this chapter is to discuss regression models for time series of counts and the associated problem of statistical inference within the context of partial and quasi likelihood estimation. In doing so, we take up the issues of model building, estimation theory, and regression diagnostics. Asymptotic theory will be omitted for the most part since it resembles closely the theory developed in earlier chapters. The statistical analysis of independent counts has been discussed in [2], [53], [190]. More recently, economic count data have been considered in [78] and [434].

[1]Many thanks to Cyprus Tourism Organization for providing the tourists arrival data.

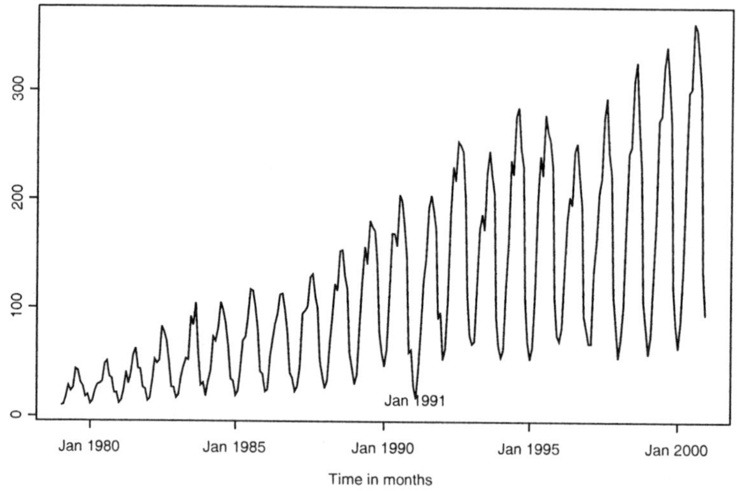

Fig. 4.1 Time series plot of monthly number of arrivals in Cyprus starting from January 1979 and ending on December 2000. The numbers are in thousands. *Source*: Cyprus Tourism Organization.

4.1 MODELING

Let $\{Y_t\}$, $t = 1, \ldots, N$ denote a time series of counts taking nonnegative integers values, and as before we think of $\{Y_t\}$ as the response process. Among several possibilities, a most natural candidate distribution model for the response process is the Poisson. Accordingly, the conditional law of $\{Y_t\}$ is specified by assuming that the conditional density of the response given the past is Poisson with mean μ_t,

$$f(y_t; \mu_t \mid \mathcal{F}_{t-1}) = \frac{\exp(-\mu_t)\mu_t{}^{y_t}}{y_t!}, \quad t = 1, \ldots, N, \tag{4.1}$$

where as before $\mathcal{F}_{t-1}$ denotes the past: the available information to the observer up to time t. For the Poisson model (4.1), the conditional expectation of the response is equal to its conditional variance,

$$\mathrm{E}[Y_t \mid \mathcal{F}_{t-1}] = \mathrm{Var}[Y_t \mid \mathcal{F}_{t-1}] = \mu_t, \quad t = 1, \ldots, N. \tag{4.2}$$

We denote by $\{\mathbf{Z}_{t-1}\}$, $t = 1, \ldots, N$ a p–dimensional vector of covariates which may include past values of the process and/or any other auxiliary information. A typical choice of $\mathbf{Z}_{t-1}$ is

$$\mathbf{Z}_{t-1} = (1, X_t, Y_{t-1})',$$

or when interactions are added

$$\mathbf{Z}_{t-1} = (1, X_t, Y_{t-1}, X_t Y_{t-1})'$$

and so on, with $\{X_t\}$ an additional process.

Following the general theory in Chapter 1, a suitable model for the analysis of count time series data is obtained by setting

$$\mu_t(\beta) = h(\mathbf{Z}'_{t-1}\beta), \quad t = 1, \ldots, N, \tag{4.3}$$

where β is a p–dimensional vector of unknown parameters and the *inverse link* $h(.)$ function maps a subset $H \subseteq R$ one–to–one onto $(0, \infty)$, since $\mu_t > 0$ for every t.

Model (4.3) and certain modifications thereof have been applied by a number of authors. For example a variation of the form

$$\mu_t(\beta) = \exp(\beta_1 X_t)\left[1 + \exp(-\beta_0 - \beta_2 Y_{t-1})\right], \quad t = 1, \ldots, N \tag{4.4}$$

has been suggested in [439], and in [213] the author applied (4.3) in a contagion model for aircraft hijacking. A multiplicative model which generalizes (4.3) in the sense that only the first and second moments are specified has been advanced in [447], and this was extended by including nonlinear functions of the parameters in [13]. An interesting modification was suggested in [394] where an ARMA model is fitted to the residuals from Poisson regression using mortality data.

Toward a more general model, Fokianos [151] assumes that the conditional distribution of the response given the past is the *doubly truncated Poisson*. Let $\{Y_t\}$, $t = 1, \ldots, N$ be a time series of counts and suppose that values below a known fixed constant c_1 and values exceeding another known fixed constant c_2 are omitted with $c_1 < c_2$. Then the *doubly truncated Poisson* conditional density is

$$f(y_t; \mu_t; c_1, c_2 \mid \mathcal{F}_{t-1}) = \frac{\mu_t^{y_t}}{y_t! \psi(c_1, c_2, \mu_t)}, \quad y_t = c_1, \ldots, c_2, \tag{4.5}$$

for $t = 1, \ldots, N$, where the function ψ is defined by

$$\psi(c_1, c_2, \mu) = \begin{cases} \sum_{y=c_1}^{c_2} \frac{\mu^y}{y!}, & \text{if } 0 \le c_1 < c_2 \\ \psi(0, c_2, \mu), & \text{otherwise.} \end{cases} \tag{4.6}$$

Clearly $\psi(0, \infty, \mu) = \exp(\mu)$. The choice $c_1 = 0$ and $c_2 = \infty$ leads to the common Poisson model (4.1) while equation (4.5) reduces to the so called *left truncated Poisson distribution* when $c_2 = \infty$. In particular, when $c_1 = 1$ and $c_2 = \infty$, (4.5) becomes

$$f(y_t; \mu_t; 1; \infty \mid \mathcal{F}_{t-1}) = \frac{{\mu_t}^{y_t}}{y_t!(\exp(\mu_t) - 1)}, \quad t = 1, \ldots, N.$$

This is the probability mass function of the *positive Poisson distribution*. Similarly, substituting $c_1 = 0$, equation (4.5) yields the right truncated Poisson distribution. The compendium [231, Ch. 4.10] is a useful source for additional information regarding the properties of these distributions.

It can be shown (see Problem 1) that the conditional mean and variance of the truncated Poisson distribution are given by

$$\mathrm{E}^{\mathrm{tr}}[Y_t; c_1; c_2 \mid \mathcal{F}_{t-1}] = \mu_t \frac{\psi(c_1 - 1, c_2 - 1, \mu_t)}{\psi(c_1, c_2, \mu_t)} \tag{4.7}$$

and

$$\text{Var}^{\text{tr}}[Y_t; c_1; c_2 \mid \mathcal{F}_{t-1}] = \frac{1}{\psi^2(c_1, c_2, \mu_t)} \Big\{ \mu_t^2 \psi(c_1 - 2, c_2 - 2, \mu_t) \psi(c_1, c_2, \mu_t)$$
$$+ \mu_t \psi(c_1 - 1, c_2 - 1, \mu_t) \left[\psi(c_1, c_2, \mu_t) - \mu_t \psi(c_1 - 1, c_2 - 1, \mu_t) \right] \Big\}, \quad (4.8)$$

respectively. Model (4.3) may also be used for regression analysis of truncated count distribution. It is important to note that for the Poisson model the conditional mean is equal to the conditional variance, however, this fact does not hold for the doubly truncated Poisson model.

We have considered the Poisson model as a candidate for regression analysis of time series of counts and certain of its generalization, the doubly truncated Poisson model. We now take up the issue of choosing the link function.

4.2 MODELS FOR TIME SERIES OF COUNTS

4.2.1 The Poisson Model

Equation (4.3) indicates that any inverse link function $h(.)$ that maps the real line to the interval $(0, \infty)$ can be employed for the analysis of time series of counts. However, in applications the most frequently used model is given by the canonical link which for the Poisson model (4.1) turns out to be the logarithm. By rewriting equation (4.1) as

$$f(y_t; \mu_t \mid \mathcal{F}_{t-1}) = \exp\left\{ (y_t \log \mu_t - \mu_t) - \log y_t! \right\}, \quad t = 1, \ldots, N,$$

the canonical link is obtained by

$$\theta_t = \log \mu_t, \quad t = 1, \ldots, N.$$

Thus, the inverse link function $h(.)$ (see (4.3)) is

$$h(\eta_t) = \exp(\eta_t) \quad t = 1, \ldots, N,$$

with $\eta_t = \mathbf{Z}'_{t-1}\beta$ and β an unknown vector of regression coefficients. Consequently, we obtain the so called *log–linear model*

$$\mu_t(\beta) = \exp(\mathbf{Z}'_{t-1}\beta), \quad t = 1, \ldots, N. \quad (4.9)$$

Another interesting inverse link that may be used in the regression modeling of time series of counts is the square root function (see Problem 6)

$$h(\eta_t) = \sqrt{\eta_t}, \quad t = 1, \ldots, N.$$

However, for the rest of our discussion we focus on model (4.9).

It should be pointed out that (4.9) is problematic when the covariates are unbounded, a case we rule out under Assumption A. To see this, consider the following vector of covariates

$$\mathbf{Z}_{t-1} = (1, X_t, Y_{t-1})',$$

with X_t a univariate auxiliary process and Y_{t-1} the first-order lagged value of the response. Expression (4.9) becomes

$$\mu_t(\boldsymbol{\beta}) = \exp(\beta_0 + \beta_1 X_t + \beta_2 Y_{t-1}), \quad t = 1, \ldots, N \tag{4.10}$$

with $\boldsymbol{\beta} = (\beta_0, \beta_1, \beta_2)'$. Suppose Y_{t-1} is unbounded. Then from (4.10), when $\beta_2 > 0$, the conditional expectation of the response given the past of the process tends to grow in an exponential rate. When there is no dependence on the covariate X_t, model (4.10) leads to a stable process only when $\beta_2 < 0$. Apparently, when Y_{t-1} is unbounded equation (4.10) accommodates only negative association but not positive association without growing exponentially fast.

Still, in most cases in practice even when a covariate such as Y_{t-1} is Poisson distributed, that is technically unbounded, the chance of it assuming very large values are extremely small and we may still use (4.10) safely with Y_{t-1} or $\log(Y_{t-1})$ and similar covariates in the regression equation. Thus, when applying the theory, one should proceed cautiously to avoid nonsensical results, but at the same time be in tune with practical considerations.

Recently Davis et al. [114] have considered model (4.9) with a covariate vector $\mathbf{Z}_{t-1} = (\mathbf{X}_t, e_{t-1}, \ldots, e_{t-p})$ where $\mathbf{X}_t$ denotes a multivariate auxiliary process and the random sequence $\{e_t\}$ is defined by

$$e_t = \frac{Y_t - \mu_t}{\mu_t^{\lambda}}, \quad t = 1, \ldots, N, \tag{4.11}$$

for $\lambda \geq 0$. The residual process (4.11) is of the type proposed by Shephard [383] and Benjamin et al. [36] (see (1.9)). Second-order moments of linear functions of e_t admit an interesting simplification for $\lambda = 1/2$ as described in Problem 3.

4.2.1.1 *Example: Incorporation of a Periodic Component:* Figure 4.2 shows some typical realizations of the log–linear model (4.10) for $N = 200$ and for different choices of the regression parameters, where $X_t = \cos(2t\pi/12)$ is a sinusoidal component. Notice that $\beta_2 < 0$ throughout. Figure 4.3 depicts the sample autocorrelation function plots of the simulated data. Apparently the first two time series are more oscillatory and, particularly from the sample autocorrelation, the sinusoidal component is evident in all the graphs.

Figure 4.4 shows realizations of length $N = 200$ from the log–linear model (4.9) with $\mathbf{Z}_{t-1} = (1, \cos(2\pi t/12), e_{t-1})$ with $e_t = (Y_t - \mu_t)/\sqrt{\mu_t}$ for different choices of the regression parameters. Apparently, the first two time series are more oscillatory than the remaining two, and from Figure 4.5, the four sample autocorrelation plots reveal the presence of a sinusoidal component.

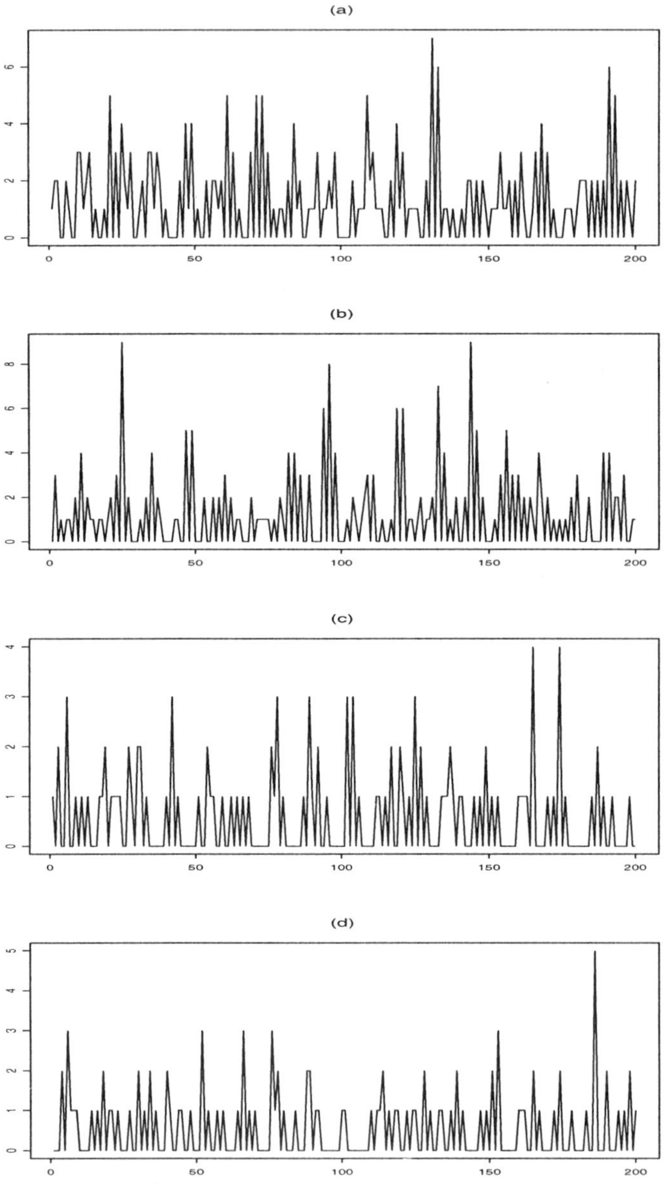

Fig. 4.2 Typical realizations of the log–linear model (4.10). Here $X_t = \cos(2t\pi/12)$ and $N = 200$. (a) $\boldsymbol{\beta} = (0.5, 1, -0.5)'$. (b) $\boldsymbol{\beta} = (0.5, 1, -1)'$. (c) $\boldsymbol{\beta} = (-0.5, -1, -0.5)'$. (d) $\boldsymbol{\beta} = (-0.5, -1, -1)'$.

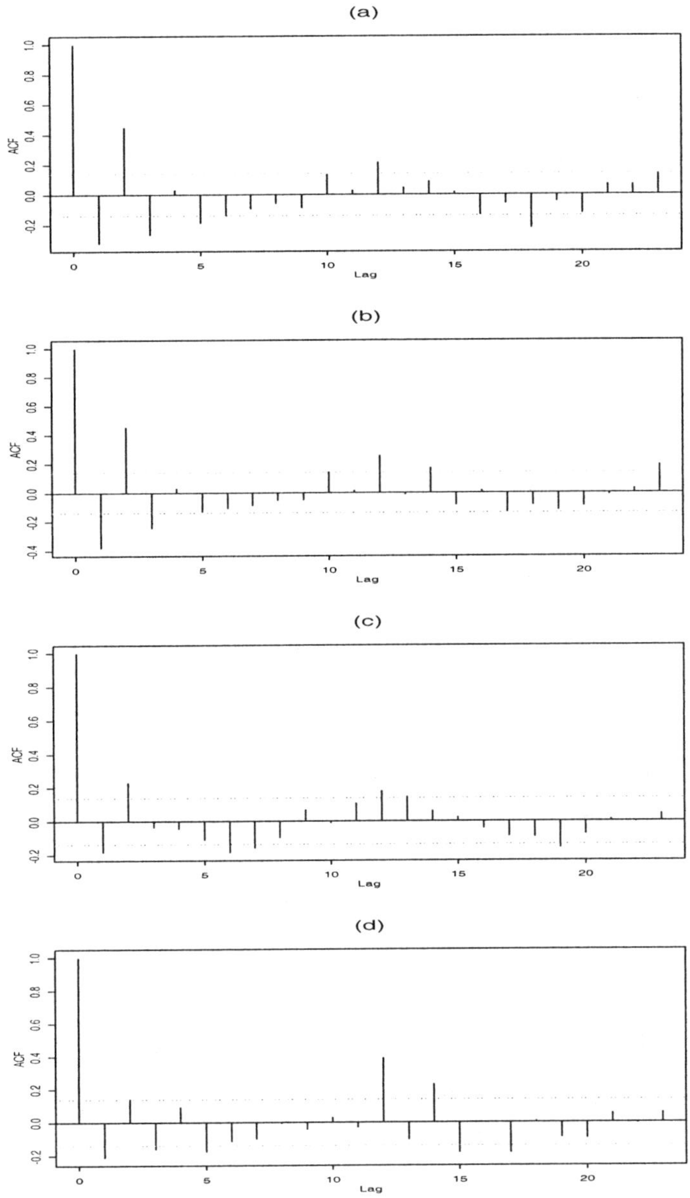

Fig. 4.3 Sample autocorrelation functions of the simulated data from Figure 4.2.

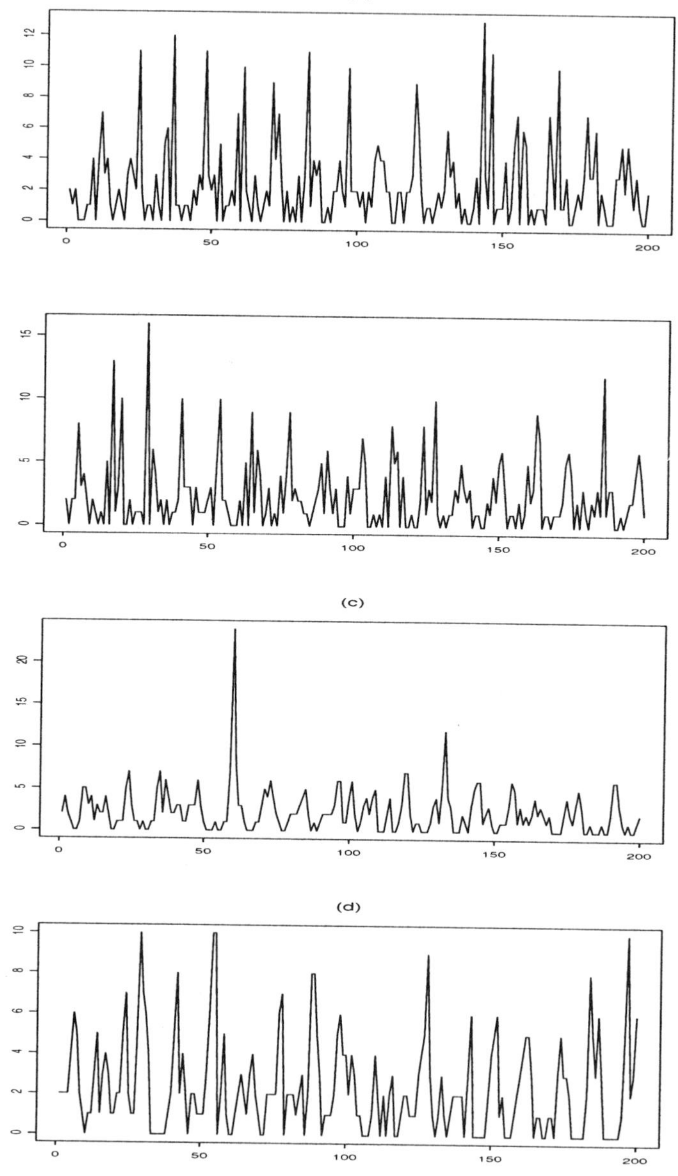

Fig. 4.4 Typical realizations of the log–linear model (4.9) with $\mathbf{Z}_{t-1} = (1, \cos(2\pi t/12), e_{t-1})$. Here $X_t = \cos(2\pi t/12)$, e_t is defined by (4.11) for $\lambda = 1/2$ and $N = 200$. (a) $\boldsymbol{\beta} = (0.5, 1, -0.5)'$. (b) $\boldsymbol{\beta} = (0.5, -1, -0.5)'$. (c) $\boldsymbol{\beta} = (0.5, 1, 0.5)'$. (d) $\boldsymbol{\beta} = (0.5, -1, 0.5)'$.

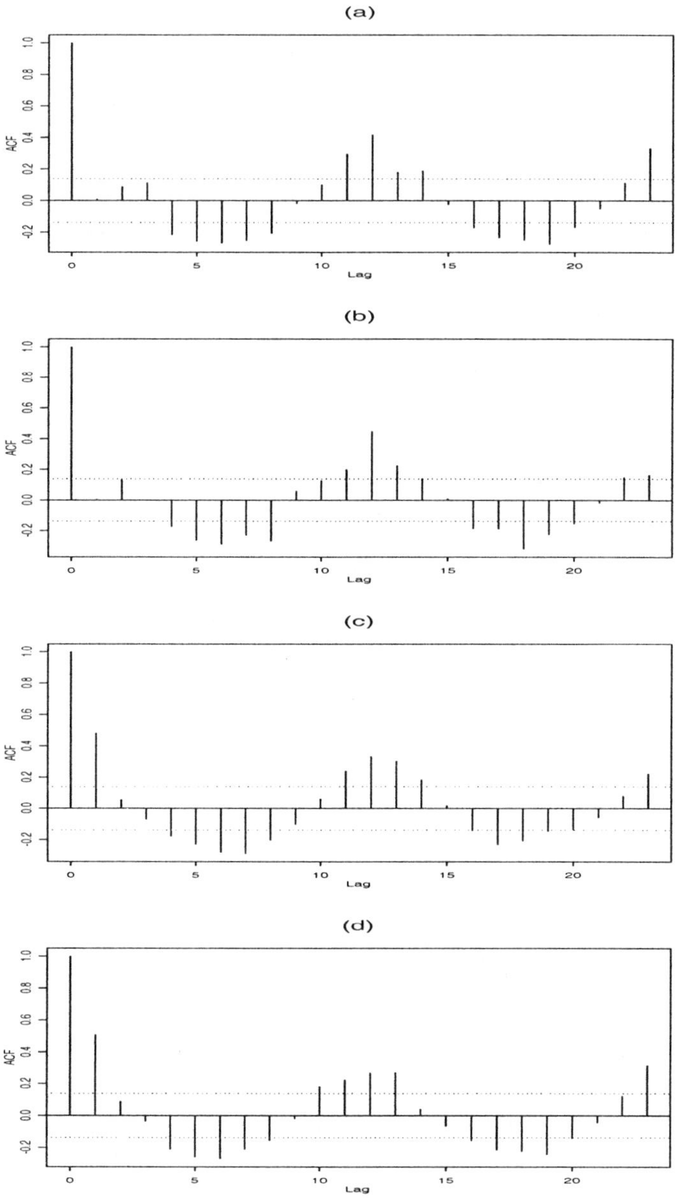

Fig. 4.5 Sample ACF of the simulated data shown in Figure 4.4.

4.2.2 The Doubly Truncated Poisson Model

We turn now to the doubly truncated Poisson model (4.5). Besides its generality, this model is more realistic in applications since in many instances the data are confined to an interval so that no data are observe outside the interval.

We notice that equation (4.5) can be rewritten for $t = 1, \ldots, N$ as

$$f(y_t; \mu_t, c_1, c_2 \mid \mathcal{F}_{t-1}) = \exp\left(y_t \log \mu_t - \log \psi(c_1, c_2, \mu_t) - \log(y_t!)\right), \quad (4.12)$$

for $y_t = c_1, \ldots, c_2$, where c_1 and c_2 are assumed known. Consequently the doubly truncated Poisson distribution is a member of the exponential family of distributions when the truncation points are known, a fact which implies that the canonical link model is the logarithm and thus the inverse link function $h(.)$ is exponential. Therefore, we obtain the log–linear model

$$\mu_t(\boldsymbol{\beta}) = \exp(\mathbf{Z}_{t-1}'\boldsymbol{\beta}), \quad t = 1, \ldots, N.$$

From (4.7) and (4.8), μ_t is not equal to the conditional expectation of the response and the conditional mean need not be equal to the conditional variance of the (truncated) response process.

4.2.2.1 *Example: Incorporation of a Periodic Component* Figures 4.6 and 4.8 display typical realizations of a truncated time series of counts of length 250 generated by the sinusoidal model, model

$$\log \mu_t(\boldsymbol{\beta}) = \beta_0 + \beta_1 \cos(\frac{2\pi t}{12}) + \beta_2 \log(Y_{t-1}). \quad (4.13)$$

Here $\boldsymbol{\beta} = (\beta_0, \beta_1, \beta_2)'$ and $\mathbf{Z}_{t-1} = (1, \cos(2\pi t/12), \log(Y_{t-1}))'$. Figure 4.6 illustrates typical realizations when the truncation points are $c_1 = 1$ and $c_2 = \infty$, that is, the case of the positive Poisson distribution. Figure 4.8 shows time series plots of the same model but now the truncation points are fixed at $c_1 = 1$ and $c_2 = 10$. For both figures the top realizations correspond to $\beta_0 = 0.50$, $\beta_1 = -1$ and $\beta_2 = -1$, while the bottom realizations correspond to $\beta_0 = -0.50$, $\beta_1 = -0.25$ and $\beta_2 = 0.75$. It is interesting to observe that the oscillation of the process depends on the sign of the coefficient of $\log(Y_{t-1})$. Negative values of the parameter β_2 produce an enhanced oscillation, a fact which is manifested also from the sample autocorrelation function. Such time series are shown in Figures 4.6(a) and 4.8(a), and the corresponding sample autocorrelation plots are displayed in Figures 4.7(a) and 4.9(a). We can deduce that successive observations tend to be located on different sides of the overall mean, and hence the increased oscillation. The sinusoidal pattern of the autocorrelation functions is quite apparent. By contrast, Figures 4.6(b) and 4.8(b) exhibit time series with less oscillation. This fact is also confirmed by the plots of their sample autocorrelation functions in Figures 4.7(b) and 4.9(b), respectively, which exhibit rather short term correlation.

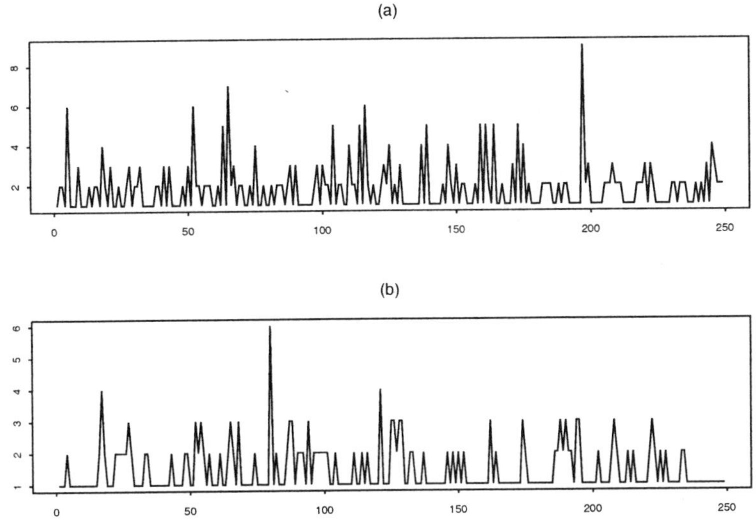

Fig. 4.6 Typical realizations of a time series $\{Y_t\}$, $t = 1, \ldots, 250$ for truncated counts. The data have been generated according to model (4.13). The truncation points are $c_1 = 1$ and $c_2 = \infty$. (a) Realizations with $\beta_0 = 0.50$, $\beta_1 = -1$ and $\beta_2 = -1$. (b) Realizations with $\beta_0 = -0.50$, $\beta_1 = -0.25$ and $\beta_2 = 0.75$.

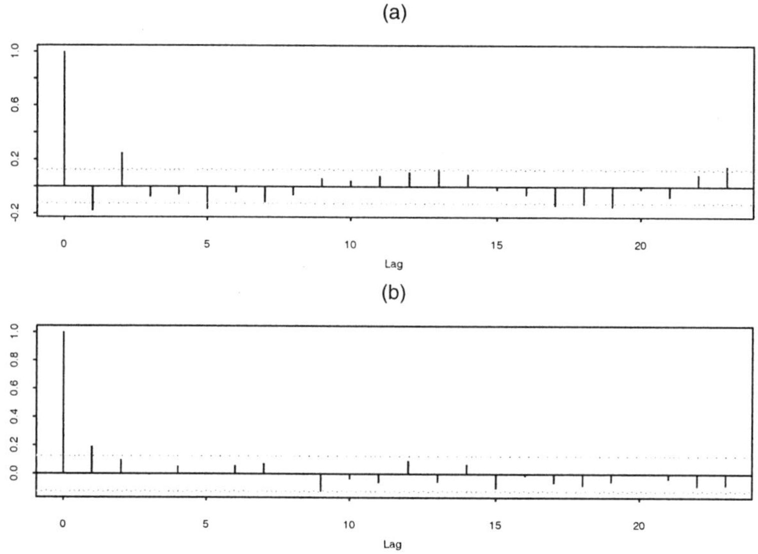

Fig. 4.7 (a) Sample autocorrelation function of the simulated time series corresponding to the top panel of Figure 4.6. (b) Sample autocorrelation function of the simulated time series corresponding to the bottom panel of Figure 4.6.

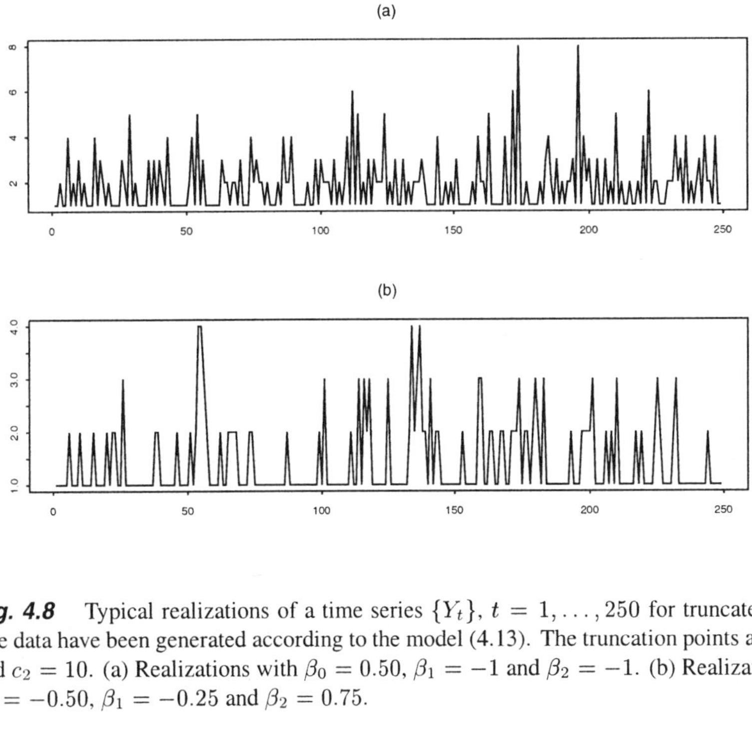

Fig. 4.8 Typical realizations of a time series $\{Y_t\}$, $t = 1, \ldots, 250$ for truncated counts. The data have been generated according to the model (4.13). The truncation points are $c_1 = 1$ and $c_2 = 10$. (a) Realizations with $\beta_0 = 0.50$, $\beta_1 = -1$ and $\beta_2 = -1$. (b) Realizations with $\beta_0 = -0.50$, $\beta_1 = -0.25$ and $\beta_2 = 0.75$.

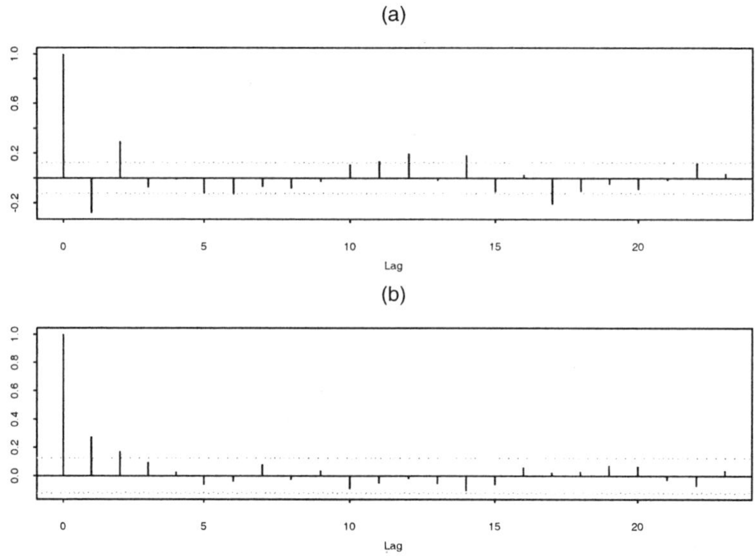

Fig. 4.9 (a) Sample autocorrelation function of the simulated time series corresponding to the top panel of Figure 4.8. (b) Sample autocorrelation function of the simulated time series corresponding to the bottom panel of Figure 4.8.

4.2.3 The Zeger–Qaqish Model

Zeger and Qaqish [447] introduce the following *multiplicative* model

$$\begin{aligned} \mu_t(\boldsymbol{\beta}) &= \exp\left(\beta_0 + \beta_1 X_t + \beta_2 \log(\tilde{Y}_{t-1})\right) \\ &= \exp(\beta_0 + \beta_1 X_t)(\tilde{Y}_{t-1})^{\beta_2}, \quad t = 1, \ldots, N, \end{aligned} \tag{4.14}$$

without specifying any distributional assumptions. Here $\mathbf{Z}_{t-1} = (1, X_t, \log(\tilde{Y}_{t-1}))'$, $\boldsymbol{\beta} = (\beta_0, \beta_1, \beta_2)'$ and $\tilde{Y}_{t-1}$ is defined either through

$$\tilde{Y}_{t-1} = \max(c, Y_{t-1}), \quad 0 < c < 1 \tag{4.15}$$

or

$$\tilde{Y}_{t-1} = Y_{t-1} + c, \quad c > 0, \tag{4.16}$$

so that $Y_{t-1} = 0$ is not an absorbing state. These empirical adjustments may influence the regression coefficients.

In addition to specifying the first moment in (4.14), it is assumed that the conditional variance of the response is given by

$$\text{Var}[Y_t \mid \mathcal{F}_{t-1}] = \phi V(\mu_t), \tag{4.17}$$

where $V(\mu_t)$ is a variance function and ϕ is an unknown dispersion parameter (see Section 1.7). That is, assumptions are made *only on first and second conditional moments* of the response.

Model (4.14) can be broaden by considering the following *multiplicative error* model

$$\mu_t(\boldsymbol{\beta}) = \exp(\beta_0 + \beta_1 X_t) \left(\frac{\tilde{Y}_{t-1}}{\exp(\beta_0 + \beta_1 X_{t-1})} \right)^{\beta_2}, \quad t = 1, \ldots, N. \tag{4.18}$$

Equation (4.18) implies that when $\beta_2 < 0$ there is an inverse relationship between $\tilde{Y}_{t-1}$ and $\mu_t(\boldsymbol{\beta})$, and when $\beta_2 > 0$, $\mu_t(\boldsymbol{\beta})$ grows with $\tilde{Y}_{t-1}$. When $\beta_2 = 0$, model (4.18) reduces to a log–linear model. Substitution of (4.16) in equation (4.18) gives

$$\mu_t(\boldsymbol{\beta}) = \exp(\beta_0 + \beta_1 X_t) \left[\frac{Y_{t-1} + c}{\exp(\beta_0 + \beta_1 X_{t-1})} \right]^{\beta_2} \tag{4.19}$$

for some $c > 0$ and $t = 1, \ldots, N$.

Equation (4.18) can be generalized further by considering the following model,

$$\mu_t(\boldsymbol{\beta}) = \exp\left[\mathbf{X}_t'\gamma + \sum_{i=1}^{q} \theta_i \left(\log \tilde{Y}_{t-i} - \mathbf{X}_{t-1}'\gamma \right) \right] \quad t = 1, \ldots, N, \tag{4.20}$$

with an $s+q$ dimensional vector $\boldsymbol{\beta} = (\gamma', \theta_1, \ldots, \theta_q)'$ containing the covariate vector, an s-dimensional vector of covariates $\{\mathbf{X}_t\}$, $t = 1, \ldots, N$, and $\{\tilde{Y}_t\}$, $t = 1, \ldots, N$, is defined through (4.15) or (4.16). For example, when $s = 2, q = 1, \gamma = (\beta_0, \beta_1)'$, $\mathbf{X}_t = (1, X_t)'$ and $\theta_1 = \beta_2$, then equation (4.20) reduces to (4.18).

Models such as (4.14) and its variants are motivated by the theory of size–dependent branching processes (see for example [187]). To see this, consider (4.14) and for simplicity assume that $\beta_1 = 0$. Set Y_t for the number of units in a population during generation t. Then

$$Y_t = \sum_{j=1}^{Y_{t-1}} Z_j(Y_{t-1}), \quad Y_{t-1} > 0,$$

where $Z_j(Y_{t-1})$ is the number of offsprings of the j'th individual at time $t-1$. When $Y_{t-1} = 0$, we assume that the population restarts to grow with Z_0 individuals. If $Z_j(Y_{t-1})$ is distributed according to the Poisson distribution with mean

$$\mathrm{E}\left[Z_j(Y_{t-1}) \mid Y_{t-1}\right] = \left(\frac{\mu}{Y_{t-1}}\right)^{1-\beta_2},$$

and

$$\mathrm{E}\left[Z_0\right] = \left(\frac{\mu}{c}\right)^{1-\beta_2}$$

with $\mu = \exp(\beta_0)$, then we arrive at model (4.14).

Various generalizations of model (4.20) have been proposed. For example Albert [13] extends this model to account for nonlinear parameters. See also [77] and [299] for applications of the model.

4.3 INFERENCE

4.3.1 Partial Likelihood Estimation for the Poisson Model

Suppose that equations (4.1) and (4.3) hold and consider the problem of estimating $\boldsymbol{\beta}$. The partial likelihood function is given by

$$\begin{aligned} \mathrm{PL}(\boldsymbol{\beta}) &= \prod_{t=1}^{N} f(y_t; \boldsymbol{\beta} \mid \mathcal{F}_{t-1}) \\ &= \prod_{t=1}^{N} \frac{\exp(-\mu_t(\boldsymbol{\beta}))\mu_t(\boldsymbol{\beta})^{y_t}}{y_t!}. \end{aligned} \tag{4.21}$$

Therefore, the partial log–likelihood is

$$\begin{aligned} l(\boldsymbol{\beta}) &\equiv \log \mathrm{PL}(\boldsymbol{\beta}) \\ &= \sum_{t=1}^{N} y_t \log \mu_t(\boldsymbol{\beta}) - \sum_{t=1}^{N} \mu_t(\boldsymbol{\beta}) - \sum_{t=1}^{N} \log(y_t!). \end{aligned} \tag{4.22}$$

Inserting expression (4.3) into (4.22) yields

$$\log \mathrm{PL}(\beta) = \sum_{t=1}^{N} y_t \log\left(h(\mathbf{Z}'_{t-1}\beta)\right) - \sum_{t=1}^{N} h(\mathbf{Z}'_{t-1}\beta) - \sum_{t=1}^{N} \log(y_t!).$$

By differentiation, we obtain the partial score function

$$\begin{aligned}
\mathbf{S}_N(\beta) &= \nabla l(\beta) = \left(\frac{\partial l(\beta)}{\partial \beta_1}, \ldots, \frac{\partial l(\beta)}{\partial \beta_p}\right)' \\
&= \sum_{t=1}^{N} \mathbf{Z}_{t-1} \frac{\partial h(\eta_t)}{\partial \eta_t} \frac{1}{\sigma_t^2(\beta)} \left(Y_t - \mu_t(\beta)\right),
\end{aligned} \tag{4.23}$$

where $\eta_t = \mathbf{Z}'_{t-1}\beta$ and $\sigma_t^2(\beta) = \mathrm{Var}[Y_t \mid \mathcal{F}_{t-1}]$. The *partial score process* is defined by the partial sums

$$\mathbf{S}_t(\beta) = \sum_{s=1}^{t} \mathbf{Z}_{s-1} \frac{\partial h(\eta_s)}{\partial \eta_s} \frac{1}{\sigma_s^2(\beta)} \left(Y_s - \mu_s(\beta)\right). \tag{4.24}$$

The solution of the score equation,

$$\mathbf{S}_N(\beta) = \nabla \log \mathrm{PL}(\beta) = \mathbf{0} \tag{4.25}$$

is the maximum partial likelihood estimator denoted by $\hat{\beta}$. The system (4.25) is nonlinear and is solved by the Fisher scoring method described in Section 1.3.2.

The cumulative conditional information matrix, $\mathbf{G}_N(\beta)$, is defined by

$$\begin{aligned}
\mathbf{G}_N(\beta) &= \sum_{t=1}^{N} \mathrm{Cov}\left[\mathbf{Z}_{t-1} \frac{\partial h(\eta_t)}{\partial \eta_t} \frac{(Y_t - \mu_t(\beta))}{\sigma_t^2(\beta)} \mid \mathcal{F}_{t-1}\right] \\
&= \sum_{t=1}^{N} \mathbf{Z}_{t-1} \left(\frac{\partial h(\eta_t)}{\partial \eta_t}\right)^2 \frac{1}{\sigma_t^2(\beta)} \mathbf{Z}'_{t-1},
\end{aligned} \tag{4.26}$$

and the unconditional information matrix is

$$\mathrm{Cov}(\mathbf{S}_N(\beta)) = \mathbf{F}_N(\beta) = \mathrm{E}[\mathbf{G}_N(\beta)]. \tag{4.27}$$

As before, let $\mathbf{H}_N(\beta)$ denote the matrix of second derivatives of the log–partial likelihood multiplied by -1,

$$\mathbf{H}_N(\beta) \equiv -\nabla\nabla\, l(\beta).$$

Then the following representation follows easily,

$$\mathbf{H}_N(\beta) = \mathbf{G}_N(\beta) - \mathbf{R}_N(\beta),$$

where

$$\mathbf{R}_N(\beta) = \sum_{t=1}^{N} \mathbf{Z}_{t-1} d_t(\beta) \mathbf{Z}'_{t-1} (Y_t - \mu_t(\beta)) \tag{4.28}$$

with $d_t(\beta) = [\partial^2 \log h(\eta_t)/\partial \eta_t^2]$.

4.3.1.1 *Inference for Log–linear Model*

Certain simplifications are obtained when we use the canonical link (see Example 1.4.2.1). For the Poisson model (4.1) with the canonical link we have

$$\mu_t(\beta) = \exp(\mathbf{Z}'_{t-1}\beta),$$

and equations (4.23) and (4.26) transform to

$$\mathbf{S}_N(\beta) = \sum_{t=1}^{N} \mathbf{Z}_{t-1}(Y_t - \mu_t(\beta))$$

and

$$\mathbf{G}_N(\beta) = \sum_{t=1}^{N} \mathbf{Z}_{t-1}\mathbf{Z}'_{t-1}\sigma_t^2.$$

Following the general theory, in the canonical link case, if a maximum partial likelihood estimator exists then it is unique.

4.3.2 Asymptotic Theory

We omit the details for proving large sample properties of the MPLE $\hat{\beta}$ since the relevant discussion in earlier chapters applies here as well. We only state a theorem that asserts the consistency and asymptotic normality of $\hat{\beta}$.

Theorem 4.3.1 Under Assumption A and the Poisson model, the probability that a locally unique maximum partial likelihood estimator exists converges to one. Moreover, there exists a sequence of maximum partial likelihood estimators $\hat{\beta}$ which is consistent and asymptotically normal:

$$\sqrt{N}(\hat{\beta} - \beta) \rightarrow \mathcal{N}_p\left(0, \mathbf{G}^{-1}(\beta)\right)$$

in distribution, as $N \rightarrow \infty$, where $\mathbf{G}$ is the limiting information matrix.

Notice that $\mathbf{G}$ is the $p \times p$ matrix

$$\mathbf{G}(\beta) = \int \mathbf{Z}\left(\frac{\partial h(\eta)}{\partial \eta}\right)^2 \frac{1}{h(\eta)} \mathbf{Z}'\nu(dZ)$$

with $\eta = \mathbf{Z}'\beta$, obtained from the limit

$$\frac{\mathbf{G}_N(\beta)}{N} \rightarrow \mathbf{G}(\beta), \tag{4.29}$$

in probability, as $N \rightarrow \infty$.

4.3.3 Prediction Intervals

Theorem 4.3.1 together with the delta method yield the following $100(1-\alpha)\%$ interval for μ_t in the Poisson model,

$$\mu_t(\beta) \doteq \mu_t(\hat{\beta}) \pm z_{\alpha/2} \frac{|h'(\mathbf{Z}'_{t-1}\beta)|}{\sqrt{N}} \sqrt{\mathbf{Z}'_{t-1}\mathbf{G}^{-1}(\beta)\mathbf{Z}_{t-1}}, \tag{4.30}$$

where h' is the derivative of h, and $z_{\alpha/2}$ is the upper $\alpha/2$ point of the standard normal distribution.

A similar result holds in the truncated case by affixing "tr" as needed [151] .

4.3.4 Inference for the Zeger–Qaqish Model

We turn now to the estimation problem in the Zeger–Qaqish model (4.20) [447].

There are two cases that need to be considered. First, assume that c (see (4.15)) is known and recall (4.17). Then estimation of the regression parameter β can be carried out by the following estimating function or *quasi–score* (see Section 1.7):

$$\mathbf{S}_N(\beta) = \sum_{t=1}^{N} \mathbf{Z}_{t-1} \frac{\partial \mu_t}{\partial \eta_t} \frac{(Y_t - \mu_t(\beta))}{\phi V(\mu_t(\beta))} \tag{4.31}$$

which clearly resembles the score equation (4.25), except that the true conditional variance is replaced by the working variance.

According to the theory in Section 1.7, the quasi maximum partial likelihood estimator $\hat{\beta}$ is asymptotically normally distributed,

$$\sqrt{N}\left(\hat{\beta} - \beta\right) \to \mathcal{N}_p\left(\mathbf{0}, \mathbf{G}^{-1}(\beta)\mathbf{G}_1(\beta)\mathbf{G}^{-1}(\beta)\right) \tag{4.32}$$

with $\mathbf{G}(\beta)$ and $\mathbf{G}_1(\beta)$ defined by the following probability limits:

$$\frac{1}{N}\sum_{t=1}^{N} \mathbf{Z}_{t-1}\left(\frac{\partial \mu_t}{\partial \eta_t}\right)^2 \frac{1}{\phi V(\mu_t(\beta))}\mathbf{Z}'_{t-1},$$

and

$$\frac{1}{N}\sum_{t=1}^{N} \mathbf{Z}_{t-1}\left(\frac{\partial \mu_t}{\partial \eta_t}\right)^2 \frac{\sigma_t^2(\beta)}{\phi^2 V^2(\mu_t(\beta))}\mathbf{Z}'_{t-1} \to \mathbf{G}_1(\beta),$$

respectively.

When the constant c is unknown, it is suggested in [282] to redefine the unknown vector of parameters β as $(\gamma, \theta_1, \ldots, \theta_q, c)'$ and then use the estimating function

$$\sum_{t=1}^{N}\left[\sum_{i=1}^{q}\left(\frac{\theta_i}{c}\right) I_{[Y_{t-i}<c]}\right] \tag{4.33}$$

in addition to (4.31). We now have a further step whereby equation (4.33) is solved for c given the current values of γ and $\theta_1, \ldots, \theta_q$. Identifiability of the parameter c is guaranteed only if $\theta_i \neq 0$ for some i [282].

The overdispersion parameter ϕ appearing in equation (4.17) can be estimated consistently by

$$\hat{\phi} = \frac{\sum_{t=1}^{N} r_t^2}{N - p}, \tag{4.34}$$

where r_t denotes the Pearson residual

$$r_t = \frac{(Y_t - \mu_t(\hat{\beta}))}{\sqrt{V(\mu_t(\hat{\beta}))}},$$

for $t = 1, \ldots, N$.

4.3.5 Hypothesis Testing

Consider the hypotheses

$$\mathrm{H}_0 : \mathbf{C}\beta = \beta_0 \text{ against } \mathrm{H}_1 : \mathbf{C}\beta \neq \beta_0, \tag{4.35}$$

where $\mathbf{C}$ is an appropriate known matrix with full rank, say $r \leq p$. Denote by $\tilde{\beta}$ the restricted partial maximum likelihood estimator under the hypothesis (4.35). Following the general theory, the most commonly used statistics for testing (4.35) in the context of Poisson regression are:

- The partial likelihood ratio statistic

$$\lambda_N = 2\{\mathrm{pl}(\hat{\beta}) - \mathrm{pl}(\tilde{\beta})\}. \tag{4.36}$$

- The Wald statistic

$$w_N = \left(\mathbf{C}\hat{\beta} - \beta_0\right)' \left(\mathbf{C}\mathbf{G}^{-1}(\hat{\beta})\mathbf{C}'\right)^{-1} \left(\mathbf{C}\hat{\beta} - \beta_0\right). \tag{4.37}$$

- The partial score statistic

$$c_N = \frac{1}{N}\mathbf{S}_N'(\tilde{\beta})\mathbf{G}^{-1}(\tilde{\beta})\mathbf{S}_N(\tilde{\beta}). \tag{4.38}$$

The following theorem states the asymptotic distribution of the last three statistics.

Theorem 4.3.2 Under assumption A, the test statistics λ_N, w_N and c_N are asymptotically equivalent. Furthermore, under hypothesis (4.35), their asymptotic distribution is chi–square with r degrees of freedom.

Theorem 4.3.2 also holds for the doubly truncated Poisson model and for the Zeger and Qaqish model (see [282] and [218].) It has been suggested that the score test performs better for small sample sizes based on simulation results [282].

4.4 GOODNESS OF FIT

4.4.1 Deviance

For count time series $\{Y_t\}$ under the Poisson log–linear model the scaled deviance takes on the form

$$D = -2\sum_{t=1}^{N}\left\{Y_t \log\left(\frac{Y_t}{\hat{\mu}_t}\right) - (Y_t - \hat{\mu}_t)\right\}, \tag{4.39}$$

and the Pearson statistic becomes

$$\mathcal{X}^2 = \sum_{t=1}^{N}\sum_{t=1}^{N}\frac{(Y_t - \hat{\mu}_t)^2}{V(\hat{\mu}_t)}, \tag{4.40}$$

with $\hat{\mu}_t = \mu_t(\hat{\beta})$, $t = 1, \ldots, N$. It can be shown that under suitable regularity conditions, the asymptotic distribution of both (4.39) and (4.40) approaches the chi–square distribution with $N - p$ degrees of freedom.

4.4.2 Residuals

Residuals in the context of Poisson regression for time series of counts are defined along the lines of Section 1.6.3. We have the *raw* residuals

$$\hat{e}_t = Y_t - \hat{\mu}_t, \quad t = 1, \ldots, N,$$

the Pearson residuals

$$\hat{r}_t = \frac{Y_t - \hat{\mu}_t}{\sqrt{V(\hat{\mu}_t)}}, \quad t = 1, \ldots, N,$$

the working residuals

$$\hat{wr}_t = \frac{Y_t - \hat{\mu}_t}{\partial\mu_t/\partial\eta_t}, \quad t = 1, \ldots, N,$$

where $\partial\mu_t/\partial\eta_t$ is evaluated at $\hat{\beta}$, and the deviance residuals given by

$$\hat{d}_t = \text{sign}(Y_t - \hat{\mu}_t)\sqrt{2\left[l_t(Y_t) - l_t(\hat{\mu}_t)\right]}, \quad t = 1, \ldots, N.$$

4.5 DATA EXAMPLES

The Poisson regression methodology is applied next in several real data problems. As before, dependence of time series of counts is accommodated by inclusion of past values of the response and/or additional covariates when available. Comparison of several competing models is carried out by considering the sum of squares of the working residuals (WR), $(Y_t - \hat{\mu}_t)/(\partial\mu_t/\partial\eta_t)$, evaluated at $\hat{\beta}$, the mean sum of squares (MSE) of the response or raw residuals, $Y_t - \hat{\mu}_t$, the sum of squares of Pearson residuals χ^2, the scaled deviance D with df $= N - p$ degrees of freedom, and the AIC and BIC computed up to an additive constant from the deviance as $D + 2p$ and $D + p\log(N)$, respectively.

4.5.1 Monthly Count of Rainy Days

To discover covariates that influence the monthly count of rainy days, a time series of $N = 401$ counts in 30-day periods was obtained from Mt. Washington, NH, for the period January 1, 1965 to December 31, 1997, together with the maximum temperature, and the mean of 65 minus the average daily temperature (heating degree), where temperature is measured in degrees Fahrenheit. Thus, associated with every period of 30 days are the count of rainy days Y_t (the response), the maximum temperature T_t, and the mean of daily heating degree H_t. Figure 4.10 shows sections of the three series Y_t, T_t, H_t. Interestingly, T_t, H_t are highly periodic while Y_t displays a more erratic oscillation.

Table 4.1 Models used in Poisson regression of the rainy days count in 30–day periods at Mt. Washington. The Mj in parentheses specify the corresponding model in hypothesis testing, where the hypothesized model is the larger model in each case. Thus, a p-value of 0.628 associated with model M9 and M3 means that Y_{t-1},Y_{t-2} are not significant, while a p-values of 0.000 associated with M3 and M2 means that T_t is very significant.

Model	Covariates	D	df	AIC	BIC	p-value
M1	Y_{t-1}	317.2	399	321.2	329.2	
M2	$Y_{t-1} + Y_{t-2}$	311.1	398	317.1	329.0	0.013 (M1)
M3	$Y_{t-1} + Y_{t-2} + T_t$	240.6	397	248.6	264.6	0.000 (M2)
M4	$Y_{t-1} + Y_{t-2} + T_t + H_t$	238.5	396	248.5	268.4	0.140 (M3)
M5	$Y_{t-1} + Y_{t-2} + H_t$	241.1	397	249.1	265.1	0.000 (M2)
M6	$\log Y_{t-1} + \log Y_{t-2}$	312.2	398	318.2	330.1	
M7	$Y_{t-1} + Y_{t-2} + \log T_t$	245.6	397	253.6	269.6	0.000 (M2)
M8	$Y_{t-1} + Y_{t-2} + T_t + T_{t-1}$	239.9	396	249.9	269.9	0.398 (M3)
M9	T_t	241.6	399	245.6	253.5	0.628 (M3)
M10	$T_t + T_{t-1}$	241.4	398	247.4	259.4	0.699 (M9)

A quick look at Table 4.1 reveals that the maximum temperature in 30-day periods, T_t, is a conspicuous covariate, responsible for a significant reduction in the scaled deviance. In fact, both the AIC and BIC are minimized at model M9 for which T_t is the sole covariate, and Figure 4.11 shows that in addition the residuals from M9 are appreciably whiter than, for example, those from M2.

The fitted model M9 is given by

$$\hat{\mu}_t = \mu_t(\hat{\beta}) = \exp(\mathbf{Z}'_{t-1}\hat{\beta}) = \exp\left\{\hat{\beta}_0 + \hat{\beta}_1 T_t\right\}, \tag{4.41}$$

where $\hat{\beta} = (\hat{\beta}_0, \hat{\beta}_1)' = (3.353, -0.0098)'$, with standard errors $(0.05, 0.001)'$.

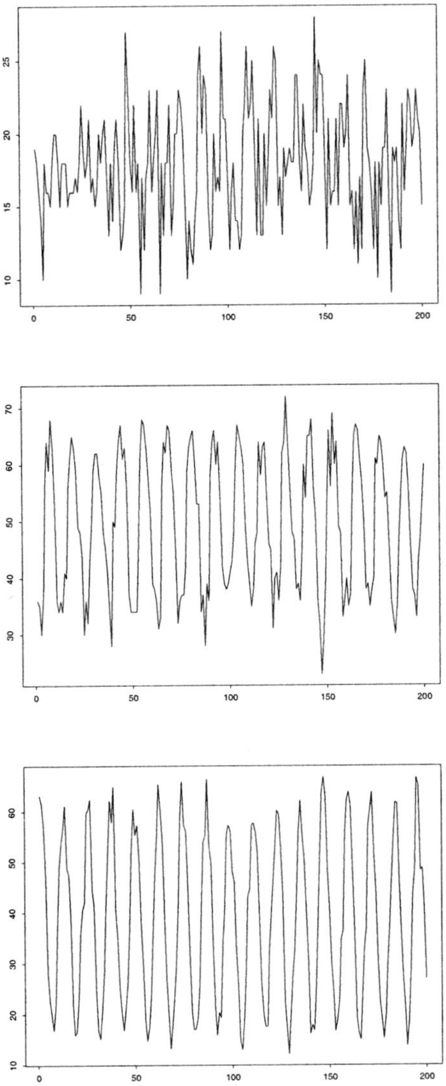

Fig. 4.10 The number of rainy days (top left), average maximum daily temperature (top right), and 65 minus average daily temperature in two hundred 30-day periods (bottom) at Mt. Washington. *Source*: `http://www.nws.noaa.gov/er/box/dailydata.htm`

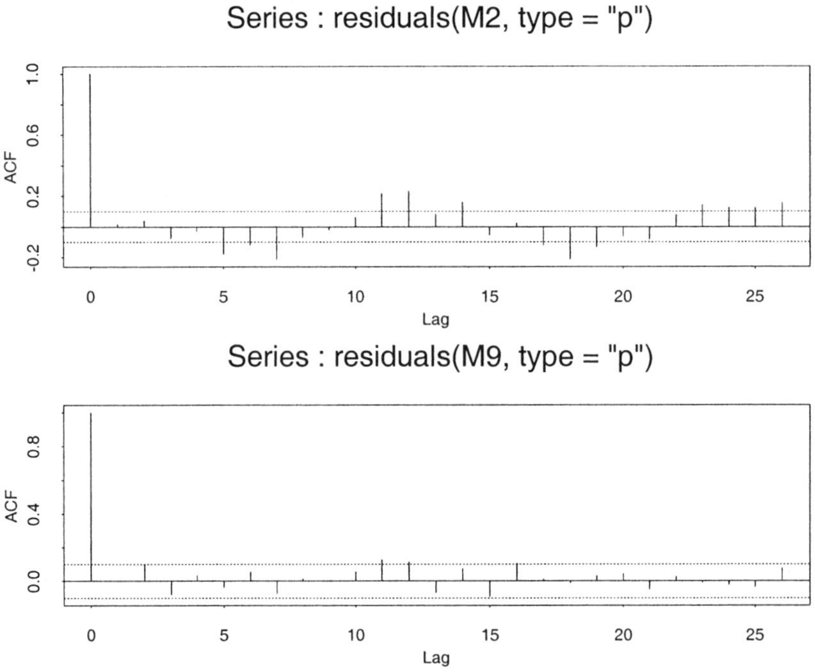

Fig. 4.11 Sample autocorrelation of the Pearson residuals from models M2 and M9.

4.5.2 Tourist Arrival Data

Table 4.3 reports the monthly number of tourist arrivals in Cyprus starting from January 1979 and ending in December 2000, a total of 264 observations in thousands. A time series plot of this time series was already displayed in Figure 4.1. The plot does not only reveal periodicity of the data but also shows the presence of an upward trend and increased variability: as time progresses so does the variability in the number of tourist arrivals. Furthermore, it is interesting to note the sudden reduction in the number of tourist arrivals during January 1991, the date that marks the beginning of the Gulf War. In what follows we consider as response Y_t the number of tourist arrivals divided by 1000 and apply the Poisson regression method to the data.

As a first step toward the analysis of these data we consider the problem of trend modeling. It is instructive to group each of the 22 years of available data and calculate both the annual sample means and the corresponding annual sample variances of the number of tourist arrivals. This information is displayed in Figure 4.12 which shows that there exists a relationship between the annual sample means and the corresponding sample variances. We see that as the annual sample mean of tourist arrivals increases so does the annual sample variances. Consequently, the underlying data trend can be taken into account by modeling the relationship between the conditional mean and the conditional variance (see (4.17)). To this end, we suggest the model

$$V(\mu_t) = \phi \mu_t^2$$

which is supported by Figure 4.12. Consequently, we use the quasi–likelihood estimation method to fit model (4.20) in light of the mean–variance relationship specification.

The second step in the analysis, is the inclusion of a covariate that takes into account the Gulf War effect. This is accomplished by considering the following *pulse* function [62],

$$I_t = \begin{cases} 1, & \text{if } t\text{=January 1991} \\ 0, & \text{otherwise.} \end{cases}$$

Third, as the data display marked well regulated cyclical oscillation, the inclusion of several sinusoidal components is sensible.

Based on all these consideration, and employing $N = 258$ observation due to lagging, we fit the following model to the tourist arrival data,

$$\begin{aligned} \log \mu_t(\boldsymbol{\beta}) &= \beta_0 + \beta_1 \cos\left(\frac{2\pi t}{12}\right) + \beta_2 \sin\left(\frac{2\pi t}{12}\right) + \beta_3 \cos\left(\frac{2\pi t}{6}\right) \\ &+ \beta_4 \sin\left(\frac{2\pi t}{6}\right) + \beta_5 \cos\left(\frac{2\pi t}{4}\right) + \beta_6 \sin\left(\frac{2\pi t}{4}\right) \\ &+ \beta_7 I_t + \beta_8 \log(Y_{t-1}) + \beta_9 \log(Y_{t-2}) + \beta_{10} \log(Y_{t-5}). \quad (4.42) \end{aligned}$$

The estimated regression coefficients for model (4.42) together with their standard errors are given in Table 4.2. We observe from the table that all the model components are significant and therefore useful in explaining the variability in the tourist arrival data. In particular, the negative sign of $\hat{\beta}_7$ points to the *negative* effect of the Gulf War on the number of tourists in Cyprus in January 1991.

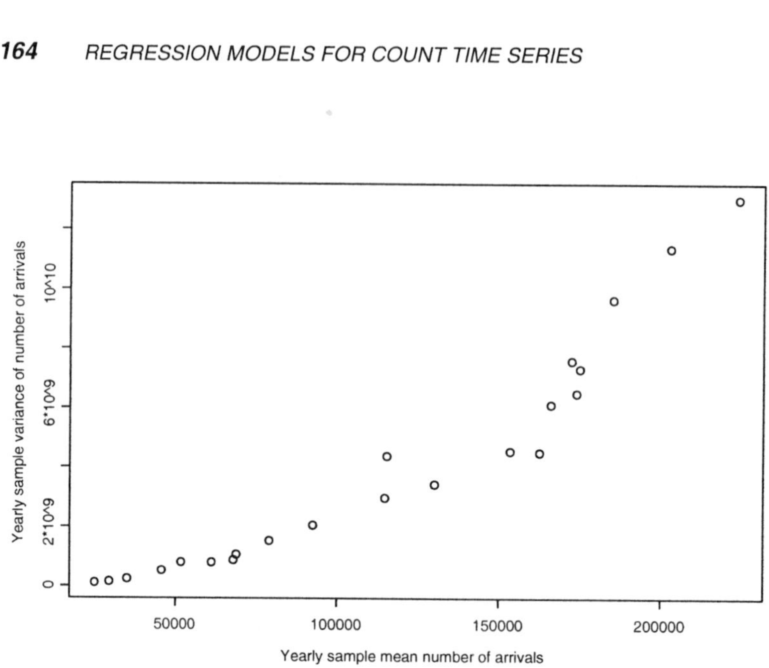

Fig. 4.12 Plot of the annual sample mean versus the annual sample variance of the number of tourist arrivals.

The fit of (4.42) resulted in $p = 11$, WR=6.24, MSE=299.35, $\mathcal{X}^2 = 6.24$, $D = 6.47$, AIC=28.47 and BIC=67.56. Notice that both the Pearson and deviance test statistics support the choice of the model because of their small magnitude. From (4.34), the parameter ϕ is estimated by $\hat{\phi} = 6.24/247 = 0.025$.

Table 4.2 Estimates of the regression coefficients with their standard errors for the tourist arrival data for model (4.42). N=258.

Regression Coefficient	$\hat{\beta}$	$\sqrt{\hat{\text{Var}}(\hat{\beta})}$	$\hat{\beta}/\sqrt{\hat{\text{Var}}(\hat{\beta})}$
β_0	0.207	0.070	2.957
β_1	-0.657	0.046	-14.282
β_2	0.074	0.041	1.804
β_3	-0.174	0.022	-7.909
β_4	-0.066	0.017	-3.882
β_5	0.068	0.015	4.533
β_6	-0.076	0.014	-5.428
β_7	-0.356	0.163	-2.184
β_8	0.376	0.055	6.836
β_9	0.353	0.059	5.983
β_{10}	0.230	0.051	4.509

Figure 4.13 displays time series plots of the raw, Pearson and deviance residuals and Figure 4.14 shows cumulative periodogram plots together with 95% confidence

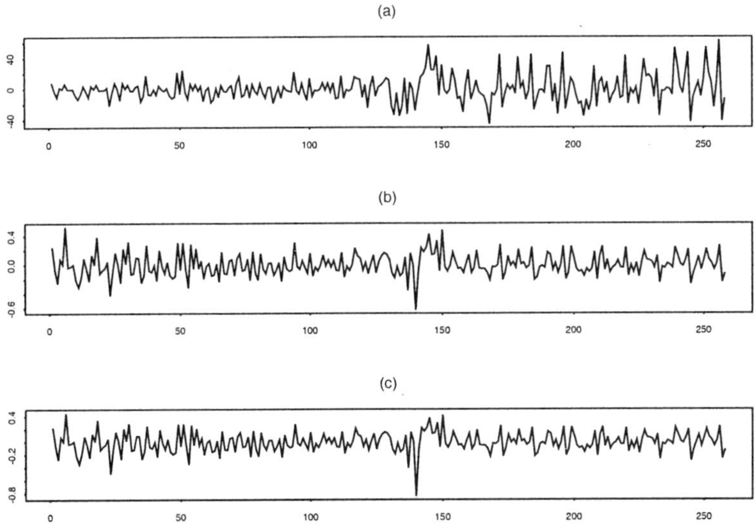

Fig. 4.13 Time series plot of the (a) raw, (b) Pearson and (c) deviance residuals from model (4.42).

limits for the raw and Pearson residuals. Figure 4.15 shows a time series plot of the predicted versus the true monthly number of arrivals. These diagnostic plots show that model (4.42) fits the data quite reasonably.

It is interesting to consider adding lagged variables of the pulse function I_t to model (4.42). It turns out that the addition of I_{t-1}, I_{t-2}, I_{t-3}, I_{t-4} and I_{t-5} does not improve the fit significantly–the scaled deviance reduces to 5.54. However the estimated regression coefficients of both I_{t-1} and I_{t-2} are negative–an indication of the negative effect of the Gulf War on the number of monthly tourist arrivals in Cyprus for some period of time after the beginning of the war. Additionally, the Wald test statistic (4.37) shows that the regression coefficient of I_{t-1} is significant as opposed to the regression coefficients of I_{t-2}, I_{t-3}, I_{t-4} and I_{t-5}. This preliminary analysis based on Poisson regression suggests that the effect of the Gulf War on the number of tourists arrivals per month had a duration of approximately a month or two.

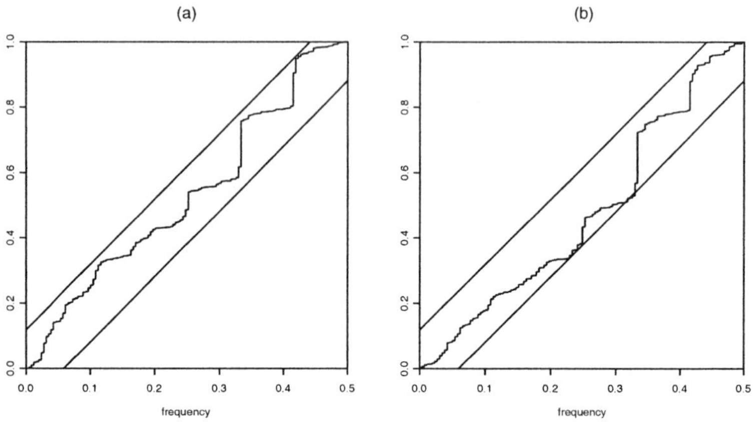

Fig. 4.14 Cumulative periodogram plot of the (a) raw and (b) Pearson residuals from model (4.42).

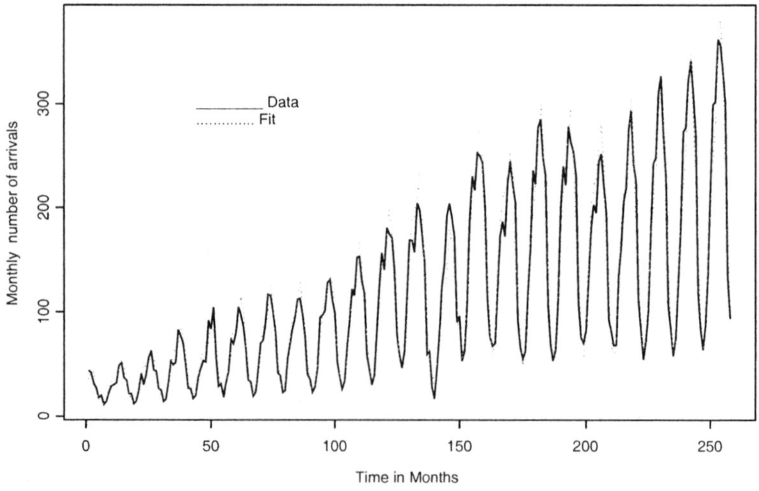

Fig. 4.15 Time series plot of predicted versus observed monthly number of tourist arrivals from model (4.42). The numbers are in thousands.

Table 4.3 Monthly arrivals of tourists in Cyprus from 1979 to 2000. *Source*: Cyprus Tourism Organization.

Year	Jan.	Feb.	Mar.	Apr.	May	June	July	Aug.	Sep.	Oct.	Nov.	Dec.
1979	9718	10889	17846	28200	22873	25835	43573	41744	31514	27253	17623	19845
1980	11248	14071	23369	28843	30274	31978	48316	50882	36629	34813	21337	21615
1981	11858	14550	24290	41040	29990	39651	56420	62542	44116	43356	26357	24963
1982	14016	16227	32511	52437	48961	51455	82715	77238	69375	50175	26715	26355
1983	16856	19691	36446	45752	53114	51953	91906	83577	104251	57536	28581	31063
1984	18198	32784	42768	73208	69210	80708	104653	97310	85411	65263	34530	32929
1985	19209	23288	44127	69653	73007	90958	116756	115581	100810	79772	41486	38960
1986	22763	24656	54721	70355	85126	94176	112056	112971	98047	77174	40399	35493
1987	23138	27564	45934	94604	97172	101356	128048	130808	111040	99158	51563	38166
1988	26018	33275	66266	92628	121848	115818	152130	153024	129486	117810	58710	44805
1989	30144	38826	83388	124812	156534	140226	180540	174672	171276	140766	77583	58869
1990	46398	62295	112452	168848	168553	157250	204337	197925	174949	147128	59253	61821
1991	27528	16748	44815	81168	124753	145217	192000	204000	190100	172300	90300	96200
1992	52700	62300	107000	187100	230300	217000	253100	249800	244000	200700	111500	75500
1993	67200	69400	119600	172500	186500	172200	224500	244800	222500	205300	91500	65000
1994	55000	62000	118000	160000	236000	223000	277000	285000	247000	231000	106000	69000
1995	53000	64000	112000	200000	240000	222000	278000	263000	254000	231000	108000	75000
1996	69900	82400	133700	180700	202800	195000	244400	252000	220900	194500	92500	81200
1997	67731	67848	134000	161356	206345	219460	275535	293887	242625	224371	111110	83732
1998	54291	71634	101575	179452	242833	248426	309983	326854	270283	228881	105773	82716
1999	57740	74041	126494	180076	273317	276879	322041	341088	309498	270732	118105	84274
2000	63553	87243	135487	221785	299355	302011	362299	356686	329964	300597	133500	93722

4.6 PROBLEMS AND COMPLEMENTS

1. Verify that the mean and the variance of the truncated Poisson distribution are given by equations (4.7) and (4.8). Using these expressions calculate the mean and variance of the positive Poisson distribution, that is $c_1 = 1$ and $c_2 = \infty$.

2. Suppose that Y is a random variable which is distributed as Poisson with parameter $\mu\Theta$ conditionally on Θ. Assume further that Θ follows the Gamma distribution with mean 1 and variance $1/\theta$. Show that the marginal distribution of Y is negative binomial with mean μ and variance $\mu + \mu^2/\theta$.

3. *Some properties of the residual process (4.11)* [114]. Assume that $\{Y_t\}$ denotes a time series of counts with $\mu_t = \mathrm{E}\left[Y_t \mid \mathcal{F}_{t-1}\right]$, $t = 1, \ldots, N$. Define

$$e_t = \frac{Y_t - \mu_t}{\mu_t^\lambda}, \quad t = 1, \ldots, N,$$

for $\lambda \geq 0$. Show that

(a)

$$\mathrm{E}\left[e_t\right] = 0.$$

and

$$\mathrm{E}\left[e_t^2\right] = \mu_t^{1-2\lambda}.$$

(b) Put $W_t = \log(\mu_t)$ and set

$$W_t = \mathbf{X}_t\beta + \sum_{i=1}^{p} \gamma_i e_{t-i}.$$

Show that

$$\mathrm{E}\left[W_t\right] = \mathbf{X}_t'\beta,$$

$$\mathrm{Var}\left[W_t\right] = \sum_{i=1}^{p} \gamma_i^2 \mu_{t-i}^{1-2\lambda},$$

and for $k > 0$

$$\mathrm{Cov}\left[W_t W_{t+k}\right] = \sum_{i=1}^{p-k} \gamma_i \gamma_{i+k} \mu_{t-i}^{1-2\lambda}.$$

Notice that for $\lambda = 0.5$ all the above expressions do not depend on t.

4. *The double Poisson distribution* [131]. The double Poisson distribution is an exponential combination of two Poisson densities.

$$f(y; \mu, \theta) = C(\mu, \theta) \left[P(\mu)\right]^\theta \left[P(y)\right]^{1-\theta},$$

where θ is a dispersion parameter, $C(\mu, \theta)$ is the normalizing constant and the notation $P(\cdot)$ corresponds to the Poisson probability mass function.

(a) Show that

$$\frac{1}{C(\mu, \theta)} \approx 1 + \frac{1-\theta}{12\theta\mu}\left(1 + \frac{1}{\theta\mu}\right),$$

and that the mean and variance of the double Poisson distribution are approximately equal to μ and μ/θ, respectively.

(b) Suppose that we observe $Y_1, \ldots, Y_n$ independent random variables from the double Poisson model with associated covariates $\mathbf{X}_1, \ldots, \mathbf{X}_n$. Assume that $\mu_i = \exp(\mathbf{X}_i'\boldsymbol{\beta}), i = 1, \ldots, n$. Discuss estimation of the vector of regression coefficients $\boldsymbol{\beta}$ and the dispersion parameter θ.

5. *The zero–inflated Poisson model* [285]. Consider the probability mass function of the zero–inflated Poisson distribution

$$\begin{aligned}
\mathrm{P}[Y_i = 0] &= \theta_i + (1-\theta_i)\exp(-\mu_i) \\
\mathrm{P}[Y_i = k] &= (1-\theta_i)\frac{\exp(-\mu_i)\mu_i^k}{k!}, \qquad k = 1, 2, \ldots,
\end{aligned}$$

which is a suitable model when the number of zeroes is excessive.

(a) Verify that the above equations define a probability mass function and calculate its mean and variance. What do you observe?

(b) Suppose that $\mu_i = \exp(\mathbf{X}_i'\boldsymbol{\beta})$ for vectors $\mathbf{X}_i$ of covariates and $\theta_i = 1/(1 + \exp(-\mathbf{Z}_i'\gamma))$, for other vectors of covariates $\mathbf{Z}_i$. Discuss joint estimation of $(\boldsymbol{\beta}, \gamma)'$ assuming parameter identifiability.

6. Suppose that $\{Y_t\}$, denotes a time series of counts observed jointly with a covariate vector process $\mathbf{Z}_{t-1}$, $t = 1, \ldots, N$. Discuss partial likelihood estimation of the vector of unknown parameters $\boldsymbol{\beta}$ when the inverse link function is given by the square root. That is,

$$h(\eta_t) = \sqrt{\eta_t}, \quad, t = 1, \ldots, N,$$

with $\eta_t = \mathbf{Z}_{t-1}'\boldsymbol{\beta}$.

7. *Maximum likelihood in misspecified models* [429]. Suppose that $Y_1, \ldots, Y_N$ is a random sample with *true* probability density function $f^\star(y;\gamma)$ where γ is a vector of parameters. Suppose that we misspecify the model by fitting to the data $f(y, \boldsymbol{\theta})$ instead. In this case

$$\mathrm{E}_\star\left[\frac{\partial \log f}{\partial \boldsymbol{\theta}}\right] \neq \mathbf{0},$$

where the notation $\mathrm{E}_\star$ denotes expectation with respect to $f^\star$. Consequently, the maximum likelihood estimator of $\boldsymbol{\theta}$, say $\hat{\boldsymbol{\theta}}$, is not consistent. In fact, the following are true:

(a) Show that

$$\hat{\boldsymbol{\theta}} \to \boldsymbol{\theta}^{\star},$$

in probability, as $N \to \infty$ where the value $\boldsymbol{\theta}^{\star}$ maximizes the following probability limit

$$\frac{1}{N}\sum_{i=1}^{N} \log f(y_i; \boldsymbol{\theta}),$$

calculated under $f^{\star}$.

(b) Furthermore, under some regularity conditions, show that

$$\sqrt{N}\left(\hat{\boldsymbol{\theta}} - \boldsymbol{\theta}\right) \to \mathcal{N}\left(\boldsymbol{\theta}, \mathbf{A}_{\star}^{-1}\mathbf{B}\mathbf{A}_{\star}^{-1}\right),$$

in distribution, as $N \to \infty$, where

$$\mathbf{A} = -\lim_{N\to\infty} \frac{1}{N} \mathrm{E}_{\star}\left[\sum_{i=1}^{N} \frac{\partial^2 \log f(y_i; \boldsymbol{\theta})}{\partial\boldsymbol{\theta}\partial\boldsymbol{\theta}'}\right]_{\boldsymbol{\theta}^{\star}},$$

and

$$\mathbf{B} = \lim_{N\to\infty} \frac{1}{N} \mathrm{E}_{\star}\left[\sum_{i=1}^{N} \frac{\partial \log f(y_i; \boldsymbol{\theta})}{\partial\boldsymbol{\theta}} \frac{\partial \log f(y_i; \boldsymbol{\theta})}{\partial\boldsymbol{\theta}'}\right]_{\boldsymbol{\theta}^{\star}}.$$

8. *GMM estimation* [197]. A widely used technique for estimation, especially in econometrics is that of *generalized method of moments* (GMM). Accordingly, suppose that $Y_1, \ldots, Y_N$ denotes the observed sample and consider the moment conditions

$$\mathrm{E}\left[\mathbf{h}_i(Y_i; \boldsymbol{\theta})\right] = \mathbf{0}, \quad i = 1, \ldots, N,$$

where $\mathbf{h}_i(.)$ is a q–dimensional vector and $\boldsymbol{\theta}$ is p–dimensional vector of unknown parameters, such that $q \geq p$. Notice that the number of moment conditions *exceeds* the number of unknown parameters. Then the so called generalized method of moments estimator, denoted by $\hat{\boldsymbol{\theta}}_{\text{gmm}}$, is the value that minimizes

$$\left\{\sum_{i=1}^{N} \mathbf{h}_i(Y_i, \boldsymbol{\theta})\right\}' \mathbf{V}_N \left\{\sum_{i=1}^{N} \mathbf{h}_i(Y_i, \boldsymbol{\theta})\right\},$$

where $\mathbf{V}_N$ is in general a $q \times q$ stochastic positive definite matrix of weights, such that it converges in probability to a non stochastic matrix $\mathbf{V}$. Assume some regularity conditions to show that

(a) The estimator $\hat{\boldsymbol{\theta}}_{\text{gmm}}$ is consistent.

(b) Moreover,

$$\sqrt{N}\left(\hat{\boldsymbol{\theta}}_{\text{gmm}} - \boldsymbol{\theta}\right) \to \mathcal{N}_p\left(\mathbf{0}, \mathbf{A}^{-1}\mathbf{B}\mathbf{A}^{-1}\right),$$

in distribution, as $N \to \infty$. Identify the matrices $\mathbf{A}$ and $\mathbf{B}$.

9. *Inference for the doubly truncated Poisson model* [151]. For the doubly truncated Poisson model specified by (4.5) and (4.9) derive the score function and the conditional information matrix of the regression coefficients.

10. *A parameter driven model for time series of counts.* [68], [115], [445]. Zeger [445] introduced a regression models for time series of counts by assuming that the observed process is driven by a latent process that is a *parameter driven* model in the terminology of [106]. To be more specific, suppose that $\{Y_t\}$, $t = 1, \ldots, N$, denotes a sequence of independent counts given an unobserved process $\{\epsilon_t\}$, $t = 1, \ldots, N$, such that

$$\mathrm{E}[Y_t \mid \epsilon_t] = \mathrm{Var}[Y_t \mid \epsilon_t] = \epsilon_t \exp(\mathbf{X}_t'\beta), \tag{4.43}$$

with $\mathbf{X}_t$ a covariate vector. If we further assume that $\{\epsilon_t\}$, $t = 1, \ldots, N$ is a stationary process with $\mathrm{E}[\epsilon_t] = 1$ and $\mathrm{Cov}[\epsilon_t, \epsilon_{t+\tau}] = \sigma^2 \rho_\epsilon(\tau)$, show that

(a) i.

$$\mu_t = \mathrm{E}[Y_t] = \exp(\mathbf{X}_t'\beta), \tag{4.44}$$

ii.

$$\mathrm{Var}[Y_t] = \mathrm{E}[Y_t] + \mathrm{E}^2[Y_t]\sigma^2, \tag{4.45}$$

iii.

$$\mathrm{Cov}[Y_t, Y_{t+\tau}] = \sigma^2 \mu_t \mu_{t+\tau} \rho_\epsilon(\tau). \tag{4.46}$$

The above formulation although similar to a Poisson-loglinear model, reveals that the observed data are overdispersed, that is,

$$\mathrm{Var}[Y_t] > \mathrm{E}[Y_t].$$

Recent work in [79] uses Zeger's model to investigate the relationship between sudden infant death syndrome and environmental temperature.

(b) Recall expressions (4.44), (4.45) and (4.46). Let $\mathbf{Y} = (Y_1, \ldots, Y_N)'$, $\boldsymbol{\mu} = (\mu_1, \ldots, \mu_N)'$ and $\mathbf{V} = \mathrm{Cov}[\mathbf{Y}]$. Show that the regression vector β can be estimated by $\hat{\beta}$ which is the solution of the estimating equations

$$\frac{\partial \boldsymbol{\mu}'}{\partial \beta} \mathbf{V}^{-1}(\mathbf{Y} - \boldsymbol{\mu}) = \mathbf{0}. \tag{4.47}$$

Notice that the matrix $\mathbf{V}$ equals to $\mathbf{A} + \sigma^2 \mathbf{A}\mathbf{R}_\epsilon \mathbf{A}$, with $A = \mathrm{diag}(\mu_i)$, an $N \times N$ diagonal matrix, and $\mathbf{R}_\epsilon$ an $N \times N$ matrix whose (j, k) element is given by $\rho_\epsilon(|j - k|)$. Equations (4.47) are solved by the Fisher scoring algorithm.

(c) Show that $\sqrt{N}(\hat{\beta} - \beta)$ converges in distribution to a normal random vector with covariance matrix

$$\mathbf{V}_{\hat{\beta}} = \lim_{N \to \infty} \left(\frac{1}{N} \frac{\partial \boldsymbol{\mu}'}{\partial \beta} \mathbf{V}^{-1} \frac{\partial \boldsymbol{\mu}}{\partial \beta} \right)^{-1}.$$

(d) However, equations (4.47) require the inversion of a $N \times N$ matrix. To overcome this problem, Zeger [445] suggests to approximate $\mathbf{V}$ by

$$\mathbf{V}_R = \mathbf{D}^{1/2}\mathbf{R}(\boldsymbol{\alpha})\mathbf{D}^{1/2},$$

where $\mathbf{D} = \text{diag}(\mu_t + \sigma^2\mu_t)$, an $N \times N$ diagonal matrix, and $\mathbf{R}(\boldsymbol{\alpha})$ the autocorrelation matrix of a stationary autoregressive process with $\boldsymbol{\alpha}$ a vector of parameters. Then (4.47) is modified according to

$$\frac{\partial \boldsymbol{\mu}'}{\partial \boldsymbol{\beta}}\mathbf{V}_R^{-1}(\mathbf{Y} - \boldsymbol{\mu}) = \mathbf{0}. \tag{4.48}$$

This approach has several advantages. First there is a great deal of simplification of calculations required to obtain $\mathbf{V}_R^{-1}$. Furthermore, the Fisher scoring algorithm still applies. Show that the asymptotic distribution of the solution of equation (4.48), $\hat{\boldsymbol{\beta}}_R$, is given by

$$\sqrt{N}(\hat{\boldsymbol{\beta}}_R - \boldsymbol{\beta}) \to \mathcal{N}_p(\mathbf{0}, \mathbf{V}_{\hat{\boldsymbol{\beta}}_R}),$$

where

$$\mathbf{V}_{\hat{\boldsymbol{\beta}}_R} = \mathbf{I}_0^{-1}\mathbf{I}_1\mathbf{I}_0^{-1},$$

with

$$\mathbf{I}_0 = \lim_{N\to\infty}\left(\frac{1}{N}\frac{\partial \boldsymbol{\mu}'}{\partial \boldsymbol{\beta}}\mathbf{V}_R^{-1}\frac{\partial \boldsymbol{\mu}}{\partial \boldsymbol{\beta}}\right)$$

and

$$\mathbf{I}_1 = \lim_{N\to\infty}\left(\frac{1}{N}\frac{\partial \boldsymbol{\mu}'}{\partial \boldsymbol{\beta}}\mathbf{V}_R^{-1}\mathbf{V}\mathbf{V}_R^{-1}\frac{\partial \boldsymbol{\mu}}{\partial \boldsymbol{\beta}}\right).$$

The author shows by a simulation that even though $\hat{\boldsymbol{\beta}}_R$ is not based on the true variance covariance matrix, its efficiency is high compared with that of $\hat{\boldsymbol{\beta}}$. Notice that this approach does not require any distributional assumptions on the latent process. An alternative estimation method based on parametric assumptions about the latent process was implemented in [88] by means of a Monte Carlo EM algorithm. More recently, in [115] the likelihood is approximated under the assumption that $\log(\epsilon_t)$ follows a Gaussian AR(p) model.

(e) Show that inference regarding the parameter σ^2 is carried out by considering the following method of moments estimator

$$\hat{\sigma^2} = \frac{\sum_{t=1}^{N}\left((Y_t - \hat{\mu}_t)^2 - \hat{\mu}_t\right)}{\sum_{t=1}^{N}\hat{\mu}_t^2},$$

while for $\rho_\epsilon(\tau)$

$$\hat{\rho}_\epsilon(\tau) = \frac{1}{\hat{\sigma^2}}\frac{\sum_{t=\tau+1}^{N}(Y_t - \hat{\mu}_t)(Y_{t+\tau} - \hat{\mu}_{t+\tau})}{\sum_{t=\tau+1}^{N}\hat{\mu}_t\hat{\mu}_{t+\tau}}.$$

For small sample sizes $\hat{\sigma}^2$ may be negative and $\hat{\rho}_\epsilon(\tau)$ may fall outside the range $(-1, 1)$ as the author points out–a fact also confirmed by [79].

Further detailed studies of Zeger's model can be found in [68], [230] and more recently in [115]. Davis et al. [115] address the problem of existence of the latent stochastic process $\{\epsilon_t\}$ and derive the asymptotic distribution of the regression coefficients when the latter exist. They also suggest adjustments for the estimators of σ^2 and for the autocovariance.

11. *Poisson AR(1)* [316]. Let $\{B_i\}$ be a sequence of i.i.d. Bernoulli(α) random variables, and $\{W_n\}$ an independent sequence of i.i.d. Poisson($\theta(1-\alpha)$) random variables. Consider the process,

$$X_n = \sum_{i=1}^{X_{n-1}} B_i + W_n, \quad n = 1, 2, ...$$

   (a) Argue that $\{X_n\}$ is a first order autoregressive process.

   (b) Show that if $X_0 \sim$ Poisson(θ), independently of $\{B_i\}$ and $\{W_n\}$, then also $X_n \sim$ Poisson(θ), $n \geq 1$.

   (c) Show that the autocorrelation function of $\{X_n\}$ is $\rho(k) = \alpha^k$, $k \geq 0$.

12. *The Galton-Watson process with immigration admits an asymptotic AR(1) representation* [210],[424]. We define a branching process with immigration as follows. Let (offspring) $\{Y_{n,i}\}$, $i, n = 1, 2, 3, \cdots$, be a family of i.i.d. zero-one random variables such that $Y_{n,i} = 1$ with probability m (i.e. independent Bernoulli(m) random variables). Let (immigration) $\{I_n\}, n = 1, 2, 3, \cdots$, be a sequence of i.i.d. Poisson(λ) random variables, independent of the $Y_{n,i}$. Recall that the possible values of I_n are $0, 1, 2, \cdots$, and that $E[I_n] = \lambda$ for all n. Put $X_0 = 0$, and $\sum_{i=1}^{0} \equiv 0$. The Galton-Watson Process with immigration $\{X_n\}$ is defined by the equation,

$$X_n = \sum_{i=1}^{X_{n-1}} Y_{n,i} + I_n, \qquad n = 1, 2, 3, \cdots \tag{4.49}$$

Clearly, this is a *Markov chain* on the nonnegative integers. Let $\mathcal{F}_n$ be the σ-field generated by $X_0, X_1, X_2, \cdots, X_n$. Think of $\mathcal{F}_{n-1}$ as representing "past information" relative to time n.

   (a) Show that $E[X_n|\mathcal{F}_{n-1}] = mX_{n-1} + \lambda$.

   (b) With $\epsilon_n \equiv X_n - E[X_n|\mathcal{F}_{n-1}]$, show that $\{X_n\}$ satisfies the *stochastic regression*,

$$X_n = mX_{n-1} + \lambda + \epsilon_n, \qquad n = 1, 2, 3, \cdots,$$

where $E[\epsilon_n|\mathcal{F}_{n-1}] = 0$ (i.e. $\{\epsilon_n\}$ is a *martingale difference*), and $E[\epsilon_n\epsilon_k] = 0, n \neq k$.

(c) Since $m < 1$, as $n \to \infty$, $\{X_n\}$ approaches a stationary regime. Suppose $\{X_n\}$ is in its stationary regime. Show that its mean is

$$E[X_n] \equiv \mu = \frac{\lambda}{1-m}$$

and that $Y_n \equiv X_n - \mu$, satisfies the AR(1) form,

$$Y_n = mY_{n-1} + \epsilon_n, \qquad n = 1, 2, 3, \cdots,$$

where $E[\epsilon_n Y_{n-k}] = 0$, $k \geq 1$. What are the possible values of Y_n?

(d) Show that $\rho_k = m^k$, the same acf as that of a Gaussian AR(1) with the same regression parameter.
(*Hint*: For $k \geq 1$, $E[\epsilon_n\epsilon_{n-k}] = E\{\epsilon_{n-k}E[\epsilon_n|\mathcal{F}_{n-1}]\}$.)

(e) Fix $m = .99$, $\lambda = 0.05$. By a computer simulation of the process (4.49), illustrate the fact that the distribution of Y_n is markedly skewed, and hence far from normal (see also [253]).

(f) Obtain the least squares estimators of m and λ. The least squares estimators converge almost surely for $m < 1$. Improved estimators are obtained by weighted least squares [424], [435].

5

Other Models and Alternative Approaches

This chapter introduces the reader to a fair number of additional regression and autoregression models appropriate for integer–valued time series, switching models, models of hidden periodicities, mixture models, and more. Some of these models have been known for a long time but their debut in the time series literature is fairly recent. We intend to provide enough useful information without delving deeply into mathematical details.

5.1 INTEGER AUTOREGRESSIVE AND MOVING AVERAGE MODELS

5.1.1 Branching Processes with Immigration

An important model for integer-valued time series is the *branching process with immigration*, also known as the *Galton-Watson* process with immigration, defined by the stochastic equation

$$X_n = \sum_{i=1}^{X_{n-1}} Y_{n,i} + I_n, \qquad n = 1, 2, 3, \ldots, \tag{5.1}$$

where the initial value X_0 is a nonnegative integer-valued random variable, and $\sum_1^0 \equiv 0$. The processes $\{Y_{n,i}\}$ and $\{I_n\}$ which drive the system are mutually independent, independent of X_0, and each consisting of independently and identically distributed (i.i.d.) random variable. This defines a Markov chain $\{X_n\}$ with nonnegative integer states, originally introduced and applied by Smoluchowski (1916) in studying the fluctuations in the number of particles contained in a small volume in connection with the second law of thermodynamics [89]. Since then, the process has been applied extensively in biological, sociological and physical branching phenomena [41], [42], [55], [64] and [157]. A good review of early work can be found in [435].

In the vernacular of branching processes, X_n is the size of the nth generation of a population, $Y_{n,1}, ..., Y_{n,X_{n-1}}$ are the offspring of the $(n-1)$st generation, and I_n is the contribution of immigration to the nth generation, that is, the number of immigrants at time n. An important role in the behavior of $\{X_n\}$ is played by the mean $m = \mathrm{E}[Y_{n,i}]$ of the offspring distribution, where the cases $m < 1, m = 1, m > 1$, are referred to as subcritical, critical, and supercritical, respectively. In the subcritical case $\{X_n\}$ has a limiting stationary distribution, while in the supercritical case $\{X_n\}$ explodes at an exponential rate. In the critical case the process is either null recurrent or transient.

The process (5.1) admits a useful autoregressive representation as follows. Let $\lambda = \mathrm{E}[I_n]$, and let $\mathcal{F}_n$ be generated by the past information $X_0, X_1, X_2, \cdots, X_n$. Then $\mathrm{E}[X_n \mid \mathcal{F}_{n-1}] = mX_{n-1} + \lambda$. Therefore, with $\epsilon_n \equiv X_n - \mathrm{E}[X_n|\mathcal{F}_{n-1}]$, the stochastic equation (5.1) is transformed into a stochastic regression model,

$$X_n = mX_{n-1} + \lambda + \epsilon_n, \qquad n = 1, 2, 3, \ldots, \tag{5.2}$$

where the noise $\{\epsilon_n\}$ is a *martingale difference*), that is, $\{\epsilon_n\}$ is $\mathcal{F}_n$–measurable and $\mathrm{E}[\epsilon_n \mid \mathcal{F}_{n-1}] = 0$. This implies that $\mathrm{E}[\epsilon_n \epsilon_k] = 0, n \neq k$, and that sums in terms of $\{\epsilon_n\}$ tend to be normally distributed under fairly general conditions. Another fact is that $\mathrm{E}[\epsilon_n^2 \mid \mathcal{F}_{n-1}] = \mathrm{Var}[Y_{n,i}]X_{n-1} + \mathrm{Var}[I_n]$ is unbounded as X_{n-1} increases.

As suggested by (5.2), the least squares estimators for m, λ are obtained by minimizing,

$$\sum_{i=1}^{n} \epsilon_i^2 = \sum_{i=1}^{n} (X_i - mX_{i-1} - \lambda)^2,$$

and are given by

$$\begin{aligned}
\tilde{m} &= \frac{\sum X_i \sum X_{i-1} - n \sum X_i X_{i-1}}{(\sum X_{i-1})^2 - n \sum X_{i-1}^2} \\
\tilde{\lambda} &= \frac{\sum X_{i-1} X_i \sum X_{i-1} - \sum X_{i-1}^2 \sum X_i}{(\sum X_{i-1})^2 - n \sum X_{i-1}^2},
\end{aligned}$$

where the summation limits are from $i = 1$ to $i = n$. It turns out that $\tilde{m}$ is consistent in all three cases, while $\tilde{\lambda}$ is not consistent in the critical and supercritical cases.

Improved estimators are obtained by weighted least squares. We write (5.2) as

$$\frac{X_n}{\sqrt{X_{n-1}+1}} = m\sqrt{X_{n-1}+1} + \frac{(\lambda - m)}{\sqrt{X_{n-1}+1}} + \frac{\epsilon_n}{\sqrt{X_{n-1}+1}}, \tag{5.3}$$

and estimate m and $\lambda - m$ by minimizing $\sum \delta_i^2$ where $\delta_i = \epsilon_i / \sqrt{X_{i-1}+1}$ to obtain [435],

$$\hat{m} = \frac{\sum X_i \sum \frac{1}{X_{i-1}+1} - n \sum \frac{X_i}{X_{i-1}+1}}{\sum (X_{i-1}+1) \sum \frac{1}{X_{i-1}+1} - n^2} \tag{5.4}$$

$$\hat{\lambda} = \frac{\sum X_{i-1} \sum \frac{X_i}{X_{i-1}+1} - \sum X_i \sum \frac{X_{i-1}}{X_{i-1}+1}}{\sum (X_{i-1}+1) \sum \frac{1}{X_{i-1}+1} - n^2}, \tag{5.5}$$

where again the summation limits are from 1 to n.

Then for $0 < m < \infty$, $\hat{m} \to m$ in probability, that is $\hat{m}$ is consistent in all cases, provided $m > 0$, and furthermore, the limiting distribution of $\hat{m}$ is normal in noncritical cases and nonnormal in the critical case. On the other hand, $\hat{\lambda}$ is consistent for $m \leq 1$, but not for $m > 1$, and is asymptotically normal when $m < 1$ or $m = 1$ and $2\lambda > \text{Var}[Y_{n,i}]$ [424],[435].

5.1.1.1 *A Stochastic Model for Rain Rate* There is ample evidence that certain rain characteristics tend to be approximately lognormally distributed in the sense of statistical goodness of fit, pertaining in particular to rainfall amounts and rates under some conditions. See [118] and the rainfall references cited in [111]. Insight into this problem can be gained from the stochastic process (5.1) and its manifestation (5.2) using a heuristic argument as follows [253].

Conditional on rain, suppose we observe at discrete time points a rain element or volume in space containing droplets of water having the following dynamics. At time $n-1$ some droplets give rise to a new generation of droplets through a complicated physical process, some droplets leave the volume while new ones, called immigrants, arrive to join the droplets of the new generation. It is really a process of replacement and immigration: each droplet in the volume is replaced by a nonnegative number of droplets where zero can be interpreted as complete departure from the volume, and the totality of these plus the immigrants give rise to a new generation of droplets at the new time step. This process can be described by (5.1), and switching now to (5.2), X_n may be interpreted as rain rate, since a multiplication by a constant throughout the equation leaves the process essentially intact.

Under the continuity assumption

$$|X_n - X_{n-1}| << X_{n-1}$$

and conditional on rain, $\delta_n \equiv (X_n - X_{n-1})/X_{n-1}$ is small, and

$$X_n = (1+\delta_n)(1+\delta_{n-1}) \cdots (1+\delta_1)X_0,$$

or for sufficiently small δ_i,

$$\log(X_n/X_0) \approx \sum_{i=1}^{n} \delta_i.$$

Thus, we arrive at the intriguing equation,

$$\log(X_n/X_0) + \sum_{i=1}^{n} [(1-m) - \lambda/X_{i-1}] \approx \sum_{i=1}^{n} \epsilon_i/X_{i-1}.$$

So, if there is any hope of seeing lognormality here, at the very least m should be close to 1 and λ should be close to 0. This however is verifiable from data by appealing to the estimators (5.4) and (5.5).

The Global Atmospheric Research Program's (GARP) Atlantic Tropical Experiment (GATE) was conducted in the summer of 1974 in the eastern Atlantic off the coast of west Africa. During roughly three triweekly periods or phases, detailed rainfall measurements were obtained from precipitation radars on an array of research vessels over a large area of about 400 km in diameter every 15 minutes. The GATE data set consists of a collection of radar reflectivity snapshots which were then converted into rain rates binned into $4 \times 4 \quad km^2$ pixels. For technical details see [344].

Twenty time series of length 1716 each of rain rate for individual $4 \times 4 \quad km^2$ pixels, and then of rain rate averaged over larger $40 \times 40 \quad km^2$ pixels have been extracted from the first phase of GATE and the parameters m, λ were estimated by the weighted least squares estimators (5.4) and (5.5). The results reported in Table 5.1 show that for area average rain rate obtained from the larger $40 \times 40 \quad km^2$ pixels, $\hat{m}$ tends to be closer to 1 and $\hat{\lambda}$ tends to be closer to 0, than the same quantities obtained from $4 \times 4 \quad km^2$ pixels. This trend is seen very well from Figure 5.1 as the pixel size increases all the way to $400 \times 400 \quad km^2$. This result suggests that, conditional on rain, a lognormal fit is apparently more appropriate for rain rate averaged over a large area.

Table 5.1 Pairs of estimates $(\hat{m}, \hat{\lambda})$ for $4 \times 4 \quad km^2$ and $40 \times 40 \quad km^2$ from 20 different time series of area average rain rate. *Source*: [253].

		$4 \times 4 \quad km^2$		
0.93, 0.40	0.94, 0.38	0.94, 0.41	0.93, 0.50	0.90, 0.68
0.85, 0.51	0.91, 0.40	0.93, 0.37	0.94, 0.38	0.94, 0.42
0.88, 0.34	0.92, 0.32	0.95, 0.22	0.94, 0.34	0.90, 0.61
0.88, 0.38	0.89, 0.39	0.94, 0.21	0.91, 0.37	0.92, 0.50
		$40 \times 40 \quad km^2$		
0.98, 0.05	0.97, 0.05	0.92, 0.09	0.97, 0.05	0.96, 0.07
0.98, 0.07	0.99, 0.05	0.98, 0.05	0.98, 0.07	0.98, 0.08
0.98, 0.08	0.99, 0.06	0.99, 0.06	0.99, 0.06	0.99, 0.07
0.98, 0.08	0.98, 0.08	0.99, 0.06	0.99, 0.08	0.99, 0.07

5.1.2 Integer Autoregressive Models of Order 1

Although the integer autoregressive model of order 1, INAR(1), is a special case of the branching process with immigration (5.1), it deserves a special consideration due to the *thinning* operation or calculus. The calculus of thinning operators provides further insight into the probabilistic structure of branching processes with immigration by using the simple device of sums of a random number of Bernoulli random variables.

The thinning operator is defined as follows [400], [417, p. 85].

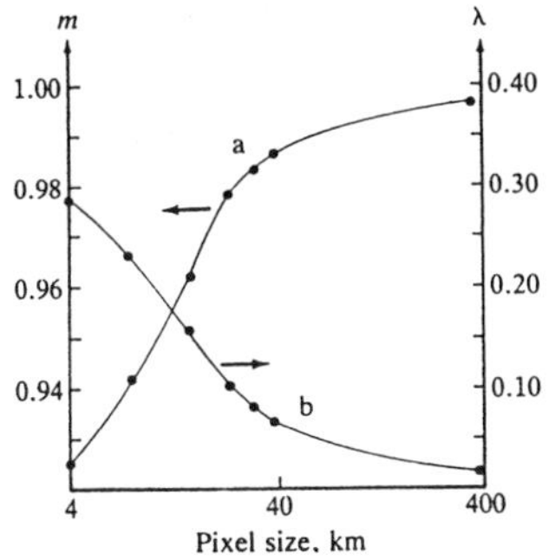

Fig. 5.1 The monotone increase in $\hat{m}$ (Curve a) and the monotone decrease in $\hat{\lambda}$ (Curve b) as a function of the square root of the area. *Source*: [253].

Definition 5.1.1 Suppose that X is a non–negative integer random variable and let $\alpha \in [0, 1]$. Then, the thinning operator, denoted by $\circ$, is defined as

$$\alpha \circ X = \sum_{i=1}^{X} Y_i,$$

where $\{Y_i\}$ is a sequence of independent and identically distributed Bernoulli random variables–independent of X–with success probability α. The sequence $\{Y_i\}$ is termed a counting series.

The random variable $\alpha \circ X$ counts the number of successes in a random number of Bernoulli trials where the probability of success α remains constant throughout the experiment so that given X, $\alpha \circ X$ is a binomial random variable with parameters X and α.

It is easy to see that $0 \circ X = 0$ and $1 \circ X = X$. In addition, the following properties hold.

$$\beta \circ (\alpha \circ X) \sim (\beta\alpha) \circ X \tag{5.6}$$

$$\mathrm{E}[\alpha \circ X \mid X] = \alpha X \tag{5.7}$$

$$\mathrm{E}[\alpha \circ X] = \alpha \mathrm{E}[X] \tag{5.8}$$

$$\mathrm{Var}[\alpha \circ X \mid X] = \alpha(1 - \alpha)X \tag{5.9}$$

$$\mathrm{Var}[\alpha \circ X] = \alpha^2 \mathrm{Var}[X] + \alpha(1 - \alpha)\mathrm{E}[X], \tag{5.10}$$

where equation (5.6) implies equality of distributions.

The notion of binomial thinning is extended to multinomial thinning. Let $\alpha_1, \ldots, \alpha_p$ be positive constants such that $\sum_{i=1}^{p} \alpha_i < 1$. Then the conditional distribution of the vector $(\alpha_1 \circ X, \ldots, \alpha_p \circ X)'$ given X is multinomial with parameters X and $(\alpha_1, \ldots, \alpha_p)$.

Integer autoregressive models imitate the structure of the common autoregressive process, discussed in Appendix A and more thoroughly in [63], in the sense that the thinning operation is applied instead of scalar multiplication [9], [15], [316], [317], [318].

Let $\alpha \in [0,1]$ and let $\{\epsilon_t\}$ be a sequence of independent and identically distributed nonnegative integer valued random variables with $\mathrm{E}[\epsilon_t] = \mu$ and $\mathrm{Var}[\epsilon_t] = \sigma^2$. The integer autoregressive process of order 1 (INAR(1)) $\{X_t\}$, $t = 1, \ldots, N$ is defined by the equation

$$X_t = \alpha \circ X_{t-1} + \epsilon_t, \tag{5.11}$$

where $\alpha \circ X_{t-1}$ is the sum of X_{t-1} Bernoulli random variables all of which are independent of X_{t-1} (recall Definition 5.1.1). It should be noted that the Bernoulli variables used in $\alpha \circ X_{t-1}$ are independent of those used in $\alpha \circ X_{t-2}$, and so on. Clearly, (5.11) is a special case of (5.1).

Figure 5.2 features realizations of 200 observations from the INAR(1) model for different values of α while the innovation variable ϵ_t has the Poisson distribution with mean equal to 1. Apparently as α grows, the process tends to be less oscillatory, a fact that is confirmed theoretically by the autocorrelation function (5.16) of INAR(1).

Distributional properties–including second order properties of the INAR(1) process–can be studied by expressing X_t in terms of present and past values of ϵ_t. By repeated substitutions and property (5.6) we obtain a "moving average" representation,

$$\begin{aligned} X_t &= \alpha \circ X_{t-1} + \epsilon_t \\ &= \alpha \circ (\alpha \circ X_{t-2} + \epsilon_{t-1}) + \epsilon_t \\ &= \alpha^2 \circ X_{t-2} + \alpha \circ \epsilon_{t-1} + \epsilon_t \\ &= \ldots \\ &= \sum_{j=0}^{\infty} \alpha^j \circ \epsilon_{t-j}. \end{aligned} \tag{5.12}$$

An important consequence of the representation (5.12) is that for $\alpha \in (0,1)$ the dependence of $\{X_t\}$ on the sequence $\{\epsilon_t\}$ decays exponentially as t grows.

The mean and variance of the INAR(1) are given by

$$\begin{aligned} \mathrm{E}[X_t] &= \alpha \mathrm{E}[X_{t-1}] + \mu \\ &= \alpha^t \mathrm{E}[X_0] + \mu \sum_{j=0}^{t-1} \alpha^j, \end{aligned} \tag{5.13}$$

$$\begin{aligned} \mathrm{Var}[X_t] &= \alpha^2 \mathrm{Var}[X_{t-1}] + \alpha(1-\alpha)\mathrm{E}[X_{t-1}] + \sigma^2 \\ &= \alpha^{2t} \mathrm{Var}[X_0] + (1-\alpha) \sum_{j=1}^{t} \alpha^{2j-1} \mathrm{E}[X_{t-j}] \\ &+ \sigma^2 \sum_{j=1}^{t} \alpha^{2(j-1)}. \end{aligned} \tag{5.14}$$

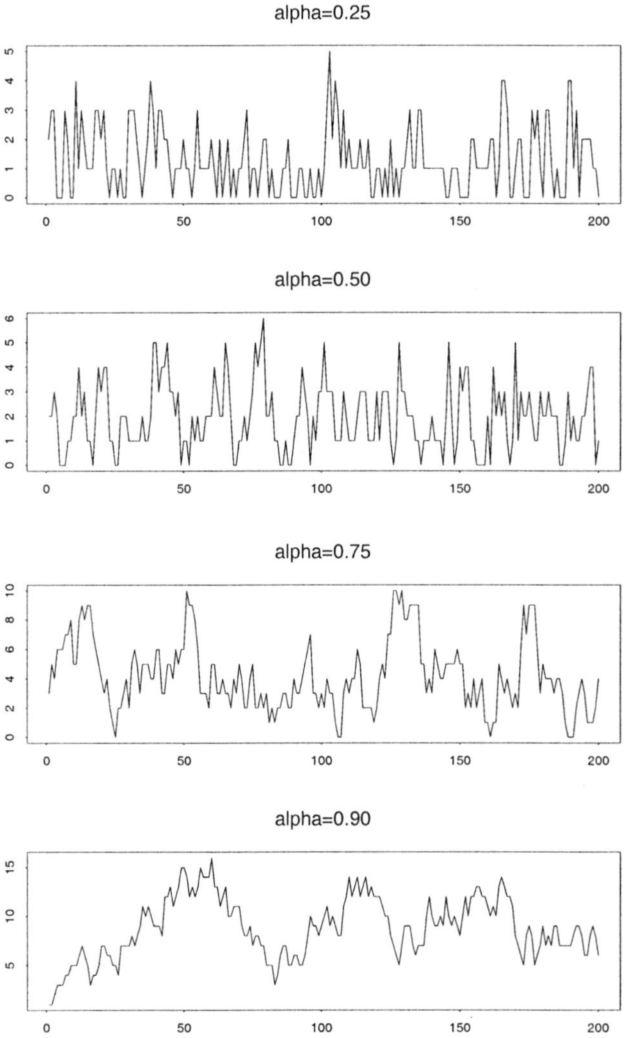

Fig. 5.2 Typical realizations of 200 observations from the INAR(1) model (5.11) for different values of α. Here ϵ_t Poisson with mean equal to 1.

Thus second-order stationarity implies that $\mathrm{E}[X_t] = \mu/(1-\alpha)$, $\mathrm{Var}[X_t] = (\alpha\mu + \sigma^2)/(1-\alpha^2)$, and the autocovariance function evaluated at lag k, $c(k)$, is given by

$$c(k) \equiv \mathrm{Cov}[X_t, X_{t-k}] = \alpha^k c(0). \tag{5.15}$$

Consequently, the autocorrelation function, $\rho(k)$, is

$$\rho(k) = \frac{c(k)}{c(0)} = \alpha^k, \tag{5.16}$$

so that $\rho(k)$ decays exponentially with the lag k as in AR(1), but unlike the autocorrelation of a stationary AR(1) process, it is always positive for $\alpha \in (0,1)$.

Under suitable conditions, it can be shown that X_t has a discrete self-decomposable distribution. This, in turn, implies unimodality properties and characterization of the distribution of X_t through ϵ_t. For example, X_t follows the Poisson distribution if and only if ϵ_t follows the Poisson distribution [9].

5.1.2.1 Poisson INAR(1) An important special case is that of Poisson INAR(1) [316] and [318]. That is, $X_t = \alpha \circ X_{t-1} + \epsilon_t$, with $\{\epsilon_t\}$ a sequence of independent and identically distributed Poisson random variables with mean μ. Then $\rho(k) = \alpha^k$,

$$\mathrm{E}[X_t \mid X_{t-1}] = \alpha X_{t-1} + \mu,$$

and

$$\mathrm{Var}[X_t \mid X_{t-1}] = \alpha(1-\alpha)X_{t-1} + \mu.$$

The conditional distribution of X_t given X_{t-1} is

$$\begin{aligned} p(y \mid x) &= \mathrm{P}[X_t = y \mid X_{t-1} = x] \\ &= x! \exp(-\mu) \sum_{k=0}^{m} \frac{\alpha^k (1-\alpha)^{x-k} \mu^{y-k}}{k!(x-k)!(y-k)!}, \quad y = 0, 1, \ldots, \end{aligned}$$

where $m = \min(x, y)$. In this case $\{X_t\}$ is a reversible Markov process with transition matrix specified by $p(y \mid x)$, with nonnegative integers x, y.

5.1.2.2 Vector INAR(1) To define the vector INAR(1) process recall the notion of multinomial thinning. That is, suppose $\boldsymbol{\alpha} = (\alpha_1, \ldots \alpha_p)'$ is a vector of nonnegative components whose sum does not exceed one. Then conditional on X, the random vector $\boldsymbol{\alpha} \circ X = (\alpha_1 \circ X, \ldots \alpha_p \circ X)'$ has a multinomial distribution with parameters X and $(\alpha_1, \ldots \alpha_p)'$. For a $p \times p$ matrix $\mathbf{A} = (\boldsymbol{\alpha}_1, \ldots, \boldsymbol{\alpha}_p)$, with $\boldsymbol{\alpha}_1, \ldots, \boldsymbol{\alpha}_p$ satisfying the same conditions as those of $\boldsymbol{\alpha}$, and a p–dimensional random vector $\mathbf{X} = (X_1, \ldots, X_p)'$, define

$$\mathbf{A} \circ \mathbf{X} = \sum_{i=1}^{p} \boldsymbol{\alpha}_i \circ X_i,$$

where each multinomial thinning operator is performed independently. Then, the vector INAR(1) process $\{\mathbf{X}_t\}$ is determined by the stationary solution of

$$\mathbf{X}_t = \mathbf{A} \circ \mathbf{X}_{t-1} + \mathbf{E}_t,$$

where $\{\mathbf{E}_t\}$ is a sequence of independent and identically distributed nonnegative integer valued random vectors.

5.1.3 Estimation for INAR(1) Process

Estimation in INAR(1) means estimation in the branching process with immigration in the subcritical case, and this has already been discussed earlier. Still it is interesting to note a few facts regarding estimation in the Poisson INAR(1).

Estimation procedures for the parameters α and μ of the INAR(1) model (5.11) assuming that the sequence $\{\epsilon_t\}$ follows the Poisson distribution has been discussed in [9]. Under the Poisson assumption, $\sigma^2 = \mu$ and, equation (5.15) yields a method of moments estimator for α given by

$$\hat{\alpha} = \frac{\sum_{t=0}^{N-1}(X_t - \bar{X})(X_{t+1} - \bar{X})}{\sum_{t=0}^{N}(X_t - \bar{X})^2},$$

while μ can be estimated by

$$\hat{\mu} = \frac{\sum_{t=1}^{N} \hat{\epsilon}_t}{N},$$

where $\hat{\epsilon}_t = X_t - \hat{\alpha}X_{t-1}$, for $t = 1, \ldots, N$.

An alternative estimation method is that of conditional least squares. Upon noticing that

$$\mathrm{E}\left[X_t \mid X_{t-1}\right] = \alpha X_{t-1} + \mu,$$

the conditional least squares of the parameters α and μ are those values that minimize

$$\sum_{t=1}^{N} \left(X_t - \alpha X_{t-1} - \mu\right)^2.$$

Asymptotic properties of the resulting estimators (see Problem 4) are deduced by using results from [269], [435]. The two methods, moments and least squares, do not require full distributional assumptions.

Maximum likelihood estimation, which requires a full distributional assumption about the innovations, has been suggested in [9]. Under the Poisson assumption, the likelihood function of a time series of $(N+1)$ observations from the INAR(1) process is

$$\left(\prod_{t=1}^{N} P_t(X_t)\right) \frac{(\mu/(1-\alpha))^{X_0}}{X_0!} \exp\left(-\mu/(1-\alpha)\right), \tag{5.17}$$

where for $t = 1, \ldots, N$,

$$P_t(y) = \exp(-\mu) \sum_{i=0}^{\min(X_t, X_{t-1})} \frac{\mu^{y-i}}{(y-i)!} \binom{X_{t-1}}{i} \alpha^i (1-\alpha)^{X_{t-1}-i}. \tag{5.18}$$

Differentiation of (5.17) gives the unconditional maximum likelihood estimates, whereas differentiation of $\left(\prod_{t=1}^{N} P_t(X_t)\right)$ yields the conditional maximum likelihood estimates of α and μ given X_0.

5.1.4 Integer Autoregressive Models of Order p

A straightforward extension to the integer autoregressive model of order 2, INAR(2), gives

$$X_t = \alpha_1 \circ X_{t-1} + \alpha_2 \circ X_{t-2} + \epsilon_t,$$

and more generally the pth order model, INAR(p), is defined as

$$X_t = \sum_{i=1}^{p} \alpha_i \circ X_{t-i} + \epsilon_t, \tag{5.19}$$

where $\{\epsilon_t\}$ is a sequence of independent and identically distributed nonnegative integer valued random variables with mean μ and variance σ^2, and for stability of the process it is required that $\sum_i \alpha_i < 1$ [15].

The conditional distribution of $(\alpha_1 \circ X_t, \ldots, \alpha_p \circ X_t)'$ given X_t is multinomial and is independent of the past of the process. In other words given X_t, $\alpha_i \circ X_t$ is independent of X_{t-k} and $\alpha_j \circ X_{t-k}$ for $i, j = 1, 2, \ldots, p$ and $k > 0$. The meaning of expression (5.19) generalizes that of (5.11) in the sense that the total size of the population at time t is equal to the number of offsprings of the last p generations $\alpha_i \circ X_{t-i}$ plus the immigration process.

Sufficient conditions for the process $\{X_t\}$ to have a limiting distribution are that the roots of

$$\lambda^p - \alpha_1 \lambda^{p-1} - \ldots - \alpha_{p-1}\lambda - \alpha_p = 0, \quad \alpha_p \neq 0, \tag{5.20}$$

are less than in 1 in absolute value and $\sum_{j=0}^{\infty}(j+1)^{-1} p_j < \infty$, where $p_j = \sum_{k=j+1}^{\infty} \mathrm{P}[\epsilon_1 = k]$ [15]. Moreover, as t grows the mean of the process is

$$\mu_x = \mathrm{E}[X_t] = \frac{\mu}{1 - \sum_{i=1}^{p} \alpha_i}, \tag{5.21}$$

while the autocovariance function satisfies

$$c(k) = \sum_{i=1}^{p} \alpha_i c(k-i) + \sum_{i=k+1}^{p} v(k-i, \alpha_i) + \delta_k(0)\sigma^2, \tag{5.22}$$

with

$$v(-k, \alpha_i) = \sum_{j=1}^{k-1} \alpha_j v(j-k, \alpha_i) + \alpha_i \left(\delta_i(k) - \alpha_k\right) \mu_x \tag{5.23}$$

and $v(k-i, \alpha_i)$ is determined from (5.23) for $k < i$, and $v(k-i, \alpha_i) = 0$ for $k \geq i$. Here $\delta_k(0) = 1$ if $k = 0$. Equation (5.22) points out that the autocovariance function of the INAR(p) process has a form similar to that of a Gaussian ARMA(p, $p-1$)

process due to dependence in $\alpha_i \circ X_{t-i}, i = 1, ..., p$ appearing in different times (see [15] and also the discussion in [128].) Existence and generalizations of INAR(p) are studied in [67], [289] and [290] while unifying work based on convolution is presented in [229].

5.1.5 Regression Analysis of Integer Autoregressive Models

The INAR(1) model (5.11) has been extended by including explanatory variables. Following [65], assume that the observed response process $\{Y_t\}$ is a realization of

$$Y_t = \alpha \circ Y_{t-1} + \epsilon_t$$

with $\{\epsilon_t\}$ a sequence of Poisson random variables with mean $\exp(\boldsymbol{\beta}'\mathbf{X}_t)$ where $\{\mathbf{X}_t\}$ is a covariate process. In addition, assume that

$$\alpha = \frac{1}{1+\exp(-\gamma)}$$

so that $0 < \alpha < 1$. Then,

$$\mathrm{E}\left[Y_t \mid \mathbf{X}_t, Y_{t-1}\right] = \left(\frac{1}{1+\exp(-\gamma)}\right) Y_{t-1} + \exp(\boldsymbol{\beta}'\mathbf{X}_t),$$

and

$$\mathrm{Var}\left[Y_t \mid \mathbf{X}_t, Y_{t-1}\right] = \left(\frac{\exp(-\gamma)}{(1+\exp(-\gamma))^2}\right) Y_{t-1} + \exp(\boldsymbol{\beta}'\mathbf{X}_t)$$

by employing (5.7) and (5.9). Estimation is based on minimizing with respect to $(\boldsymbol{\beta}, \gamma)$ the unweighted sum of squares

$$\sum_t \left\{ Y_t - \left(\frac{1}{1+\exp(-\gamma)}\right) Y_{t-1} - \exp(\boldsymbol{\beta}'\mathbf{X}_t) \right\}^2 .$$

The corresponding standard errors need to be adjusted though to allow for heteroscedasticity. An alternative method is that of weighted least squares which leads to more efficient estimators. The above specification can be extended by introducing the dynamic parameterization $\alpha_t = 1/(1+\exp(-\boldsymbol{\gamma}'\mathbf{Z}_t))$ with $\mathbf{Z}_t$ another covariate process. Again, least squares or weighted least squares can be used in the estimation of $(\boldsymbol{\beta}, \gamma)$.

5.1.6 Integer Moving Average Models

In the spirit of the INAR(p) process, the integer moving average model of order q, abbreviated INMA(q), is defined by the equation [8], [318],

$$X_t = \beta_0 \circ \epsilon_t + \beta_1 \circ \epsilon_{t-1} + \ldots + \beta_q \circ \epsilon_{t-q}, \tag{5.24}$$

where $\{\epsilon_t\}$ is a sequence of independent and identically distributed non negative integer–valued random variables with mean μ and variance σ^2, $\beta_0 = 1$ and β_i belong to $[0,1]$, $i = 1, \ldots, q$, and that all thinning operations are performed independently. Figure 5.3 illustrates typical realizations of the INMA(1) process,

$$X_t = \epsilon_t + \beta_1 \circ \epsilon_{t-1}, \tag{5.25}$$

where the ϵ_t follows the Poisson distribution with mean 1 and the parameter β_1 assumes the values 0.25, 0.50, 0.75 and 0.90. As β_1 grows, the process becomes less oscillatory since the degree of positive correlation between successive observations increases.

It is instructive to consider some properties of the INMA(1) model. Recall the properties of the thinning operator (5.8) and (5.10). Then it is easy to see that

$$\begin{aligned} \mathrm{E}[X_t] &= \mathrm{E}[\epsilon_t + \beta_1 \circ \epsilon_{t-1}] \\ &= (1+\beta_1)\mu \end{aligned} \tag{5.26}$$

and

$$\begin{aligned} \mathrm{Var}[X_t] &= \mathrm{Var}[\epsilon_t + \beta_1 \circ \epsilon_{t-1}] \\ &= (1+\beta_1^2)\sigma^2 + \beta_1(1-\beta_1)\mu. \end{aligned} \tag{5.27}$$

By a conditional argument and using the independence of the thinning operations we also obtain

$$c(k) = \beta_1 \sigma^2 \tag{5.28}$$

for $k = 1$. When $k > 1$, then $\mathrm{Cov}[X_t, X_{t-k}] = 0$. The autocorrelation function of the INMA(1) process is then

$$\rho(k) = \frac{c(k)}{c(0)} = \frac{\sigma^2 \beta_1}{(1+\beta_1^2)\sigma^2 + \beta_1(1-\beta_1)\mu}, \tag{5.29}$$

for $k = 1$, and is equal to 0 otherwise, similar to the autocorrelation function of the standard MA(1) model. For Poisson distributed errors (5.29) reduces to [318],

$$\rho(k) = \begin{cases} \beta_1/(1+\beta_1), & k = 1, \\ 0, & k > 1. \end{cases}$$

It is not hard to generalize formulas (5.26)–(5.29) for the case of INMA(q) model (5.24). We have

$$\mathrm{E}[X_t] = \mu \sum_{i=0}^{q} \beta_i, \tag{5.30}$$

$$\mathrm{Var}[X_t] = \sigma^2 \left(\sum_{i=0}^{q} \beta_i^2 \right) + \mu \left(\sum_{i=0}^{q} \beta_i (1-\beta_i) \right), \tag{5.31}$$

$$c(k) = \sigma^2 \sum_{i=0}^{q-k} \beta_i \beta_{i+k} + \mu \sum_{i=0}^{q-k} \beta_i \left(\beta_k - \beta_{i+k} \right), \tag{5.32}$$

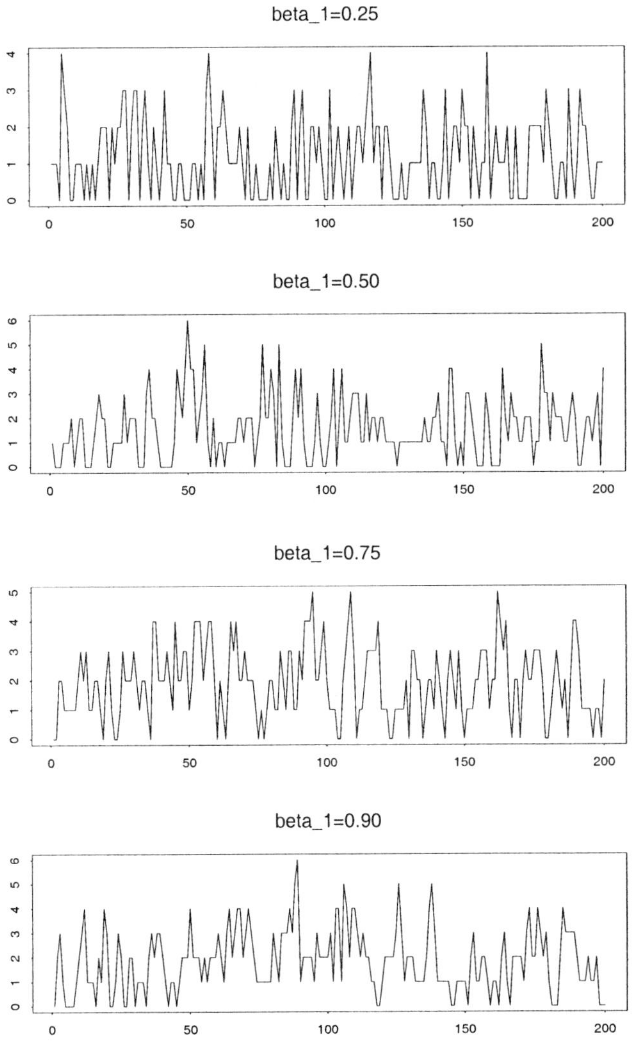

Fig. 5.3 Typical realizations of 200 observations from the INMA(1) model (5.11) for different values of β_1. Here ϵ_t Poisson with mean equal to 1.

for $k = 1, 2, \ldots, q$ and 0 otherwise,

$$\rho(k) = \frac{c(k)}{c(0)} = \frac{\sigma^2 \sum_{i=0}^{q-k} \beta_i \beta_{i+k} + \mu \sum_{i=0}^{q-k} \beta_i \left(\beta_k - \beta_{i+k}\right)}{\sigma^2 \left(\sum_{i=0}^{q} \beta_i^2\right) + \mu \left(\sum_{i=0}^{q} \beta_i (1 - \beta_i)\right)} \tag{5.33}$$

for $k = 1, 2, \ldots, q$ and 0 otherwise. Problem 7 asks the reader to verify all these moment calculations. An extension of the INMA(q) model and the accompanying estimation problem are discussed in [66].

5.1.7 Extensions and Modifications

Among the early works on autoregressive processes with prescribed marginals is that of Gaver and Lewis [165], where it is shown that there exists an innovation sequence $\{\epsilon_t\}$ such that X_n from the AR(1) sequence $X_n = \rho X_{n-1} + \epsilon_n, \quad n = 0, \pm 1, \pm 2, \ldots$, has a gamma distribution for $0 \leq \rho < 1$. In particular, there is a sequence $\{\epsilon_t\}$ of i.i.d. random variables such that the X_n have an exponential distribution and the resulting process is called exponential autoregressive process of order 1, EAR(1). Related works on time series having prescribed marginals such as exponential MA, EMA(1), and exponential ARMA, EARMA(p,q), includes that of [291], [292], [295] [310].

In connection with AR(1) having gamma marginals we must mention that under some conditions, in the branching process with immigration (5.1), which we know admits the AR(1) representation (5.2), $(1 - m)X(m)$ converges in distribution as $m \to 1_-$ to a random variable with a gamma distribution, where $X(m)$ has the limiting stationary distribution of X_n [360].

The INAR(p) and INMA(q) models can be combined to form the integer autoregressive moving average model of order (p, q), INARMA(p,q). The distributional properties of INAR, INMA, and INARMA depend on the assumptions regarding the innovation process. Besides the Poisson assumption, also binomial, geometric and negative binomial ϵ_t's have been proposed [316], [317], [318]. A representative example is the autoregressive negative binomial model, INAR(1)-NB,

$$Y_t = \Pi \circ Y_{t-1} + I_t,$$

where Π is a beta random variable and Y_0, I_t are independent negative binomial random variables. The INAR(1)-NB model has been applied to the analysis of count data from a panel diary study about the relationship between personality factors and emotional experiences in [55]. In the same vein, the definition of the binomial thinning operator is extended to the so called hypergeometric thinning operator for the construction of binomial and generalized Poisson models in [10], [16].

A generalization of binomial thinning is given in [229],

$$X_t = A_t(X_{t-1}; \alpha) + \epsilon_t, \quad t = 1, 2, \ldots.,$$

where A_t is a random transformation, and $A_t(X_{t-1}; \alpha)$ and ϵ_t are independent. Based on this general thinning, a class of stationary moving average processes with margins in the class of infinitely divisible exponential dispersion models was introduced in [237].

Another recent generalization is that of the first order conditional linear autoregressive process, CLAR(1),

$$m(X_{t-1}) = \phi X_{t-1} + \lambda,$$

where $m(X_{t-1}) = \mathrm{E}[X_t|X_{t-1}]$, and ϕ, λ are real numbers. The CLAR(1) class includes many of the non-Gaussian AR(1) models proposed in the literature and allows various generalizations of previous results [185]. Interestingly, For $|\phi| < 1$, the autocorrelation is $\rho(k) = \phi^k$, $k = 1, 2, ...$, as in other first-order autoregressive processes including the branching process with immigration (5.1).

5.2 DISCRETE AUTOREGRESSIVE MOVING AVERAGE MODELS

An early attempt to form autoregressive and moving average models for discrete-valued time series data was made by the introduction in [221] and [222] of the so called discrete autoregressive moving average models, DARMA, defined as follows. Let $\{Y_t\}$ be a sequence of independent discrete random variables each having an arbitrary distribution π such as Poisson or geometric. By definition, $\mathrm{P}[Y_t = i] = \pi_i$ for i in some countable set. Let $\{U_t\}$ and $\{V_t\}$ be independent sequences of Bernoulli random variables such that $\mathrm{P}[U_t = 1] = \beta$, $0 \leq \beta \leq 1$, and $\mathrm{P}[V_t = 1] = \rho$, $0 \leq \rho < 1$, and denote by $\{S_t\}$ a sequence of independent and identically distributed random variables supported in $\{0, 1, \ldots, N\}$ with distribution F.

We call the process $\{X_t\}$ discrete autoregressive moving average process of order $(1, N+1)$, DARMA(1, $N+1$), if

$$X_t = U_t Y_{t-S_t} + (1 - U_t) A_{t-(N+1)}, \quad t = 1, 2, \ldots, \tag{5.34}$$

where

$$A_t = V_t A_{t-1} + (1 - V_t) Y_t, \quad ,t = -N, -N+1, \ldots. \tag{5.35}$$

The last expression (5.35) is termed as discrete autoregressive process of order 1, where A_t equals A_{t-1} with probability ρ or A_t coincides with Y_t with probability $1 - \rho$. Similarly, X_t is equal to one of the $Y_t, Y_{t-1}, \ldots, Y_{t-N}$ with probability β or $X_t = A_{t-(N+1)}$ with probability $1 - \beta$.

Properties of the DARMA(1, $N + 1$) process are discussed in detail in [221]. When the process starts with $A_{-(N+1)}$ having the distribution π independently of Y_t for $t \geq -N$, $\{U_t\}$, $\{V_t\}$, and $\{S_t\}$, then $\{X_t\}$, $t = 1, 2, \ldots$ is stationary with marginal distribution π and an autocorrelation structure that depends on ρ, β and F,

$$\begin{aligned} \rho(k) &= \beta^2 \sum_{j=0}^{N-k} F(j)F(j+k) \\ &+ \beta(1-\beta)\left((1-\rho)\rho^{-(N+1-k)} \sum_{j=N+1-k}^{N} \rho^j F(j)\right) \\ &+ (1-\beta)^2 \rho^k, \end{aligned}$$

for $1 \leq k \leq N$, and

$$\rho(k) = \beta(1-\beta)\rho^{k-(N+1)} \sum_{j=0}^{N} \rho^j (1-\rho) F(j) + (1-\beta)^2 \rho^k,$$

for $k > N$.

It can be shown that $\{X_t\}$ is, in general, not Markovian but it is so for $\beta = 0$, and that, in general, it is not time reversible in the sense that $\{X_1, ..., X_k\}$ does not, in general, have the same distribution as $\{X_{-k}, ..., X_{-1}\}$. For more on the DARMA model and its asymptotic properties regarding estimates of moments, percentiles and quantiles, and a goodness of fit test for the marginal distribution π see [222]. Applications of DARMA models related to meteorological problems are discussed in [90], [91], [92] and [119].

An extension of (5.34) and (5.35) has been considered in [223] where the enlarged discrete autoregressive moving average model of order $(p, N+1)$ is considered in addition to other models. Keeping the same notation as in (5.34) and (5.35), the DARMA$(p, N+1)$ model is given by

$$X_t = U_t Y_{t-S_t} + (1 - U_t) A_{t-(N+1)}, \quad t = 1, 2, \ldots$$

and

$$A_t = V_t A_{t-D_t} + (1 - V_t) Y_t, \quad , t = -N, -N+1, \ldots,$$

where $\{D_t\}$ is a sequence of independent identically distributed random variables from a distribution G, taking values $1, 2, \ldots, p$. It can be shown that the correlation structure of this process is similar to that of the ARMA(p, q) model, and that some ad hoc nonparametric estimation methods perform reasonably well compared with maximum likelihood estimators [223].

A related theory for bivariate exponential and geometric autoregressive moving average models has been developed in [54] and [287] where the concept of positive dependence is used to show that all these models consist of associated random variables.

5.3 THE MIXTURE TRANSITION DISTRIBUTION MODEL

The *mixture transition distribution* model has been introduced by Raftery [362] extending previous work by Pegram [345] as a parsimonious approach to the analysis of higher order Markov chains.

Suppose the process $\{X_t\}$ takes values in $\{1, 2, \ldots, m\}$ satisfying

$$\mathrm{P}[X_t \mid X_{t-1}, X_{t-2}, \ldots] = \mathrm{P}[X_t \mid X_{t-1}, X_{t-2}, \ldots, X_{t-p}], \qquad (5.36)$$

for some $p \geq 2$. As indicated in Chapter 3, as p and m grow the number of parameters increases exponentially according to the formula $m^p(m-1)$. The mixture transition distribution model–abbreviated MTD–bypasses this problem by specifying the conditional probability of observing $X_t = i_0$ given the past as a linear combination of contributions from $X_{t-1}, \ldots, X_{t-p}$. More precisely it is assumed that

$$\mathrm{P}[X_t = i_0 \mid X_{t-1} = i_1, \ldots, X_{t-p} = i_p] = \sum_{j=1}^{p} \lambda_j \mathrm{P}[X_t = i_0 \mid X_{t-j} = i_j]$$

$$= \sum_{j=1}^{p} \lambda_j q_{i_j i_0}, \tag{5.37}$$

where $i_0, \ldots, i_p$ belong to $\{1, 2, \ldots, m\}$, $q_{i_j i_0}$ are elements of the $m \times m$ transition matrix $\mathbf{Q}$ and the vector of lag parameters $\boldsymbol{\lambda} = (\lambda_1, \ldots, \lambda_p)'$ satisfies

$$\sum_{j=1}^{p} \lambda_j = 1, \quad \lambda_j \geq 0,$$

so that the right-hand side of (5.37) is between 0 and 1. An alternative set of restrictions for $\boldsymbol{\lambda}$ is given in [365].

Besides reducing considerably the number of parameters to $m(m-1) + (p-1)$, model (5.37) has several more useful properties. It can be shown that the limiting behavior of MTD model is the same as the full parameterized higher order Markov chain (see [362] and [1]). Furthermore, defining the $m \times m$ matrix $\mathbf{B}(k)$ whose elements are

$$b_{ij}(k) = \mathrm{P}\left[X_t = i, X_{t+k} = j\right], \quad i, j = 1, 2, \ldots, m,$$

for k integer, then

$$\mathbf{B}(k) = \sum_{j=1}^{p} \lambda_j \mathbf{Q} \mathbf{B}(k-j). \tag{5.38}$$

As a result, if Y_t is a random variable whose distribution is given by

$$\mathrm{P}\left[Y_t = i \mid X_t = j\right] = q_{ji}$$

and $\rho(k)$, $\tilde{\rho}(k)$ denote the correlations between X_{t+k} and X_t and between Y_{t+k} and X_t, respectively, then

$$\rho(k) = \sum_{j=1}^{p} \lambda_j \tilde{\rho}_{k-j}.$$

That is, the autocorrelation satisfies a system similar to the Yule–Walker equations. Various generalizations of the MTD model have been proposed. For example, Raftery [363] considers the multi–matrix mixture transition distribution model called MTDg. The MTDg model uses a different transition matrix for each lag as follows:

$$\mathrm{P}\left[X_t = i_0 \mid X_{t-1} = i_1, \ldots, X_{t-p} = i_p\right] = \sum_{j=1}^{p} \lambda_j q_{i_j i_0}^{(j)}. \tag{5.39}$$

Model (5.39) is less parsimonious than model (5.37) in the sense that it requires $m(m-1)+1$ additional parameters for each lag. However it accommodates a dynamic relation between each lag and time period.

The work in [293] and more recently that in [436] and [438] extend definition (5.37) to arbitrary state space models. The spatial MTD model is investigated in [38] and [364], and the double chain Markov model is studied in [37].

5.3.1 Estimation in MTD Models

Estimation of the parameters $\boldsymbol{\lambda}$ and q_{ij} of the mixture transition model (5.37) is accomplished by maximizing the log-likelihood [38], [365],

$$\sum_{i_0,\ldots,i_p=1}^{m} n_{i_0,\ldots,i_p} \log\left(\sum_{j=1}^{p} \lambda_j q_{i_j i_0}\right)$$

subject to constraints on $\boldsymbol{\lambda}$. Here $n_{i_0,\ldots,i_p}$ denotes the number of sequences satisfying $\{X_t = i_0, \ldots, X_{t-p} = i_p\}$. Alternative estimation methods include the minimum χ^2 estimation ([365]) and E–M algorithm ([293]). Software (programs MTD and GMTD) for fitting the mixture transition model can be obtained from the web address `http://lib.stat.cmu.edu/general`. A thorough review of the mixture transition distribution model for higher order Markov chains and non-Gaussian time series can be found in [39].

5.3.2 Old Faithful Data Revisited

Recall Example 2.5.2 regarding modeling successive eruptions of the Old Faithful geyser in Yellowstone National Park, Wyoming. The analysis, based on regression models for binary time series, suggested that a second order model fits the data reasonably well. This fact is reconfirmed by Table 5.2 which reports the deviance, AIC and BIC from models (5.36), (5.37) and (5.39) fitted to the Old Faithful data. The second column of Table 5.2 lists the number of estimated parameters in each model. For example, the first order Markov chain model consists only of one free parameter for the Old Faithful data since 0 is never followed by 0. Therefore the transition probability from 0 to 1 is equal to 1 and the only free parameter is the transition probability from 1 to 1. In a similar manner we obtain the rest of this column's entries. These results are based on the first 259 observations for a fair comparison with Table 2.7. The first five lines of Table 5.2 show that there is a close agreement between the Markov chain models and the corresponding regression models pointing again to the second order Markov chain model as the best candidate for these data.

The estimation results for the MTD model (5.37) are reported only for $p = 1$, since there was no any further improvement in the deviance, AIC and BIC for higher order models. It is seen that by the AIC result, the MTDg of order 2 is somewhat preferable to the MTD model, but by the BIC criterion there is no advantage to any of the MTDg models. Among the MTDg models (5.39), the AIC selects a second-order MTDg model while the BIC points to a first-order MTDg model. Overall, from all the cases considered, the full second-order Markov chain is selected as the preferable model by both AIC and BIC.

Table 5.2 Results from Markov chain and MTD models fitted to the Old Faithful Data. $N = 259$.

Model	Number of Parameters	D	AIC	BIC
Independence	1	331.12	333.12	336.66
Markov chain of order 1	1	227.37	229.37	232.92
Markov chain of order 2	2	215.52	219.52	226.61
Markov chain of order 3	3	215.07	221.07	231.69
Markov chain of order 4	5	213.95	223.95	241.65
MTD of order 1	1	227.37	229.37	232.92
MTDg of order 1	1	227.37	229.37	232.92
MTDg of order 2	5	215.81	225.81	243.52
MTDg of order 3	6	215.52	227.52	248.77
MTDg of order 4	11	215.61	237.61	276.55

5.3.3 Explanatory Analysis of DNA Sequence Data Revisited

Consider Example 3.5.1 concerning DNA sequence data of the gene BNRF1 of the Epstein–Barr virus. Table 5.3 reports the results obtained by fitting models (5.36), (5.37) and (5.39) and it should be compared with Table 3.4.

The first line of Table 5.3 reports results under independence, the selected model under the BIC criterion. The next four rows show the analysis based on full Markov chain modeling, that is, model (5.36). We see that a Markov chain of order 4 leads to the smallest AIC with 321 parameters, while the BIC selects the first-order model with 12 parameters.

MTD fitting points to the first-order model. Indeed, an MTD model of order 1 is simply a Markov chain of order 1. It can be seen though that higher order MTD models do not affect the fit considerably since the changes in deviance are rather small. Compared with the output of the multinomial logits model (3.4) for the DNA data, the MTD models reduce both the AIC and BIC criteria. In addition the number of parameters that need to be estimated is appreciably less than the number of parameters that need to be estimated for both the multinomial logits model and the full Markov chain. Note that, as in orders 2 and 3, the equality between the number of parameters for some models may not leave degrees of freedom for testing certain hypotheses.

The multi–lag MTDg model points to the first order Markov chain. Notice again that, as with MTD of order 1, an MTDg model of order 1 is simply a Markov chain of order 1. Regarding the fitted MTDg models, here the number of parameters becomes large compared with those of the MTD model and this leads to an increase in both the AIC and BIC values. Compared with the multinomial logits fit, the optimal AIC and BIC values from the MTDg models are fairly similar.

5.3.4 Soccer Forecasting Data Revisited

The final example in this section concerns the fits of models (5.36), (5.37) and (5.39) to the soccer categorical data (see Section 3.5.2). Table 5.4 reports the results of this analysis only for the games played in the first position. We see that the proportional

Table 5.3 Results from Markov chain and MTD models fitted to gene BNRF1 of the Epstein-Barr virus DNA data. $N = 996$.

Model	Number of Parameters	D	AIC	BIC
Independence	3	2711.31	2717.31	2732.02
Markov chain of order 1	12	2677.75	2701.75	2760.60
Markov chain of order 2	48	2627.68	2723.68	2959.06
Markov chain of order 3	179	2463.22	2821.22	3698.99
Markov chain of order 4	321	1808.33	2450.33	4024.44
MTD of order 1	12	2677.75	2701.75	2760.60
MTD of order 2	13	2677.75	2703.75	2767.51
MTD of order 3	13	2677.11	2703.11	2766.86
MTD of order 4	14	2676.18	2704.18	2772.83
MTDg of order 1	12	2677.75	2701.75	2760.60
MTDg of order 2	25	2664.27	2714.27	2836.87
MTDg of order 3	36	2647.27	2719.27	2895.80
MTDg of order 4	46	2631.12	2723.12	2948.70

odds model performs better than all the alternatives considered in the table in the sense of minimizing both the AIC and BIC. To explain this notice the relatively small number of parameters required when fitting a proportional odds model. The final result is consistent with the previous analysis. That is, the model of independence fits the soccer data quite well, leading once more to the conclusion that the soccer forecasting game is fair.

Table 5.4 Results from Markov chain and MTD models applied to the Soccer Forecasting Data for the first position. $N = 289$.

Model	Number of Parameters	D	AIC	BIC
Independence	2	562.43	566.43	573.74
Markov chain of order 1	6	558.69	570.69	592.64
Markov chain of order 2	18	549.21	585.21	651.08
MTD of order 1	6	558.69	570.69	592.64
MTD of order 2	7	558.68	572.68	598.29
MTDg of order 1	6	558.69	570.69	592.64
MTDg of order 2	12	557.84	581.84	625.75

5.4 HIDDEN MARKOV MODELS

Hidden Markov models specify that the observed process is driven by some unobserved process which is assumed to be a Markov chain. To be more specific, assume that $\{X_t\}, t = 1, \ldots, N$ is an observed non-negative integer time series and that $\{A_t\}$ is an unobserved or "hidden" irreducible homogeneous Markov chain taking values on $\{1, 2, \ldots, m\}$ with transition probability matrix $\mathbf{Q}$. That is, the (i, j)-element of

Q is

$$Q_{ij} = \mathrm{P}\left[A_t = j \mid A_{t-1} = i\right], \quad i, j = 1, 2, \ldots, m. \tag{5.40}$$

Set

$$\boldsymbol{\pi} = (\pi_1, \ldots, \pi_m)' \tag{5.41}$$

for the stationary distribution of $\{A_t\}$–such a distribution exists because the chain is assumed irreducible. In addition, suppose that conditional on $\{A_t\}$, the random variables $\{X_t\}$, $t = 1, \ldots, N$ are mutually independent such that

$$p_{xa} = \mathrm{P}\left[X_t = x \mid A_t = a\right], \quad x = 0, 1, \ldots, \quad a = 1, 2, \ldots, m. \tag{5.42}$$

In general, the probabilities p_{xa} may depend on t. However for our limited exposition we prefer to drop that notation. Examples (5.42) include Poisson, binomial and multinomial distributions. Hidden Markov models were introduced in [31], [32], [33], [34] and consequently found numerous applications in engineering, speech processing, genetics, econometrics, biochemistry, environmetrics and so on. We do not attempt a comprehensive study of these models and the reader is referred to the texts [132], [188], [306] and the recent review article by [280] for more references and further information on their probabilistic properties and existing estimation methods.

It is instructive to consider a Poisson hidden Markov model in which case (5.42) becomes

$$p_{xa} = \frac{\exp(-\lambda_a)\lambda_a^x}{x!}. \tag{5.43}$$

Equation (5.43) implies that the observed process is Poisson with mean λ_a given that the unobserved process is in state a. It follows that

$$\mathrm{E}\left[X_t \mid A_t = a\right] = \lambda_a,$$

and upon recalling (5.41) and defining $\boldsymbol{\lambda} = (\lambda_1, \ldots, \lambda_m)'$ we have

$$\begin{aligned} \mathrm{E}\left[X_t\right] &= \sum_{a=1}^{m} \mathrm{E}\left[X_t \mid A_t = a\right] \mathrm{P}\left[A_t = a\right] \\ & \sum_{a=1}^{m} \lambda_a \pi_a = \boldsymbol{\lambda}'\boldsymbol{\pi}. \end{aligned}$$

The second-order moments are,

$$\begin{aligned} \mathrm{E}\left[X_t^2\right] &= \sum_{a=1}^{m} \mathrm{E}\left[X_t^2 \mid A_t = a\right] \mathrm{P}\left[A_t = a\right] \\ &= \sum_{a=1}^{m} \left(\lambda_a^2 + \lambda_a\right) \pi_a \end{aligned}$$

so that

$$\begin{aligned} \mathrm{Var}\left[X_t\right] &= \sum_{a=1}^{m} \left(\lambda_a^2 + \lambda_a\right) \pi_a - \left(\boldsymbol{\lambda}'\boldsymbol{\pi}\right)^2 \\ &= \boldsymbol{\pi}'\boldsymbol{\Lambda}\boldsymbol{\lambda} + \boldsymbol{\lambda}'\boldsymbol{\pi} - \left(\boldsymbol{\lambda}'\boldsymbol{\pi}\right)^2, \end{aligned}$$

where $\mathbf{\Lambda} = \text{diag}\,(\lambda_1, \ldots, \lambda_m)$. The autocovariance function of the Poisson hidden Markov model is

$$\begin{aligned} c(k) &= \mathrm{E}\,[X_t X_{t+k}] - \mathrm{E}\,[X_t]\,\mathrm{E}\,[X_{t+k}] \\ &= \sum_{a=1}^{m}\sum_{b=1}^{m} \lambda_a \lambda_b \pi_a q_{ab}^{(k)} - \left(\boldsymbol{\lambda}'\boldsymbol{\pi}\right)^2, \end{aligned} \tag{5.44}$$

where $q_{ab}^{(k)}$ is the (a, b)-element of $\mathbf{Q}^k$ for positive k. The last equation holds since

$$\begin{aligned} \mathrm{E}\,[X_t X_{t+k}] &= \sum_{a=1}^{m}\sum_{b=1}^{m} \mathrm{E}\,[X_t X_{t+k} \mid X_t = a, X_{t+k} = b]\,\mathrm{P}\,[X_t = a, X_{t+k} = b] \\ &= \sum_{a=1}^{m}\sum_{b=1}^{m} \lambda_a \lambda_b \pi_a q_{ab}^{(k)}. \end{aligned}$$

Thus, (5.44) becomes

$$c(k) = \boldsymbol{\pi}'\mathbf{\Lambda}\mathbf{Q}^k\boldsymbol{\lambda} - \left(\boldsymbol{\lambda}'\boldsymbol{\pi}\right)^2 \tag{5.45}$$

and consequently the autocorrelation function is

$$\rho(k) = \frac{\boldsymbol{\pi}'\mathbf{\Lambda}\mathbf{Q}^k\boldsymbol{\lambda} - \left(\boldsymbol{\lambda}'\boldsymbol{\pi}\right)^2}{\boldsymbol{\pi}'\mathbf{\Lambda}\boldsymbol{\lambda} + \boldsymbol{\lambda}'\boldsymbol{\pi} - \left(\boldsymbol{\lambda}'\boldsymbol{\pi}\right)^2}. \tag{5.46}$$

Similar results are obtained by specifying (5.42) to be the probability mass function of the binomial, multinomial or any other probability distribution.

From a statistical point of view, an important problem is the estimation of model parameters based on the observed data $\{X_t\}$. In this case the unknown parameters consist of the transition probabilities (5.40) of the hidden Markov chain and any other parameters introduced by (5.42). Observe that (5.41) depends on (5.40) since $\mathbf{Q}\boldsymbol{\pi} = \boldsymbol{\pi}$. The Poisson example shows that the unknown parameters in the model are all the elements of the transition matrix plus $\lambda_1, \lambda_2, \ldots, \lambda_m$.

To estimate the parameters of a hidden Markov model, consider the likelihood of the observed data $\{X_t\}$, for $t = 1, 2, \ldots, N$

$$L = \mathrm{P}\,[X_1 = x_1, X_2 = x_2, \ldots, X_N = x_N]\,.$$

Conditioning on the event $\{A_1 = a_1, \ldots, A_N = a_N\}$, for the a_j's in $\{1, 2, \ldots, m\}$, leads to

$$\begin{aligned} L &= \sum_{a_1=1}^{m} \cdots \sum_{a_N=1}^{m} \mathrm{P}\,[X_1 = x_1, \ldots, X_N = x_N \mid A_1 = a_1, \ldots, A_N = a_N] \\ &\qquad \times \mathrm{P}\,[A_1 = a_1, \ldots, A_N = a_N] \\ &= \sum_{a_1=1}^{m} \cdots \sum_{a_N=1}^{m} \left(p_{x_1 a_1} \cdots p_{x_N a_N}\right)\left(\pi_{a_1} q_{a_1 a_2} \cdots q_{a_{N-1} a_N}\right) \end{aligned}$$

$$\begin{aligned}
&= \sum_{a_1=1}^{m} \cdots \sum_{a_N=1}^{m} \pi_{a_1} p_{x_1 a_1} q_{a_1 a_2} \cdots p_{x_N a_N} q_{a_{N-1} a_N} \\
&= \pi' \mathbf{V}_1 \mathbf{Q} \mathbf{V}_2 \ldots \mathbf{Q} \mathbf{V}_N \mathbf{1} \\
&= \pi' \left\{ \prod_{t=1}^{N} \mathbf{V}_t \mathbf{Q} \right\} \mathbf{1}, \qquad (5.47)
\end{aligned}$$

where $\mathbf{V}_t = \operatorname{diag}(p_{t1}, p_{t2}, \ldots, p_{tm})$, for $t = 1, \ldots, N$. To derive (5.47), we note that the second equality is a simple consequence of the conditional independence while the other equalities are obtained by rearranging terms expressing them in matrix notation–see, for example [451]. Computational methods for maximum likelihood estimation as well as numerical complications are discussed in [306, Section 2.7] where the authors also provide a set of Fortran routines for fitting hidden Markov models at `http://www.statoek.wiso.uni-goettingen.de/links`. Asymptotic properties of maximum likelihood estimators have been recently studied in [48], [49], [294] [374]. In Chapter 6 we consider more general state space modeling where the unobserved states are continuous.

5.5 VARIABLE MIXTURE MODELS

A broad class of models is formed by switching between models where the switching mechanism is controlled by a hidden process, a hidden Bernoulli process in the case of two components. The resulting marginal probability distribution of the observed process is a mixture of the component distributions, and the problem is to decide which distribution is applicable (the regime probability) through a regression structure. Such models are called *variable mixture models* because the parameters of the observed process keep changing over time due to switching as opposed to, for example, ARMA processes where the parameters are fixed and do not change with time [227]. The variable mixture models may be used as alternatives to threshold models where the threshold is either fixed or random, and to hidden Markov models. Useful references on mixture models include [134], [319], [320], [411],[413].

To define the variable mixture models suppose that $\{Y_t\}$ denotes a univariate time series and $\{\mathbf{Z}_{t-1}\}$, $t = 1, \ldots, N$ is a vector of random time dependent covariates. Assume that $\{\mathbf{W}_{t-1}, X_{t0}, X_{t1}\}$ are covariates composed of known functions of $\mathbf{Z}_{t-1}$. Let $\{I_t\}$ be an *unobserved* Bernoulli process taking the values 0,1, and suppose that Y_t can obey two different regimes/models where Y_t is generated by the regime 1 distribution if $I_t = 1$ and Y_t is generated by the regime 0 distribution if $I_t = 0$, where

$$\mathrm{P}_{\gamma}\left[I_t = 1 \mid \mathcal{F}_{t-1}\right] = F(\mathbf{W}_{t-1}'\gamma) \qquad (5.48)$$

and γ represents an unknown vector of regression parameters. Typical choices for $F(\cdot)$ include the logistic (logistic mixture) and standard normal (probit mixture) cdf's. The main ideas of variable mixture models can be put forth in terms of mixtures of two components only.

The previous description can be summarized as follows in the case of two components. The conditional density of Y_t given the indicator I_t, the past $\mathcal{F}_{t-1}$, and regime specific covariates and parameters α_i, denoted by $f(y_t; \boldsymbol{\theta} \mid \mathcal{F}_{t-1})$, is expressed as

$$\begin{array}{lll} g(y_t; \alpha_1 \mid x_{t1}) & \text{if} & I_t = 1, \\ g(y_t; \alpha_0 \mid x_{t0}) & \text{if} & I_t = 0, \end{array} \tag{5.49}$$

where the functions $g(y_t; \alpha_1 \mid x_{t1})$ and $g(y_t; \alpha_0 \mid x_{t0})$ are probability densities. The components of model (5.49) may be generalized linear models and in particular autoregressive processes [227], [437] and [444].

Variable mixture models have been considered previously in [122],[146], where logistic regression is used within the transition matrix of a hidden Markov model, and in [279] and [288] where logistic mixtures are employed in the mixing of hazard rates. Asymptotic results and testing for the presence of a mixture have been addressed at length in [227] for logistic mixtures.

5.5.1 Threshold Models

A closely related idea is that of threshold models developed in [414], [415]. In the case of two components, the two state self exciting threshold autoregressive (SETAR) model is defined by the switching mechanism,

$$Y_t = \begin{cases} \xi_{01} + \xi_{11} Y_{t-1} + \ldots + \xi_{p1} Y_{t-p} + \sigma_1 \cdot \epsilon_t \text{ if } Y_{t-d} > \tau \\ \xi_{00} + \xi_{10} Y_{t-1} + \ldots + \xi_{p0} Y_{t-p} + \sigma_0 \cdot \epsilon_t \text{ if } Y_{t-d} \leq \tau, \end{cases}$$

where d, p, τ are the delay, lag length, and threshold parameters, respectively. These are unknown parameters that can be estimated by conditional least squares. The SETAR model may include more components (states) and also covariates with coefficients that change depending upon whether or not the threshold is exceeded. SETAR goes under the rubric of nonlinear models.

5.5.2 Partial Likelihood Inference

Write $\boldsymbol{\theta} = (\alpha_1, \alpha_0, \gamma)'$. Then the conditional distribution of Y_t given the observed past, $\mathcal{F}_{t-1}$, is given by

$$\begin{aligned} f(y_t; \boldsymbol{\theta} \mid \mathcal{F}_{t-1}) &= g(y_t; \alpha_1 \mid x_{t1}) \mathrm{P}_{\boldsymbol{\gamma}} \left[I_t = 1 \mid \mathcal{F}_{t-1} \right] \\ &+ g(y_t; \alpha_0 \mid x_{t0}) \mathrm{P}_{\boldsymbol{\gamma}} \left[I_t = 0 \mid \mathcal{F}_{t-1} \right]. \end{aligned} \tag{5.50}$$

Therefore, the partial likelihood function evaluated at $\boldsymbol{\theta}$ is given by

$$\prod f(y_t; \boldsymbol{\theta} \mid \mathcal{F}_{t-1}).$$

The parameter θ is estimated by the EM algorithm (see [120], [321] or [440]). The specific algorithm for logistic mixtures is discussed in detail in [227].

5.5.3 Comparison with the Threshold Model

It has been noted in [227] that simple threshold models may yield inconsistent results if the threshold is measured with error. An example of this is the model

$$Y_t = \begin{cases} .6Y_{t-1} + .5 * \epsilon_t & \text{if } W_{t-1} < .1 \\ .1Y_{t-1} + .3 * \epsilon_t & \text{if } W_{t-1} \geq .1, \end{cases}$$

where $W_t = Y_t + .8\eta_t$ and both η_t and ϵ_t are unobserved i.i.d. $\mathcal{N}(0,1)$, and ϵ_t is independent of η_s for all s. That is W_{t-1} is not observed, and only Y_{t-1} is available. Not knowing that the threshold is noisy, the statistician might fit the SETAR model

$$Y_t = \begin{cases} \beta_1 Y_{t-1} + \sigma_1 * \epsilon_t & \text{if } Y_{t-1} < \tau \\ \beta_0 Y_{t-1} + \sigma_0 * \epsilon_t & \text{if } Y_{t-1} \geq \tau\,. \end{cases} \tag{5.51}$$

An alternative model is the logistic or probit mixture model for the same data. Accordingly, let $\alpha_i = (\beta_i, \sigma_i)$, $i = 0, 1$, and write

$$Y_t = \begin{cases} \beta_1 Y_{t-1} + \sigma_1 * \epsilon_t & \text{if } I_t = 1 \\ \beta_0 Y_{t-1} + \sigma_0 * \epsilon_t & \text{otherwise,} \end{cases} \tag{5.52}$$

with $\mathrm{P}_\gamma\left[I_t = 1 \mid Y_{t-1}\right] = F(\gamma_0 + \gamma_1 Y_{t-1})$, and with $F(\cdot)$ either the logistic or the standard normal cdf and $\gamma = (\gamma_1, \gamma_2)'$. In a simulation study the estimates derived under (5.52) were more precise than those obtained under (5.51) [227]. Notice that the SETAR model is a limiting case of variable mixtures when $\gamma_1 \to -\infty$ and $\gamma_0 = -\tau\gamma_1$.

5.6 ARCH MODELS

Autoregressive conditionally heteroscedastic (ARCH) models were introduced by Engle [133] to account for changes in volatility, or variability, in time series data. They have been found useful in numerous applications, especially in the context of financial time series which often exhibit large variability.

5.6.1 The ARCH(1) Model

Suppose that $\{Y_t\}$, $t = 1, \ldots, N$ denotes the observed response time series. The ARCH(1) model is specified by the following equations:

$$Y_t = \sigma_t \epsilon_t, \tag{5.53}$$

and

$$\sigma_t^2 = \beta_0 + \beta_1 Y_{t-1}^2, \tag{5.54}$$

where the coefficient β_1 is assumed positive and $\{\epsilon_t\}$ is sequence of i.i.d. standard normal random variables. A straightforward consequence of (5.53) and (5.54) is that the conditional distribution of Y_t given $Y_{t-1} = y_{t-1}$ is normal with mean 0 and variance $\beta_0 + \beta_1 y_{t-1}^2$. Moreover, subtracting (5.54) from the square of (5.53), we obtain

$$Y_t^2 - \left(\beta_0 + \beta_1 Y_{t-1}^2\right) = \sigma_t^2 \left(\epsilon_t^2 - 1\right)$$

or equivalently

$$Y_t^2 = \beta_0 + \beta_1 Y_{t-1}^2 + u_t, \tag{5.55}$$

where $u_t = \sigma_t^2 \left(\epsilon_t^2 - 1\right)$, a scaled $\mathcal{X}_1^2$ random variable with shifted mean equal to zero. By this construction both processes $\{Y_t\}$ and $\{u_t\}$ are martingale differences.

Representation (5.55) shows that the sequence $\{Y_t^2\}$, $t = 1, \ldots, N$ follows an AR(1) process, a fact which immediately points to the following:

$$\mathrm{E}\left[Y_t^2\right] = \mathrm{Var}\left[Y_t\right] = \frac{\beta_0}{1-\beta_1},$$

$$\mathrm{Var}\left[Y_t^2\right] = \mathrm{E}\left[Y_t^4\right] = \frac{3\beta_0^2}{(1-\beta_1)^2}\frac{1-\beta_1^2}{1-3\beta_1^2},$$

provided that $0 \leq \beta_1 < 1$, $3\beta_1^2 < 1$ and the variance of $\{u_t\}$, $t = 1, \ldots, N$ is finite. Thus the marginal distribution of $\{Y_t\}$ is *leptokurtic* ("fat tailed") since the *kurtosis* is equal to

$$\frac{\mathrm{E}\left[Y_t^4\right]}{\left[\mathrm{E}\left[Y_t^2\right]\right]^2} = 3\frac{1-\beta_1^2}{1-3\beta_1^2}.$$

If $3\beta_1^2 \geq 1$, then the process $\{Y_t^2\}$ is strictly stationary with infinite variance.

5.6.2 Maximum Likelihood Estimation

Estimation of the parameter vector $\boldsymbol{\beta} = (\beta_0, \beta_1)'$ is carried out by maximizing the conditional log-likelihood

$$l(\boldsymbol{\beta}) = -\frac{1}{2}\sum_{t=1}^{N}\left\{\log(\beta_0 + \beta_1 y_{t-1}^2) + \frac{y_t^2}{\beta_0 + \beta_1 y_{t-1}^2}\right\},$$

derived by (5.53) and (5.54). For more on estimation see [381].

5.6.3 Extensions of ARCH Models

There are several extensions of the ARCH(1) model. For example, the so called ARCH(p) model is specified by equation (5.53) and the following equation:

$$\sigma_t^2 = \beta_0 + \beta_1 Y_{t-1}^2 + \ldots + \beta_p Y_{t-p}^2.$$

It is also feasible to postulate an ARMA model for the mean and an ARCH model for the errors. Moreover, the so called GARCH (generalized ARCH) models enlarge the class of ARCH models by stipulating an autoregressive process for σ_t^2 [56]. Specifically, the GARCH(1,1) model stipulates that

$$\sigma_t^2 = \beta_0 + \beta_1 Y_{t-1}^2 + \gamma_1 \sigma_{t-1}^2,$$

in addition to (5.53). Other extensions include the exponential GARCH (EGARCH) and the integrated GARCH models (IGARCH), to mention a few. The reader is referred to [57], [181] and [381] for more details.

5.7 SINUSOIDAL REGRESSION MODEL

The classical problem of estimating the frequencies $\omega_1, ..., \omega_p$ in the sinusoidal regression model,

$$Z_t = \sum_{j=1}^{p} (A_j \cos(\omega_j t) + B_j \sin(\omega_j t)) + \zeta_t, \tag{5.56}$$

where $t = 0, \pm 1, \pm 2, \ldots$, has been studied from many different angles for over one hundred years starting with the introduction of Schuster's [379] celebrated *periodogram* in the late 1890s [20], [252], [357], [361]. The problem may be formulated as a regression problem with fixed amplitudes in which case the frequencies appear in the mean of $\{Z_t\}$. An alternative is to treat (5.56) as a stationary process with zero-mean uncorrelated random amplitudes, in which case the frequencies are part of the autocorrelation [252, Ch. 7].

The amplitude and frequency estimates may be obtained by nonlinear least squares provided the initial guess of the frequencies in a Newton-Raphson type algorithm is sufficiently precise as explained below. In many cases the frequencies are located "far apart" from each other so that it is sufficient, due to suitable linear operations, to concentrate on the special case of a "single frequency in noise,"

$$Y_t = \beta \cos(\omega_1 t + \phi) + \epsilon_t, \tag{5.57}$$

where β is a positive constant, $\omega_1 \in (0, \pi)$, ϕ uniformly distributed in $(-\pi, \pi]$, and where $\{\epsilon_t\}$ is a sequence of i.i.d. random variables with mean 0 and variance σ_ϵ^2, independent of ϕ.

In addition to nonlinear least squares, another well known method in frequency detection and estimation is based on the *periodogram*. Denoted by $I_N(\omega)$, the periodogram is defined by the transformation,

$$I_N(\omega) = \frac{2}{N} \left| \sum_{t=1}^{N} Y_t \exp(-i\omega t) \right|^2, \qquad \omega \in [-\pi, \pi], \tag{5.58}$$

where $Y_1, ..., Y_N$ is the time series to be transformed. Periodogram analysis consists of the search for significant peaks in $I_N(\omega)$ by treating ω as a continuous variable or, much more often, as a discrete variable taking the values in $[0, \pi]$,

$$\frac{2\pi k}{N}, \qquad k = 0, 1, \cdots, \left[\frac{N}{2}\right],$$

known as Fourier frequencies. In the latter case the periodogram can be computed very rapidly by the *fast Fourier transform*, FFT, a great computational device [103]. When a time series has a significant sinusoidal component with frequency $\omega_0 \in [0, \pi]$, then the periodogram exhibits a peak at that frequency with a high probability. In the single frequency regression model (5.57) with a fixed frequency and phase, the periodogram, apart from a constant, acts as the sample correlation between the observed series Y and the fitted data $\hat{Y}$ obtained from a linear least squares fit. The optimal frequency is then the one that maximizes this sample correlation, or the periodogram.

Given a time series $Y_1, ..., Y_N$, the more traditional methods of nonlinear least squares and periodogram maximization result in efficient estimates with precision $O(N^{-3/2})$, provided the optimization is initialized with precision $o(N^{-1})$ [369]. This requirement of a very precise initial guess of the frequencies may not be feasible in practice, and alternative methods are called for. The contraction mapping (CM) method which estimates the frequency by fixed point iterations relaxes the requirement of a very precise initial guess. With this method, under some conditions guesses precise to order $O(1)$ can still lead to estimates arbitrarily close to being efficient, and at the same time the method is computationally simple and fast [252], [296], [297], [398], [442].

The contraction mapping method (CM) is an iterative procedure involving linear parametric filters satisfying certain properties, and whose parameter takes values in $(-1, 1)$. The name of the algorithm is derived from the fact that cosines of frequencies in sinusoidal models such as (5.56) constitute *fixed points* of a certain correlation contraction to be described next [205], [252, Ch.7], [442].

Suppose we have such a parametric filter $\mathcal{L}_\theta$, $\theta \in \Theta$. The algorithm starts by equating the parameter of the filter with an initial guess of the cosine of the frequency to be detected, and it consists of the following two alternating steps. At the kth iteration, the parametric filter is applied to the data. A new parameter is then chosen as the lag–one autocorrelation of the resulting filtered time series. The series is then filtered again with the new filter, a new parameter is chosen as the lag–one autocorrelation of the filtered series, and so on. The sequence of parameters obtained in this way converges to the cosine of the frequency to be detected, a fixed point.

The contraction factor of the mapping just described can be reduced at each iteration to enhance the speed of convergence [442]. In the AR(2) family of filters used below, the contraction factor is controlled by an extra regularization parameter $\eta \in (0, 1)$, as η increases the contraction factor decreases. In general, the initial guess of the frequency is not as critical as in the case of nonlinear least squares, and under some conditions starting values of order $O(1)$ suffice [297], [398].

With $\alpha \in (-1, 1)$, define $Y_t(\alpha)$ by operating on Y_t,

$$Y_t(\alpha) = (1 + \eta^2)\alpha Y_{t-1}(\alpha) - \eta^2 Y_{t-2}(\alpha) + Y_t, \tag{5.59}$$

where $\eta \in (0, 1)$ is the *contraction parameter*. Put $Y_{-1}(\alpha) = Y_0(\alpha) = 0$. Define the sample autocorrelation

$$\hat{\rho}_1(\alpha) = \frac{\sum_{t=1}^{N-1} Y_t(\alpha)Y_{t-1}(\alpha)}{\sum_{t=0}^{N-1} Y_t^2(\alpha)}. \tag{5.60}$$

Then the CM algorithm is given by

$$\alpha_{k+1} = \hat{\rho}_1(\alpha_k), \qquad k = 0, 1, 2, \cdots \tag{5.61}$$

For an initial guess θ_0 of ω_1, the algorithm starts with $\alpha_0 = \cos(\theta_0)$, and η close to 1, for example $\eta = 0.98$. We obtain $\{Y_t(\alpha_0)\}$ from (5.59), then $\hat{\rho}_1(\alpha_0)$ from (5.60), then get a new α_1 from (5.61), $\alpha_1 = \hat{\rho}_1(\alpha_0)$, we increase η slightly and start anew with α_1 and a new η. This iterative scheme gives a sequence $\alpha_0, \alpha_1, \alpha_2 ...$, which converges to α^*. Under certain conditions, $\alpha^* = \alpha_N^*$ converges almost surely to $\cos(\omega_1)$ as N increases. The desired estimator is given by

$$\hat{\omega}_1 = \arccos(\alpha^*).$$

It has been shown recently in [398] under regularity conditions that if η is chosen such that $(1 - \eta)^2 N \to 0$ as $N \to \infty$, then $(1 - \eta)^{-1/2} N(\hat{\omega}_1 - \omega_1)$ converges in distribution to $\mathcal{N}(0, \gamma^{-1})$ as $N \to \infty$, where $\gamma = \frac{1}{2}\beta^2/\sigma_\epsilon^2$. The implication of this is that by a judicious choice of η, the precision of the CM estimate can be made arbitrarily close to that achieved by periodogram maximization and nonlinear least squares. Based on this and other results the authors suggest a two-step procedure for the implementation of the CM algorithm. See [252] and [398] for many more important related references from the statistical and engineering literature.

The following S-Plus function `KY.AR2()` gives the code for the CM algorithm. The arguments of `KY.AR2()` are $z, \theta_0, \eta, inc, niter$, where z is a vector containing the data, θ_0 is the initial guess of the frequency of interest, η is the initial value of the contraction parameter, inc is the increment of η at each iteration, and $niter$ is the number of iterations of the algorithm. The output includes the sequence of estimates, and the sample variance of the filtered process. An appreciable increase in the variance accompanies detection. The algorithm makes use of the S-Plus functions `acf(),filter()`.

```
KY.AR2 <- function(z,theta0,eta,inc,niter){
y <- rep(0,length(z))
r <- rep(0,niter); OMEGA <- rep(0,niter)
r [1] <- cos(theta0); OMEGA[1] <- theta0
cat(c("Initial frequency guess is", OMEGA[1]),fill=T)
cat(c("eta","            r(k)","           Omega(k)",
"           Var(y)"), fill=T)
for(k in 2:niter){# eta increments by inc
eta <- eta+inc
if((eta < 0)|(eta >1))
```

```
  stop("eta must be between 0 and 1")
FiltCoeff <- c((1+eta^2)*r[k-1],-(eta^2))
y <- filter(z,FiltCoeff, "rec")
# CM Iterations-------------------
rrr <- acf(y)        # motif() must be on
r[k] <- rrr$acf[2] # Gives acf(1)
# -------------------------------
OMEGA[k] <- acos(r[k])
cat(c(eta,r[k],OMEGA[k],var(y)),fill=T)}}
```

The algorithm can be used also for multiple frequencies by starting the algorithm with different initial guesses θ_0 of ω_1, that is, centering the filter at different regions in $(0, \pi)$.

As an example, consider the sum of two sinusoids plus noise with frequencies $\omega_1 = 0.513, \omega_2 = 0.771$,

$$Y_t = 0.5\cos(0.513t + \phi_1) + \cos(0.771t + \phi_2) + 2.2\epsilon_t,$$

where the ϵ_t are i.i.d. $\mathcal{N}(0, 1)$ random variables, and $t = 1, ..., 1500$. The signal to noise ratio defined as $10\log_{10}((.5^2/2 + 1^2/2)/2.2^2) = -8.890$ is relatively low.

Storing the data in a vector z and calling `KY.AR2(z,0.48,0.98,0.0015,10)` gives Table 5.5. The first iteration is not shown, only the last nine. Similarly, calling `KY.AR2(z,0.88,0.98,0.001,10)` starting at $\theta_0 = 0.88$ and changing the increment to 0.001 gives Table 5.6. The estimates are 0.5135, 0.7709, respectively, so that in both cases the error is of order 10^{-4}. The FFT evaluated at Fourier frequencies gives 0.5152, 0.7749, respectively, or errors of order 10^{-3}.

It is interesting to observe in Tables 5.5 and 5.6 the steady increase in the variance of the filtered series $Y_t(\alpha)$ as the CM algorithm builds up the power of the frequency to be detected.

Table 5.5 Nine iterations of the CM algorithm in the estimation of $\omega_1 = 0.513$. Convergence of $\hat{\omega}$ starting at $\theta_0 = 0.48$, and $\eta = 0.98$ increasing by 0.0015. The final estimate is $\hat{\omega} = 0.5135$. Error: 0.0005.

η	$\alpha(k)$	$\omega(k)$	Var($Y_t(\alpha)$)
0.9815	0.8807	0.4932	425.958
0.9830	0.8755	0.5042	574.342
0.9845	0.8723	0.5107	856.165
0.9860	0.8711	0.5131	1134.483
0.9875	0.8708	0.5138	1365.735
0.9890	0.8707	0.5139	1666.432
0.9905	0.8708	0.5138	2106.870
0.9920	0.8709	0.5136	2783.643
0.9935	0.8710	0.5135	3892.713

Table 5.6 Nine iterations of the CM algorithm in the estimation of $\omega_2 = 0.771$. Convergence of $\hat{\omega}$ starting at $\theta_0 = 0.88$, and $\eta = 0.98$ increasing by 0.001. The final estimate is $\hat{\omega} = 0.7709$. Error: 0.0001.

η	$\alpha(k)$	$\omega(k)$	Var($Y_t(\alpha)$)
0.981	0.6518	0.8607	102.987
0.982	0.6672	0.8403	128.128
0.983	0.6822	0.8199	162.341
0.984	0.6973	0.7990	215.022
0.985	0.7104	0.7806	371.580
0.986	0.7162	0.7723	988.001
0.987	0.7171	0.7710	1555.238
0.988	0.7172	0.7708	1817.274
0.989	0.7172	0.7709	2130.545

5.8 MIXED MODELS FOR LONGITUDINAL DATA

The topic of linear mixed models for longitudinal data, mentioned earlier in Chapter 1 Section 1.7.1 and Problem 13, deals with a certain class of regression models appropriate for many independent short time series. Examples of this are encountered in numerous medical and epidemiological studies where observations are made on many individuals over time for the purpose of detecting sequential changes in an individual while taking into consideration differences among individuals. In this setting the variation in each measurement may be attributed to a baseline common to all individuals as well as to subject–specific variation [123], [284] [419], [420]. Having to handle both variation within and among different time series, we are faced with some nonstandard challenging regression problems.

To present some basic facts in mixed models, the ensuing discussion focuses on mixed models for normal data. A more general treatment of random effects in generalized linear models for longitudinal data can be found in [123, Ch. 9] and [144, Ch. 7].

Although the problem is general, it is convenient to think in terms of individuals. Let Y_{ij} be the response of the ith individual at time t_{ij}, or simply think of Y_{ij} as the jth response of the ith individual, $i = 1, ..., N$, $j = 1, ..., n_i$. In vector notation, denote by $\mathbf{Y}_i$ the n_i measurements taken on the ith subject, $\mathbf{Y}_i = (Y_{i1}, Y_{i2}, ..., Y_{in_i})'$. In many applications $\mathbf{Y}_i$ contains a short time series while the number of such sequences, N, is large.

There are several ways to proceed from here. Let us first restrict attention to the model,

$$\mathbf{Y}_i = \mathbf{X}_i\beta + \mathbf{Z}_i\mathbf{b}_i + \epsilon_i, \tag{5.62}$$

where $\mathbf{X}_i$ and $\mathbf{Z}_i$ are nonrandom $n_i \times p$ and $n_i \times q$ design or covariate matrices, respectively, β is a p–dimensional parameter of *fixed effects* common to all subjects, $\mathbf{b}_i$ is q–dimensional subject–specific (nonrandom) parameter considered nuisance, and $\epsilon_i \sim \mathcal{N}_{n_i}(\mathbf{0}, \sigma^2\mathbf{I})$ is the n_i–dimensional vector representing noise or measurement error. Notice that the covariance matrix $\sigma^2\mathbf{I} = \sigma^2\mathbf{I}_{n_i}$ is $n_i \times n_i$ and its elements do not depend on i. Under these conditions the longitudinal measurements on an individual are normal and independent, $\mathbf{Y}_i \sim \mathcal{N}_{n_i}(\mathbf{X}_i\beta + \mathbf{Z}_i\mathbf{b}_i, \sigma^2\mathbf{I})$. Denote by $f_i(\mathbf{y}_i|\mathbf{b}_i, \beta, \sigma^2)$ the density of $\mathbf{Y}_i$.

Consider the case when the $\mathbf{b}_i$ are treated as (fixed) nuisance parameters, and the main objective is to estimate β and σ^2. Then, as illustrated in Problem 15, maximizing the likelihood $\prod_{i=1}^N f_i(\mathbf{y}_i|\mathbf{b}_i, \beta, \sigma^2)$ with respect to β, σ^2 and $\mathbf{b}_1, ..., \mathbf{b}_N$, does not guarantee consistent estimators because the number of $\mathbf{b}_i$ increases with N. A way out of this predicament is to employ sufficient statistics for the $\mathbf{b}_i$ in order to eliminate the latter from the likelihood [420]. Indeed, since $\mathbf{Z}_i'\mathbf{Y}_i$ is sufficient for $\mathbf{b}_i$, the conditional distribution of $\mathbf{Y}_i$ given $\mathbf{Z}_i'\mathbf{Y}_i$ is free of $\mathbf{b}_i$,

$$f_i(\mathbf{y}_i|\mathbf{Z}_i'\mathbf{y}_i, \beta, \sigma^2) = \frac{|\mathbf{Z}_i'\mathbf{Z}_i|^{\frac{1}{2}}}{(2\pi\sigma^2)^{\frac{n_i-q}{2}}} \exp\left\{-\frac{1}{2\sigma^2}(\mathbf{y}_i - \mathbf{X}_i\beta)'\mathbf{M}_i(\mathbf{y}_i - \mathbf{X}_i\beta)\right\}, \tag{5.63}$$

where $\mathbf{M}_i = \mathbf{I}_{n_i} - \mathbf{Z}_i(\mathbf{Z}_i'\mathbf{Z}_i)^{-1}\mathbf{Z}_i$ is symmetric idempotent. The density (5.63) is obtained from the ratio of the density of $\mathbf{Y}_i$ divided by that of $\mathbf{Z}_i'\mathbf{Y}_i$. The estimates for β, σ^2 are now obtained by maximizing the conditional likelihood

$$\prod_{i=1}^N f_i(\mathbf{y}_i|\mathbf{Z}_i'\mathbf{y}_i, \beta, \sigma^2),$$

which no longer depends on the $\mathbf{b}_i$.

Although quite clever, the *conditional inference* we have just described is somewhat problematic. A closer look at the quadratic form in (5.63) reveals that when the rows of $\mathbf{M}_i$ are orthogonal to the columns of $\mathbf{X}_i$ for all i, the corresponding components of β vanish from the likelihood and therefore cannot be estimated. This and the fact the likelihood contains no information about the $\mathbf{b}_i$ after conditioning by the sufficient statistics prompts us to look for alternative approaches when information about the subject–specific parameters is needed. Interestingly, this is done by assuming that the $\mathbf{b}_i$ are random, in which case they are referred to as *random effects*. When the subject–specific parameters are assumed random, inference can be carried out about their distribution, parametrically as well as nonparametrically.

The so called *mixture inference* calls for independent $\mathbf{b}_i$ each having the same *mixing distribution* $Q(\cdot)$ used in forming the mixture density,

$$f_i(\mathbf{y}_i|Q, \beta, \sigma^2) = \int f_i(\mathbf{y}_i|\mathbf{b}_i, \beta, \sigma^2)dQ(\mathbf{b}_i). \tag{5.64}$$

By maximizing the mixture likelihood $\prod_{i=1}^N f_i(\mathbf{y}_i|Q, \beta, \sigma^2)$ with respect to the triple (Q, β, σ^2) we obtain the desired maximum likelihood estimates of β, σ^2 and Q. When $Q(\cdot)$ is parametric we optimize with respect to its parameters, but when no parametric form is assumed, an estimate for $Q(\cdot)$ is derived by nonparametric maximum likelihood. We refer the reader to [420] for a discussion of this and for a literature survey.

A simpler alternative assumes that the random effects are normally distributed in which case the model is referred to as the *linear mixed–effects model*. The linear mixed–effects model with fixed effects β and random effects $\mathbf{b}_i$ is defined by the equation [284], [419, p. 24],

$$\mathbf{Y}_i = \mathbf{X}_i\beta + \mathbf{Z}_i\mathbf{b}_i + \epsilon_i, \tag{5.65}$$

where $\mathbf{X}_i$, $\mathbf{Z}_i$, and β are as above, but $\mathbf{b}_i \sim \mathcal{N}_q(\mathbf{0}, \mathbf{D})$, and $\epsilon_i \sim \mathcal{N}_{n_i}(\mathbf{0}, \mathbf{\Sigma})$ is the n_i–dimensional error. Here $\mathbf{\Sigma} = \mathbf{\Sigma}_i$ is $n_i \times n_i$ covariance matrix whose components do not depend on i. It is assumed that the ϵ_i and $\mathbf{b}_i$ are independent. Therefore the $\mathbf{Y}_i$ are independent and normal,

$$\mathbf{Y}_i \sim \mathcal{N}_{n_i}(\mathbf{X}_i\beta, \mathbf{Z}_i\mathbf{D}\mathbf{Z}_i' + \mathbf{\Sigma}). \tag{5.66}$$

For simplicity we introduce $\mathbf{V}_i(\boldsymbol{\alpha}) = \mathbf{Z}_i\mathbf{D}\mathbf{Z}_i' + \mathbf{\Sigma}$, where $\boldsymbol{\alpha}$ contains all the unknown variances and covariances in $\mathbf{D}$ and $\mathbf{\Sigma}$, a set of parameters traditionally called *variance components*. The objective is to estimate β and $\boldsymbol{\alpha}$ by maximizing the likelihood,

$$\prod_{i=1}^{N} \frac{1}{(2\pi)^{\frac{n_i}{2}} |\mathbf{V}_i(\boldsymbol{\alpha})|^{\frac{1}{2}}} \exp\left\{-\frac{1}{2}(\mathbf{Y}_i - \mathbf{X}_i\beta)'\mathbf{V}_i^{-1}(\boldsymbol{\alpha})(\mathbf{Y}_i - \mathbf{X}_i\beta)\right\}. \tag{5.67}$$

This gives

$$\hat{\beta} = \left\{\sum_{i=1}^{N} \mathbf{X}_i'\mathbf{V}_i^{-1}(\hat{\boldsymbol{\alpha}})\mathbf{X}_i\right\}^{-1} \sum_{i=1}^{N} \mathbf{X}_i'\mathbf{V}_i^{-1}(\hat{\boldsymbol{\alpha}})\mathbf{y}_i, \tag{5.68}$$

where $\hat{\boldsymbol{\alpha}}$ is the maximum likelihood of $\boldsymbol{\alpha}$ [284]. Another method for estimating $\boldsymbol{\alpha}$ is by restricted maximum likelihood which requires a linear transformation of the response variables. This is discussed in detail in [419]. The distribution of $\hat{\beta}$ is approximated by a normal distribution with mean β and covariance matrix

$$\left\{\sum_{i=1}^{N} \mathbf{X}_i'\mathbf{V}_i^{-1}(\hat{\boldsymbol{\alpha}})\mathbf{X}_i\right\}^{-1},$$

a fact that can be used in testing hypotheses about β. Thus in testing $\mathrm{H}_0 : \mathbf{C}\beta = \beta_0$ versus $\mathrm{H}_0 : \mathbf{C}\beta \neq \beta_0$ we can use the Wald statistic,

$$W = (\mathbf{C}(\hat{\beta} - \beta_0))' \left(\mathbf{C}\left\{\sum_{i=1}^{N} \mathbf{X}_i'\mathbf{V}_i^{-1}(\hat{\boldsymbol{\alpha}})\mathbf{X}_i\right\}^{-1} \mathbf{C}'\right)^{-1} \mathbf{C}(\hat{\beta} - \beta_0),$$

which is asymptotically chi-square with $\mathrm{rank}(\mathbf{C})$ degrees of freedom, and reject H_0 for large values of W.

Example: A linear mixed–effects model applied in a longitudinal hearing study of 680 males was considered in [420]. With hearing threshold (in dB) measured at 500 Hz and using the left ear only as response (Y_{ij}), and age at which the individual entered the study ($\mathrm{age_i}$), time point of at which measurements were taken (t_{ij}), and an indicator taking the value 1 at the first measurement of the subject and 0 otherwise (I_{ij}), as covariates, the model is

$$Y_{ij} = (\beta_1 + \beta_2 \mathrm{age_i} + \beta_3 \mathrm{age_i^2} + \mathrm{b_{i1}}) + (\beta_4 + \beta_5 \mathrm{age_i} + \mathrm{b_{i2}})\mathrm{t_{ij}} + \beta_6 \mathrm{I_{ij}} + \epsilon_{\mathrm{ij}}.$$

5.9 PROBLEMS AND COMPLEMENTS

1. (a) Fix m, λ and simulate the process (5.1), for $n = 1, ..., N$, where $N = 100, 200, ..., 2000$. Obtain estimates for m, λ by minimizing $\sum_{i=1}^{N} (\epsilon_n^*)^2$ in

$$\frac{X_n}{\sqrt{X_{n-1}+1}} = m\frac{X_{n-1}}{\sqrt{X_{n-1}+1}} + \frac{\lambda}{\sqrt{X_{n-1}+1}} + \epsilon_n^*,$$

where $\epsilon_n^* = \epsilon_n / \sqrt{X_{n-1}+1}$.

(b) Consider the design matrix,

$$\mathbf{X}_N = \begin{pmatrix} X_1/\sqrt{X_1+1} & 1/\sqrt{X_1+1} \\ X_2/\sqrt{X_2+1} & 1/\sqrt{X_2+1} \\ \vdots & \vdots \\ X_N/\sqrt{X_N+1} & 1/\sqrt{X_N+1} \end{pmatrix}$$

and define $\mathbf{A} \equiv \mathbf{X}_N' \mathbf{X}_N$. Let $\lambda_{min}(N)$ and $\lambda_{max}(N)$ be the smallest and largest eigenvalues of $\mathbf{A}$, respectively. Verify that $[\log \lambda_{max}(N)]/\lambda_{min}(N)$ decreases as N increases, and interpret the results relative to the convergence of your least squares estimates [283].

2. *More on the estimation of m in (5.1)* [210]. By a simulation study compare the estimate of m given by (5.4) with the estimator

$$1 - \frac{1}{2}\frac{\sum_{i=1}^{N}(X_{i-1} - X_i)^2}{\sum_{i=1}^{N}(X_{i-1} - \bar{X}_N)^2},$$

$\bar{X}_N = (X_1 + \cdots + X_N)/N$, assuming the immigration distribution is Poisson, $m < 1$, and that the process is stationary.

3. Prove equations (5.6)–(5.10).

4. Prove the following expressions:

(a)

$$\mathrm{E}\left[\alpha \circ X\right]^2 = \alpha(1-\alpha)\mathrm{E}\left[X\right] + \alpha^2 \mathrm{E}\left[X^2\right].$$

(b)

$$\mathrm{E}\left[\alpha \circ X - \alpha \circ Z\right]^2 = \alpha(1-\alpha)\mathrm{E} \mid X - Z \mid + \alpha^2 \mathrm{E}\left[X - Z\right]^2,$$

with $\alpha \circ X = \sum_{i=1}^{X} Y_i$ and $\alpha \circ Z = \sum_{i=1}^{Z} Y_i$.

(c)

$$\mathrm{E}\left[(\alpha \circ X)(\beta \circ Z)\right] = \alpha\beta \mathrm{E}\left[XZ\right],$$

where $\alpha \circ X = \sum_{i=1}^{X} Y_i$, $\beta \circ Z = \sum_{i=1}^{Z} Y_i^\star$, $\{Y_i\}$ is independent of $\{Y_i^\star\}$ and (X, Z) is independent of $\{Y_i\}$ and $\{Y_i^\star\}$.

5. Derive expression (5.15) by using the representation (5.12).

6. *Estimation for INAR(1) process* [9]. Recall the INAR(1) process (5.11) and assume that the innovations follow the Poisson distribution with mean μ.

   (a) By differentiating

$$\sum_{t=1}^{N} (X_t - \alpha X_{t-1} - \mu)^2,$$

   show that the conditional least squares estimators of α and μ are given by

$$\hat{\alpha} = \frac{\sum_{t=1}^{N} X_t X_{t-1} - \left(\sum_{t=1}^{N} X_t \sum_{t=1}^{N} X_{t-1}\right)/N}{\sum_{t=1}^{N} X_{t-1}^2 - \left(\sum_{t=1}^{N} X_{t-1}\right)^2/N}$$

   and

$$\hat{\mu} = \frac{1}{N}\left(\sum_{t=1}^{N} X_t - \hat{\alpha}\sum_{t=1}^{N} X_{t-1}\right),$$

   respectively.

   (b) Verify the conditions of [269, Theorem 3.2] to conclude that the conditional least squares derived above, are asymptotically normally distributed. In addition calculate the asymptotic covariance matrix.

7. *Moments for the INMA(q) process* [318]. For the integer moving average model of order q (5.24), show the following:

   (a) Verify equations (5.30)–(5.33).

   (b) Assume that the sequence $\{\epsilon_t\}$ is Poisson with mean $\mu/(\sum_{i=0}^{q} \beta_i)$. Then the autocorrelation function is given by

$$\rho(k) = \begin{cases} \sum_{i=0}^{q-k} \beta_i \beta_{i+k} / \sum_{i=0}^{q} \beta_i, & k = 0, 1, \ldots, q \\ 0, & k > q. \end{cases}$$

   (c) If $\{\epsilon_t\}$ is Poisson with mean $\mu/\sum_{i=0}^{q} \beta_i$, then X_t is Poisson with mean μ.

8. *INARMA(1,q) model* [317]. By combining the INAR(1) model and the INMA(q) we obtain an autoregressive moving average process of order $(1, q)$. Let ϵ_t be a sequence of independent and identically distributed Poisson random variables with mean $(1-\alpha)\mu$, and put

$$\begin{aligned} Y_t &= \alpha \circ Y_{t-1} + \epsilon_t, & (5.69)\\ X_t &= Y_{t-q} + \sum_{k=1}^{q} \beta_k \circ \epsilon_{t+1-k}, & (5.70) \end{aligned}$$

with $0 < \alpha < 1$, $0 < \beta_k < 1$ for $k = 1, \ldots, q$ and $\sum_{k=1}^{q} \beta_k < 1$. In addition assume that Y_0 is a Poisson random variable with mean μ which is independent of $\{\epsilon_t\}$. Show that the autocorrelation function of the process specified by (5.69) and (5.70) is given by

$$\rho(k) = \begin{cases} \dfrac{\left\{\alpha^k + (1-\alpha)\left(\sum_{i=1}^{q-k} \beta_i\beta_{i+k} + \sum_{i=q-k+1}^{q} \beta_i \alpha^{k-1+i-q}\right)\right\}}{\left(1+(1-\alpha)\sum_k \beta_k\right)}, & k = 1, \ldots, q \\ \dfrac{\left(\alpha^q + (1-\alpha)\sum_{i=1}^{q} \beta_i \alpha^{i-1}\right)\alpha^{k-q}}{\left(1+(1-\alpha)\sum_k \beta_k\right)}, & k > q. \end{cases}$$

9. *Properties of DAR(1) process* [223]. The discrete autoregressive process of order 1 has the following form

$$A_t = V_t A_{t-1} + (1 - V_t) Y_t,$$

with the same notation as in (5.35).

   (a) Simulate a realization of the process $\{A_t\}$ of length 200. Use the Poisson distribution for π.

   (b) Show that the autocorrelation function of the process $\{A_t\}$ is ρ^k.

   (c) Suppose that for a realization of length N from $\{A_t\}$, we denote by m_{ij} the number of times that the process moves from state i to j. Show that the log–likelihood function for a DAR(1) process is given by

$$\begin{aligned} L &= \sum_{i=0}^{\infty}\sum_{j\neq i} m_{ij} \log\left[(1-\rho)\pi_j\right] + \sum_{i=0}^{\infty} m_{ii} \log\left[1 - (1-\rho)(1-\pi_i)\right] \\ &+ \sum_{i=0}^{\infty} I(X_1 = i) \log \pi_i, \end{aligned}$$

   where $I(B)$ is the indicator function of the event B. Hence, taking derivative with respect to $x = 1 - \rho$, the maximum likelihood estimator of x, if it exists, is obtained from the equation

$$1 - \frac{1}{N-1}\sum_{i=0}^{\infty} m_{ii} \frac{1}{1 - x[1-\pi_i]} = 0.$$

10. *Properties of the MTD model* [362]. Consider the mixture transition model (5.37) and prove the following.

    (a) Model (5.37) satisfies the system of equations (5.38).

    (b) If $l = 2$, then equations (5.38) have unique solution when $0 < \lambda_1 \leq 1$.

11. *Gaussian MTD model* [293]. Suppose that $\{Y_t\}, t = 1, \ldots, N$ is a time series taking values in an arbitrary space. The Gaussian mixture transition distribution model (GMTD) is given by

$$F(y_t \mid y_1, \ldots y_{t-1}) = \alpha_0 \Phi\left(\frac{y_t - \sum_{j=1}^{p} \phi_{0j} y_{t-j}}{\sigma_0}\right) + \sum_{i=1}^{p} \alpha_i \Phi\left(\frac{y_t - \phi_i y_{t-i}}{\sigma_i}\right), \tag{5.71}$$

where F is the conditional distribution function of Y_t given the past of the process, $\alpha_i \geq 0, i = 0, \ldots, p$ such that $\sum_{i=0}^{p} \alpha_i = 1$ and Φ is the cumulative distribution function of a Gaussian random variable. Assuming that Y_t is second order stationary, prove the following:

    (a) If $\rho(l)$ denotes the lag-l autocorrelation, then

$$\rho(l) = \sum_i (\alpha_0 \phi_{0i} + \alpha_i \phi_i)\, \rho(\mid l - i \mid),$$

    for $l = 1, \ldots, p$. These equations, for different values of l, resemble the Yule–Walker equations for the AR(p) model upon noticing that the coefficients $\alpha_0 \phi_{0i} + \alpha_i \phi_i$ are replaced by the lag-i coefficient of the AR(p) process.

    (b) Suppose that $\alpha_0 = 0$ and $p = 2$. Derive the admissible region for the autocorrelations $\rho(1)$ and $\rho(2)$.

12. For a hidden Markov model satisfying the relations (5.40)–(5.42), show that

$$\mathrm{E}\left[g(X_t)\right] = \sum_{a=1}^{m} \mathrm{E}\left[g(X_t) \mid A_t = a\right] \pi_a,$$

and

$$\mathrm{E}\left[g(X_t, X_{t+k})\right] = \sum_{a=1}^{m} \sum_{b=1}^{m} \mathrm{E}\left[g(X_t, X_{t+k}) \mid A_t = a, A_{t+k} = b\right] \pi q_{ij}^{(k)},$$

provided that all expectations exist.

13. Define the binomial hidden Markov model by specifying p_{xa} of (5.42) to be the binomial distribution with parameters n and ζ_a $(0 < \zeta_a < 1)$. Calculate the

mean, variance, autocovariance and autocorrelation functions of the binomial hidden Markov Model.

14. Based on (5.47), show the following.

    (a) The marginal distribution of X_t is given by

    $$\mathrm{P}\left[X_t = x\right] = \boldsymbol{\pi}\mathbf{V}_x\mathbf{1}'.$$

    (b) The joint distribution of (X_t, X_{t+1}) is

    $$\mathrm{P}\left[X_t = u, X_{t+1} = v\right] = \boldsymbol{\pi}\mathbf{V}_u\mathbf{Q}\mathbf{V}_v\mathbf{1}'.$$

    (c) The one–step–ahead distribution is

    $$\mathrm{P}\left[X_{t+1} = x_{t+1} \mid X_1 = x_1, \ldots, X_t = x_t\right] = \frac{\boldsymbol{\pi}\mathbf{V}_1\mathbf{Q}\mathbf{V}_2\ldots\mathbf{Q}\mathbf{V}_{t+1}\mathbf{1}'}{\boldsymbol{\pi}\mathbf{V}_1\mathbf{Q}\mathbf{V}_2\ldots\mathbf{Q}\mathbf{V}_t\mathbf{1}'}.$$

15. Inconsistent MLE [337]. Let Y_{ij}, $i = 1, ..., N$, $j = 1, ..., n$, be independent random variables. For each fixed i, Y_{ij} are distributed as $\mathcal{N}(\theta_i, \sigma^2)$, $j = 1, ..., n$. Let $\hat{\theta}_i$ and $\hat{\sigma^2}$ be the MLE's of θ_i and σ^2, respectively.

    (a) Show that $\hat{\theta}_i = \bar{Y}_i$, $i = 1, ..., N$, and

    $$\hat{\sigma^2} = \frac{1}{nN}\sum_{i=1}^{N}\sum_{j=1}^{n}(Y_{ij} - \bar{Y}_i)^2.$$

    (b) Show that $\hat{\sigma^2}$ is not consistent. That is, show that as $N \to \infty$, $\hat{\sigma^2}$ converges in probability to $(n-1)\sigma^2/n$.

6

State Space Models

6.1 INTRODUCTION

To a large degree, the origin of statistical state space models can be traced to dynamical systems in engineering branches including automatic control, communications, robotics, and aerospace systems such as spacecraft attitude control. If $\mathbf{U}(t)$, $\mathbf{Y}(t)$, $\mathbf{X}(t)$, represent input, output, and state vectors, respectively, general state-space equations that govern the relationship between these variables are the nonlinear equations [21], [166],

$$\begin{aligned} \mathbf{Y}(t) &= \mathbf{G}(\mathbf{X}(t), \mathbf{U}(t), t) \\ \tfrac{d}{dt}\mathbf{X}(t) &= \mathbf{F}(\mathbf{X}(t), \mathbf{U}(t), t). \end{aligned} \tag{6.1}$$

The corresponding important special discrete-time linear case is

$$\begin{aligned} \mathbf{Y}(t) &= \mathbf{A}(t)\mathbf{X}(t) + \mathbf{B}(t)\mathbf{U}(t) \\ \mathbf{X}(t+1) &= \mathbf{C}(t)\mathbf{X}(t) + \mathbf{D}(t)\mathbf{U}(t) \end{aligned} \tag{6.2}$$

where the state variables refer to memory variables. A simple example is that of a container into which water flows at a rate $u(t)$ and from which water flows out at rate $y(t)$, and $x(t)$ is the accumulated water. Then for some g, $y = g(x)$ and $dx/dt = u - g(x)$.

The statistical adaptation of equations (6.1) and (6.2) are widely used discrete time regression-like models made of two interconnected equations, the *observation equation* and the *system equation*, which may assume various linear and nonlinear forms and commonly referred to as *state space models*. This chapter discusses linear and nonlinear state space models and their application in prediction, filtering, and smoothing or interpolation.

6.1.1 Historical Note

Early work on linear state space models by R. E. Kalman and others appeared in the late 1950s, however the models owe their widespread use and popularity to NASA's Apollo space program, designed to achieve preeminence in space for the United States including landing humans on the moon and bringing them safely back to Earth [217], [377]. In March of 1960, Kalman [242] published a seminal paper in which he developed the "Kalman filter" that gives the recursion formulas for filtering and prediction using the linear state space model in discrete time, thus extending the Wiener–Kolmogorov theory of filtering and prediction for stationary time series set forth in the 1940s. As related in [377], in the fall of that year Kalman presented his paper to scientists and engineers at the Ames Research Center (ARC) of NASA. The audience, due to notation and conceptual problems, had great difficulty at first understanding Kalman's work, but past that stage the value of the state space approach to nonlinear navigation (state estimation) became apparent and a simulation study for validation of the method and in particular the "extended Kalman filter" (linearization about the best state estimate) took place. By early 1961 it was established that on-board optical measurements combined with the equation of motion could yield adequate estimates for navigation and guidance problems, the breakthrough that the NASA scientists were hoping for. Subsequently an early (perhaps the first) Kalman filter application was made circa 1961 during feasibility studies for the Apollo space program at the Instrumentation Laboratory of MIT.

Since then the Kalman filter has been widely used in navigation and guidance systems and in many other control systems. In particular, nowadays Kalman filters are used routinely in inertial navigation systems installed on transoceanic airliners, submarines, aircraft carriers, ballistic missiles, and certain spacecraft, for the initial alignment and calibration and mid-course update [26].

It seems that the application of a Kalman filter is a simple matter for it appears that once the problem is formulated in terms of equations (6.2), the standard Kalman filter algorithm can be applied in a straightforward manner. This is only ostensibly so on two accounts. First, casting the problem in the right form, especially when the models are nonlinear, is not an easy task, and second, most of the systems are not fully observable ([179], [180]) and thus there are various difficulties in the successful application of the algorithm. Yet, in spite of these difficulties, nowadays the code of the Kalman filter algorithm is a central component in the software of many sophisticated systems[1].

The continuous time analog of the linear state space model and the Kalman filter have been studied in 1961 in another celebrated paper by Kalman and Bucy [243]. In that paper the authors combined and streamlined their ideas developed independently in previous works in the late 1950s. A very similar line of work during roughly the same period was also pursued in the former USSR by the Russian physicist R. L. Stratonovich ([404], [405]), who studied a recursive algorithm for nonlinear least square estimates of the states of nonlinear dynamical systems driven by white noise.

[1]The authors are grateful to I. Y. Bar-Itzhack for the last three references and clarifying remarks on the use of Kalman filters in navigation.

A historical account of this and other much related work in the 1940s and 1950s can be found in [241].

State space models started to permeate the statistical literature in the 1960s and 1970s through the work of individuals interested in forecasting and in particular Bayesian forecasting of nonstationary processes–where the assumption of constant coefficients is quite onerous–as is apparent from the accounts in [198] and [427]. Another reason for the interest in state space models by statisticians is the fact that general state space modeling based on recursive relations of probability densities and their integrals are useful for non-Gaussian time series with abrupt discontinuities and/or outliers [265]. Comprehensive treatments of state space models and their statistical applications can be found in [72], [144], [200], [265], [388], and [427].

6.2 LINEAR GAUSSIAN STATE SPACE MODELS

Let $Y_1, Y_2, ...$ be a sequence of (scalar) observations or responses and $X_1, X_2, ...$ the corresponding covariate sequence, and as before , let $\mathcal{F}_t$ represent the available information to the observer at time t. It is convenient to adopt the convention that

$$\mathcal{F}_0 = \emptyset, \quad \mathcal{F}_t = \{Y_1, \ldots, Y_{t-1}, Y_t\} = \{\mathcal{F}_{t-1}, Y_t\},$$

while the dependence on the covariates $\{X_1, \ldots, X_t\}$ is kept in the background in the sense that the results are interpreted as conditional on the covariates. We have,

$$\text{Observation equation}: \quad Y_t = \mathbf{z}_t' \boldsymbol{\beta}_t + v_t, \quad v_t \sim \mathcal{N}(0, V_t) \tag{6.3}$$

$$\text{System equation}: \quad \boldsymbol{\beta}_t = \mathbf{F}_t \boldsymbol{\beta}_{t-1} + \mathbf{w}_t, \quad \mathbf{w}_t \sim \mathcal{N}_p(\mathbf{0}, \mathbf{W}_t) \tag{6.4}$$

$$\text{Initial information}: \quad \boldsymbol{\beta}_0 \sim \mathcal{N}_p(\mathbf{b}_0, \mathbf{W}_0), \tag{6.5}$$

where $\mathbf{z}_t$ is a design vector of covariates, such as past observations, supposed known at time t, where $\mathbf{b}_0, \mathbf{W}_0$ are likewise assumed known, and where we first take $\mathbf{F}_t, V_t, \mathbf{W}_t$ as known. In addition, we assume that $\{v_t\}$ and $\{\mathbf{w}_t\}$ each consists of independent random variables and that $\boldsymbol{\beta}_0, \{v_t\}, \{\mathbf{w}_t\}$ are mutually independent. The main departure from the previous GLM models is that the *state* $\boldsymbol{\beta}_t$, a vector of dimension p, is time dependent and random (reminiscent of random effects), and satisfies the autoregression equation (6.4) by means of the time dependent *transition matrix* $\mathbf{F}_t$. Notice that the joint distributions of the observations and the states are determined by the distributions of the initial state $\boldsymbol{\beta}_0$ and of the error sequences $\{v_t\}, \{\mathbf{w}_t\}$. The system of equations (6.3), (6.4), (6.5) is basically a regression model called *linear state space model* or *dynamic linear model*.

Given the observations $Y_1, ..., Y_N$, the linear state space system (6.3)–(6.5) is used in three types of estimation problems at time t referred to as *prediction or forecasting, filtering*, and *smoothing or interpolation* pending on the relationship between t and N: **prediction** for $t > N$, **filtering** for $t = N$, and **smoothing for** $t < N$. Thus, the estimation of the state $\boldsymbol{\beta}_t$ or its conditional distribution $f(\boldsymbol{\beta}_t \mid \mathcal{F}_N)$ is called prediction if $t > N$, filtering if $t = N$, and smoothing when $t < N$.

6.2.1 Examples of Linear State Space Models

We next illustrate by means of some simple examples the useful fact that, at the price of some redundancy, many linear models and in particular autoregressive moving average models admit state space representations, implying that the estimation theory for linear state space models dubbed filtering, smoothing, and prediction, is quite comprehensive and can cater to stationary as well as to nonstationary data.

A simple example. Consider the linear system,

$$\begin{aligned} Y_t &= X_t + v_t, \quad v_t \sim \mathcal{N}(0, \sigma_v^2) \\ X_t &= \phi_1 X_{t-1} + \phi_2 X_{t-2} + u_t, \quad u_t \sim \mathcal{N}(0, \sigma_u^2) \end{aligned} \tag{6.6}$$

and let,

$$\mathbf{z}_t = \begin{pmatrix} 1 \\ 0 \end{pmatrix}, \quad \boldsymbol{\beta}_t = \begin{pmatrix} X_t \\ X_{t-1} \end{pmatrix}, \quad \mathbf{F}_t \equiv \mathbf{F} = \begin{pmatrix} \phi_1 & \phi_2 \\ 1 & 0 \end{pmatrix}, \quad \mathbf{w}_t = \begin{pmatrix} u_t \\ 0 \end{pmatrix}.$$

Then clearly $Y_t = \mathbf{z}_t' \boldsymbol{\beta}_t + v_t$ and

$$\begin{pmatrix} X_t \\ X_{t-1} \end{pmatrix} = \begin{pmatrix} \phi_1 & \phi_2 \\ 1 & 0 \end{pmatrix} \begin{pmatrix} X_{t-1} \\ X_{t-2} \end{pmatrix} + \begin{pmatrix} u_t \\ 0 \end{pmatrix}$$

or, $\boldsymbol{\beta}_t = \mathbf{F}\boldsymbol{\beta}_{t-1} + \mathbf{w}_t$.

Structural time series. A slight extension of (6.6) is the structural time series model

$$\begin{aligned} Y_t &= T_t + S_t + v_t \\ T_t &= \phi T_{t-1} + w_{t1} \\ S_t &= \psi S_{t-1} - S_{t-2} + w_{t2}, \quad \psi = 2\cos(\omega), \end{aligned} \tag{6.7}$$

where T_t and S_t represent trend and seasonal components, respectively. Notice that when $w_{t2} = 0$ then the solution of $S_t = \psi S_{t-1} - S_{t-2}$ is a sinusoid with frequency ω. The structural model (6.7) can be expressed in terms of (6.3) and (6.4) by observing that

$$\boldsymbol{\beta}_t = \begin{pmatrix} T_t \\ S_t \\ S_{t-1} \end{pmatrix} = \begin{pmatrix} \phi & 0 & 0 \\ 0 & \psi & -1 \\ 0 & 1 & 0 \end{pmatrix} \begin{pmatrix} T_{t-1} \\ S_{t-1} \\ S_{t-2} \end{pmatrix} + \begin{pmatrix} w_{t1} \\ w_{t2} \\ 0 \end{pmatrix}$$

and $Y_t = (1, 1, 0)\boldsymbol{\beta}_t + v_t$.

State space representation for AR(p). Let X_t be an autoregressive process of order p, not necessarily stationary,

$$X_t = \phi_1 X_{t-1} + \cdots + \phi_p X_{t-p} + \epsilon_t.$$

Then the same equation can be expressed in matrix form to give the state equation,

$$\boldsymbol{\beta}_t = \begin{pmatrix} X_{t-p+1} \\ \vdots \\ X_{t-1} \\ X_t \end{pmatrix} = \begin{pmatrix} 0 & 1 & \cdots & 0 \\ \vdots & \vdots & \ddots & \vdots \\ 0 & 0 & \cdots & 1 \\ \phi_p & \phi_{p-1} & \cdots & \phi_1 \end{pmatrix} \begin{pmatrix} X_{t-p} \\ \vdots \\ X_{t-2} \\ X_{t-1} \end{pmatrix} + \begin{pmatrix} 0 \\ \vdots \\ 0 \\ \epsilon_t \end{pmatrix}$$

or with obvious notation $\boldsymbol{\beta}_t = \mathbf{F}\boldsymbol{\beta}_{t-1} + \mathbf{w}_t$ and $Y_t = (0, 0, \cdots, 0, 1)\boldsymbol{\beta}_t$.

State space representation for ARMA(p, q). There are several state space representations for autoregressive moving averages [72], [265]. To gain insight into a particular representation–for an alternative see Problem 1–consider the polynomials in the backward shift operator B, $BX_t \equiv X_{t-1}$,

$$\phi(B) = 1 - \phi_1 B - \cdots - \phi_p B^p$$

$$\theta(B) = 1 + \theta_1 B + \cdots + \theta_q B^q,$$

where $\phi(B)$ has its roots outside the unit circle, the so called stationarity condition. Obtaining a state equation for $\phi(B)X_t = v_t$ using matrices as was done in the previous example and then an observation equation $Y_t = \theta(B)X_t$, then formally

$$Y_t = \theta(B)X_t = \theta(B)\phi^{-1}(B)v_t$$

and $\phi(B)Y_t = \theta(B)v_t$ is ARMA(p, q). Notice that the role of the *state component* X_t is implicit in the ARMA representation but explicit in the state space representation. This argument gives the necessary clue for going in the reverse direction where we start with a given ARMA(p, q).

Consider the special case where $p = 3, q = 1$,

$$Y_t = \phi_1 Y_{t-1} + \phi_2 Y_{t-2} + \phi_3 Y_{t-3} + v_t + \theta_1 v_{t-1}.$$

We first obtain a representation for $\phi(B)X_t = v_t$ or $X_t = \phi_1 X_{t-1} + \phi_2 X_{t-2} + \phi_3 X_{t-3} + v_t$,

$$\begin{pmatrix} X_{t-2} \\ X_{t-1} \\ X_t \end{pmatrix} = \begin{pmatrix} 0 & 1 & 0 \\ 0 & 0 & 1 \\ \phi_3 & \phi_2 & \phi_1 \end{pmatrix} \begin{pmatrix} X_{t-3} \\ X_{t-2} \\ X_{t-1} \end{pmatrix} + \begin{pmatrix} 0 \\ 0 \\ v_t \end{pmatrix}.$$

Next we express $Y_t = \theta(B)X_t$ in matrix form letting $\theta_0 = 1$ and adding $\theta_2 = 0$,

$$Y_t = (\theta_2, \theta_1, \theta_0) \begin{pmatrix} X_{t-2} \\ X_{t-1} \\ X_t \end{pmatrix}.$$

It is easy to check that the last two equations represent the above ARMA(3,1) time series.

The same argument holds for the general stationary ARMA(p, q) by augmenting $\phi(B)$ or $\theta(B)$ with higher powers of B with zero coefficients as needed pending on whether $p > q$ or $p \leq q$, respectively. Thus with $r = \max(p, q+1)$, $\theta_0 = 1$, and adding as needed $\phi_j = 0, j > p, \theta_j = 0, j > q$, define X_t by $Y_t = \theta(B)X_t$ and $\phi(B)X_t = v_t$, and write $\mathbf{X}_t = (X_{t-r+1}, ..., X_t)'$. Then the observation equation is $Y_t = \theta(B)X_t = (\theta_{r-1}, ..., \theta_0)\mathbf{X}_t$, and the state equation is obtained by expressing $\phi(B)X_t = v_t$ as

$$\mathbf{X}_t = \begin{pmatrix} 0 & 1 & 0 & \cdots & 0 \\ 0 & 0 & 1 & \cdots & 0 \\ \vdots & \vdots & \vdots & \ddots & \vdots \\ 0 & 0 & 0 & \cdots & 1 \\ \phi_r & \phi_{r-1} & \phi_{r-2} & \cdots & \phi_1 \end{pmatrix} \mathbf{X}_{t-1} + \begin{pmatrix} 0 \\ 0 \\ \vdots \\ 0 \\ v_t \end{pmatrix}.$$

6.2.2 Estimation by Kalman Filtering and Smoothing

Consider the state space system (6.3)–(6.5) for $t = 1, ..., N$, and let

$$\beta_{t|s} = \mathrm{E}[\beta_t \mid \mathcal{F}_s] \tag{6.8}$$

$$\mathbf{P}_{t|s} = \mathrm{E}[(\beta_t - \beta_{t|s})(\beta_t - \beta_{t|s})'] \tag{6.9}$$

be the conditional mean and its precision matrix. Observe that the covariance matrix between $\beta_t - \beta_{t|s}$ and $Y_1, ..., Y_s$ is zero for all t and s. Therefore, by the normal assumption $\beta_t - \beta_{t|s}$ is also independent of $Y_1, ..., Y_s$ for all t and s which implies that $\mathbf{P}_{t|s}$ is also the conditional covariance matrix of $\beta_{t|s}$. Let $\beta_{0|0} = \mathbf{b}_0, \mathbf{P}_{0|0} = \mathbf{W}_0$, and assume the initial condition $\beta_0 \mid \mathcal{F}_0 \sim \mathcal{N}_p(\beta_{0|0}, \mathbf{P}_{0|0})$. Then we have

Kalman Prediction

$$\begin{aligned} \beta_{t|t-1} &= \mathbf{F}_t\beta_{t-1|t-1} \\ \mathbf{P}_{t|t-1} &= \mathbf{F}_t\mathbf{P}_{t-1|t-1}\mathbf{F}_t' + \mathbf{W}_t \end{aligned} \tag{6.10}$$

Kalman Filtering

$$\begin{aligned} \beta_{t|t} &= \beta_{t|t-1} + \mathbf{K}_t(Y_t - \mathbf{z}_t'\beta_{t|t-1}) \\ \mathbf{P}_{t|t} &= [\mathbf{I} - \mathbf{K}_t\mathbf{z}_t']\mathbf{P}_{t|t-1} \end{aligned} \tag{6.11}$$

where the so called *Kalman Gain* $\mathbf{K}_t$ is given by

$$\mathbf{K}_t = \mathbf{P}_{t|t-1}\mathbf{z}_t[\mathbf{z}_t'\mathbf{P}_{t|t-1}\mathbf{z}_t + V_t]^{-1}. \tag{6.12}$$

Proof. The prediction equations (6.10) follow from (6.4),

$$\beta_{t|t-1} \doteq \mathrm{E}[\beta_t \mid \mathcal{F}_{t-1}] = \mathrm{E}[\mathbf{F}_t\beta_{t-1} + \mathbf{w}_t \mid \mathcal{F}_{t-1}] = \mathbf{F}_t\beta_{t-1|t-1}$$

and

$$\begin{aligned} \mathbf{P}_{t|t-1} &= \mathrm{E}[(\beta_t - \beta_{t|t-1})(\beta_t - \beta_{t|t-1})'] \\ &= \mathrm{E}\{[\mathbf{F}_t(\beta_{t-1} - \beta_{t-1|t-1}) + \mathbf{w}_t][\mathbf{F}_t(\beta_{t-1} - \beta_{t-1|t-1}) + \mathbf{w}_t]'\} \\ &= \mathbf{F}_t\mathbf{P}_{t-1|t-1}\mathbf{F}_t' + \mathbf{W}_t. \end{aligned}$$

To obtain (6.11), recall the initial condition $\beta_0 \mid \mathcal{F}_0 \sim \mathcal{N}_p(\beta_{0|0}, \mathbf{P}_{0|0})$, and write

$$\beta_{t-1} \mid \mathcal{F}_{t-1} \sim \mathcal{N}_p(\beta_{t-1|t-1}, \mathbf{P}_{t-1|t-1}).$$

Then

$$\beta_t \mid \mathcal{F}_{t-1} = \mathbf{F}_t\beta_{t-1} + \mathbf{w}_t \mid \mathcal{F}_{t-1} \sim \mathcal{N}_p(\beta_{t|t-1}, \mathbf{P}_{t|t-1}),$$

from which

$$Y_t \mid \mathcal{F}_{t-1} \sim \mathcal{N}(\mathbf{z}_t'\beta_{t|t-1}, \mathbf{z}_t'\mathbf{P}_{t|t-1}\mathbf{z}_t + V_t)$$

and

$$\mathrm{Cov}(\beta_t, Y_t \mid \mathcal{F}_{t-1}) = \mathbf{P}_{t|t-1}\mathbf{z}_t$$

and hence,

$$\begin{pmatrix} \beta_t \\ Y_t \end{pmatrix} \Bigg| \mathcal{F}_{t-1} \sim \mathcal{N}_{p+1} \left[\begin{pmatrix} \beta_{t|t-1} \\ \mathbf{z}_t'\beta_{t|t-1} \end{pmatrix}, \begin{pmatrix} \mathbf{P}_{t|t-1} & \mathbf{P}_{t|t-1}\mathbf{z}_t \\ \mathbf{z}_t'\mathbf{P}_{t|t-1} & \mathbf{z}_t'\mathbf{P}_{t|t-1}\mathbf{z}_t + V_t \end{pmatrix} \right].$$

Therefore, from the conditional multivariate normal distribution (see (6.14) below) and after some algebra,

$$\beta_t \mid Y_t, \mathcal{F}_{t-1} \sim \beta_t \mid \mathcal{F}_t \sim \mathcal{N}_p(\beta_{t|t-1} + \mathbf{K}_t(Y_t - \mathbf{z}_t'\beta_{t|t-1}), (\mathbf{I} - \mathbf{K}_t\mathbf{z}_t')\mathbf{P}_{t|t-1}),$$

where $\mathbf{K}_t$ is given in (6.12). This completes the proof.

In the above proof we have made use of the important fact that if the p-vector $(\mathbf{x}_1', \mathbf{x}_2')'$ has a multivariate normal distribution with corresponding means $(\boldsymbol{\mu}_1', \boldsymbol{\mu}_2')'$ and covariance matrix partitioned compatibly, $\boldsymbol{\Sigma} = (\boldsymbol{\Sigma}_{ij}), i, j = 1, 2$,

$$\begin{pmatrix} \mathbf{x}_1 \\ \mathbf{x}_2 \end{pmatrix} \sim \mathcal{N}_p \left[\begin{pmatrix} \boldsymbol{\mu}_1 \\ \boldsymbol{\mu}_2 \end{pmatrix}, \begin{pmatrix} \boldsymbol{\Sigma}_{11} & \boldsymbol{\Sigma}_{12} \\ \boldsymbol{\Sigma}_{21} & \boldsymbol{\Sigma}_{22} \end{pmatrix} \right],$$

then $\mathbf{x}_i$ has a multivariate normal distribution with mean $\boldsymbol{\mu}_i$ and covariance matrix $\boldsymbol{\Sigma}_{ii}$, $i = 1, 2$, and the conditional distribution of $\mathbf{x}_2$ given $\mathbf{x}_1$ is again multivariate normal with mean

$$\mathrm{E}[\mathbf{x}_2 \mid \mathbf{x}_1] = \boldsymbol{\mu}_2 + \boldsymbol{\Sigma}_{21}\boldsymbol{\Sigma}_{11}^{-1}(\mathbf{x}_1 - \boldsymbol{\mu}_1), \tag{6.13}$$

and covariance matrix

$$\mathrm{cov}[\mathbf{x}_2 \mid \mathbf{x}_1] = \boldsymbol{\Sigma}_{22} - \boldsymbol{\Sigma}_{21}\boldsymbol{\Sigma}_{11}^{-1}\boldsymbol{\Sigma}_{12}. \tag{6.14}$$

The smoother or interpolator for obtaining $\beta_{t-1|N}$ and its covariance matrix $\mathbf{P}_{t-1|N}$, for $t = N, N-1, ..., 1$, under normality and the initial filtering conditions $\beta_{N|N}, \mathbf{P}_{N|N}$, is given by the following recursions.

Kalman Smoothing

$$\begin{aligned} \beta_{t-1|N} &= \beta_{t-1|t-1} + \mathbf{B}_t(\beta_{t|N} - \beta_{t|t-1}) \\ \mathbf{P}_{t-1|N} &= \mathbf{P}_{t-1|t-1} + \mathbf{B}_t(\mathbf{P}_{t|N} - \mathbf{P}_{t|t-1})\mathbf{B}_t' \\ \mathbf{B}_t &\equiv \mathbf{P}_{t-1|t-1}\mathbf{F}_t'\mathbf{P}_{t|t-1}^{-1}. \end{aligned} \tag{6.15}$$

Starting at $t = N$ and going backward in time, the smoothing estimate $\beta_{t-1|N}$ is obtained by adjusting the filtering estimate $\beta_{t-1|t-1}$, adding to it a weighted difference between a smoothing estimate $\beta_{t|N}$ and a prediction estimate $\beta_{t|t-1}$.

Proof. The proof of the smoothing recursions is apparently more complicated than that of the Kalman filtering recursions as was already noted by Kalman (1960). Of the several possible lines of attack, including a proof based on projections, we follow that of maximum likelihood as suggested in [367] and elaborated on in [388]. The idea is to maximize with respect to β_{t-1}, β_t the conditional Gaussian density

$$f(\beta_{t-1}, \beta_t \mid \mathcal{F}_N), \quad t \leq N, \tag{6.16}$$

upon noting that the values of β_{t-1}, β_t that maximize (6.16) are the respective conditional means $\beta_{t-1|N}, \beta_{t|N}$. To maximize (6.16), observe that from (6.3) and (6.4)

$$\begin{aligned} f(\beta_{t-1}, \beta_t \mid \mathcal{F}_N) &\propto f(\beta_{t-1}, \beta_t, \mathcal{F}_{t-1}, Y_t, ...Y_N) \qquad (6.17) \\ &= f(\mathcal{F}_{t-1}) f(\beta_{t-1}, \beta_t \mid \mathcal{F}_{t-1}) f(Y_t, ...Y_N \mid \beta_{t-1}, \beta_t, \mathcal{F}_{t-1}) \\ &= f(\mathcal{F}_{t-1}) f(\beta_{t-1} \mid \mathcal{F}_{t-1}) f(\beta_t \mid \beta_{t-1}) f(Y_t, ..., Y_N \mid \beta_t), \end{aligned}$$

where $f(\beta_{t-1} \mid \mathcal{F}_{t-1})$ is the density of $\mathcal{N}_p(\beta_{t-1|t-1}, \mathbf{P}_{t-1|t-1})$ and $f(\beta_t \mid \beta_{t-1})$ is the density of $\mathcal{N}_p(\mathbf{F}_t\beta_{t-1}, \mathbf{W}_t)$. Assume now that $\beta_{t|N}$ has already been obtained. Then $\beta_{t-1|N}$ is obtained by *minimizing* $-2\log f(\beta_{t-1}, \beta_{t|N} \mid \mathcal{F}_N)$ with respect to β_{t-1}. This is equivalent to minimizing

$$\begin{aligned} &(\beta_{t-1} - \beta_{t-1|t-1})' \mathbf{P}_{t-1|t-1}^{-1} (\beta_{t-1} - \beta_{t-1|t-1}) \\ + &(\beta_{t|N} - \mathbf{F}_t\beta_{t-1})' \mathbf{W}_t^{-1} (\beta_{t|N} - \mathbf{F}_t\beta_{t-1}) \end{aligned}$$

by differentiating with respect to β_{t-1} and equating the derivative to zero. The solution is

$$\beta_{t-1|N} = \left(\mathbf{P}_{t-1|t-1}^{-1} + \mathbf{F}_t'\mathbf{W}_t^{-1}\mathbf{F}_t\right)^{-1} \left(\mathbf{P}_{t-1|t-1}^{-1}\beta_{t-1|t-1} + \mathbf{F}_t'\mathbf{W}_t^{-1}\beta_{t|N}\right).$$

This can be simplified using the matrix relations (Problem 5),

$$\left(P^{-1} + F'W^{-1}F\right)^{-1} = P - PF'\left(FPF' + W\right)^{-1} FP$$

$$\left(P^{-1} + F'W^{-1}F\right)^{-1} F'W^{-1} = PF'\left(FPF' + W\right)^{-1},$$

where P, F, W stand for $\mathbf{P}_{t-1|t-1}, \mathbf{F}_t, \mathbf{W}_t$, respectively. This and the prediction expressions (6.10) give the desired smoother/interpolator,

$$\begin{aligned} &\beta_{t-1|N} \\ &= \beta_{t-1|t-1} + \mathbf{P}_{t-1|t-1}\mathbf{F}_t'\left(\mathbf{F}_t\mathbf{P}_{t-1|t-1}\mathbf{F}_t' + \mathbf{W}_t\right)^{-1}(\beta_{t|N} - \mathbf{F}_t\beta_{t-1|t-1}) \\ &= \beta_{t-1|t-1} + \mathbf{P}_{t-1|t-1}\mathbf{F}_t'\mathbf{P}_{t|t-1}^{-1}(\beta_{t|N} - \beta_{t|t-1}) \\ &= \beta_{t-1|t-1} + \mathbf{B}_t(\beta_{t|N} - \beta_{t|t-1}). \end{aligned}$$

To obtain $\mathbf{P}_{t-1|N} = \mathrm{E}[(\beta_{t-1} - \beta_{t-1|N})(\beta_{t-1} - \beta_{t-1|N})']$, note that

$$\beta_{t-1} - \beta_{t-1|N} = \beta_{t-1} - \beta_{t-1|t-1} - \mathbf{B}_t(\beta_{t|N} - \beta_{t|t-1})$$

or by rearranging terms,

$$(\beta_{t-1} - \beta_{t-1|N}) + \mathbf{B}_t\beta_{t|N} = (\beta_{t-1} - \beta_{t-1|t-1}) + \mathbf{B}_t\mathbf{F}_t\beta_{t-1|t-1}, \qquad (6.18)$$

and that the following cross terms vanish

$$\mathrm{E}(\beta_{t-1} - \beta_{t-1|N})(\mathbf{B}_t\beta_{t|N})' = \mathrm{E}(\beta_{t-1} - \beta_{t-1|t-1})(\mathbf{B}_t\mathbf{F}_t\beta_{t-1|t-1})' = \mathbf{0},$$

and

$$\mathrm{E}(\boldsymbol{\beta}_{t|s}\boldsymbol{\beta}'_{t|s}) = \mathrm{E}(\boldsymbol{\beta}_t\boldsymbol{\beta}'_t) - \mathbf{P}_{t|s} = \mathbf{F}_t\mathrm{E}(\boldsymbol{\beta}_{t-1}\boldsymbol{\beta}'_{t-1})\mathbf{F}'_t + \mathbf{W}_t - \mathbf{P}_{t|s}.$$

Thus from (6.18),

$$\mathbf{P}_{t-1|N} = \mathbf{P}_{t-1|t-1} + \mathbf{B}_t(\mathbf{P}_{t|N} - \mathbf{F}_t\mathbf{P}_{t-1|t-1}\mathbf{F}'_t - \mathbf{W}_t)\mathbf{B}'_t,$$

which, together with (6.10), completes the proof.

6.2.3 Estimation in the Linear Gaussian Model

Estimation of parameters in the linear Gaussian system (6.4)–(6.5) can be carried out by the method of maximum likelihood under a certain parametrization assumption often met in practice. We shall assume that the parameters $\mathbf{b}_0, \mathbf{W}_0, \mathbf{F}_t, V_t, \mathbf{W}_t$ depend completely or in part on a vector $\boldsymbol{\theta}$ of *hyperparameters* which do not depend on t. In this case we write

$$\mathbf{b}_0 = \mathbf{b}_0(\boldsymbol{\theta}),\ \mathbf{W}_0 = \mathbf{W}_0(\boldsymbol{\theta}),\ \mathbf{F}_t = \mathbf{F}_t(\boldsymbol{\theta}),\ V_t = V_t(\boldsymbol{\theta}),\ \mathbf{W}_t = \mathbf{W}_t(\boldsymbol{\theta}),$$

and base the inference on the joint distribution of the observations $Y_1, ..., Y_N$, or equivalently the likelihood of $\boldsymbol{\theta}$. Parametrizations of correlation functions of this type is also used in the next chapter on prediction. Similarly, we may assume that $\mathbf{F}_t, V_t, \mathbf{W}_t$ in the system (6.4)–(6.5) do not depend on t and estimate $\boldsymbol{\theta} = (\mathbf{b}_0, \mathbf{W}_0, \mathbf{F}, V, \mathbf{W})$ by maximum likelihood. In either case, the likelihood is obtained from the joint distribution of the *one-step prediction errors* or *innovations*

$$\epsilon_t = Y_t - \mathrm{E}[Y_t \mid \mathcal{F}_{t-1}] = Y_t - \mathbf{z}'_t\boldsymbol{\beta}_{t|t-1} = Y_t - \mathbf{z}'_t\mathbf{F}_t\boldsymbol{\beta}_{t-1|t-1}, \quad t = 1, ..., N,$$

which are independent normal random variables with mean zero and variance

$$\sigma_t^2(\boldsymbol{\theta}) = \mathbf{z}'_t\mathbf{P}_{t|t-1}\mathbf{z}_t + v_t = \mathbf{z}'_t\{\mathbf{F}_t\mathbf{P}_{t-1|t-1}\mathbf{F}'_t + \mathbf{W}_t\}\mathbf{z}_t + v_t.$$

The Gaussian assumption implies that $\epsilon_1, ..., \epsilon_N$ is a one-to-one linear transformation of $Y_1, ..., Y_N$ so that up to a constant the log-likelihood of $\boldsymbol{\theta}$ based on $Y_1, ..., Y_N$ is given by

$$\log L_y(\boldsymbol{\theta}) = -\frac{1}{2}\sum_{t=1}^{N} \log\sigma_t^2(\boldsymbol{\theta}) - \frac{1}{2}\sum_{t=1}^{N} \epsilon_t^2/\sigma_t^2(\boldsymbol{\theta}).$$

Maximizing the likelihood with respect to $\boldsymbol{\theta}$ is sometime referred to as the *direct* method. See [200] for more details.

The *indirect* method is based on the joint distribution of both the observed time series and the unobserved states and uses the EM algorithm to maximize the resulting likelihood. From (6.4)–(6.5),

$$f(y_t \mid \boldsymbol{\beta}_t, \mathcal{F}_{t-1}; \boldsymbol{\theta}) = f(y_t \mid \boldsymbol{\beta}_t; \boldsymbol{\theta}) = f_v(y_t - \mathbf{z}'_t\boldsymbol{\beta}_t)$$

and

$$f(\boldsymbol{\beta}_t \mid \boldsymbol{\beta}_{t-1}, \boldsymbol{\beta}_{t-2}, ..., \boldsymbol{\beta}_0; \boldsymbol{\theta}) = f(\boldsymbol{\beta}_t \mid \boldsymbol{\beta}_{t-1}; \boldsymbol{\theta}) = f_w(\boldsymbol{\beta}_t - \mathbf{F}_t \boldsymbol{\beta}_{t-1}),$$

where $v_t \sim f_v$ and $\mathbf{w}_t \sim f_w$. The likelihood is then

$$\begin{aligned} L_{y,\beta}(\boldsymbol{\theta}) &= f(\boldsymbol{\beta}_0, \boldsymbol{\beta}_1, ..., \boldsymbol{\beta}_N, Y_1, ..., Y_N; \boldsymbol{\theta}) \\ &= f(\boldsymbol{\beta}_0) \prod_{t=1}^{N} f_w(\boldsymbol{\beta}_t - \mathbf{F}_t \boldsymbol{\beta}_{t-1}) f_v(y_t - \mathbf{z}_t' \boldsymbol{\beta}_t). \end{aligned} \tag{6.19}$$

The application of the EM algorithm to maximize this likelihood in the presence of unobserved states is described in [144], [388].

A general Bayesian method for the estimation of $\boldsymbol{\theta}$ in dynamic models using a Markov chain Monte Carlo method called permutation sampling is studied in [160]. Other estimation methods and tools in dynamic models are discussed in [144], [200], and [388]. In a related problem, Gerencsér [169] discusses recursive estimation in parameter dependent stochastic systems of the form $Y_n(\theta) = C(\theta)X_n(\theta)$, $X_{n+1}(\theta) = A(\theta)X_n + B(\theta)e_n$.

6.2.3.1 Example of Estimation and Filtering in a Structural Model

We apply Kalman filtering and prediction to a monthly time series, recorded in Table 7.3, of the number of unemployed women older than 20 years of age from 1997 to 2001. The time series plot in Figure 6.1 points to a slowly decreasing trend and an additional periodic component very characteristic of *structural time series*. It seems therefore that a slight extension of the structural model (6.7) discussed earlier,

$$\begin{aligned} Y_t &= T_t + S_t + v_t \\ T_t &= \phi T_{t-1} + w_{t1} \\ S_t &= \psi_1 S_{t-1} + \psi_2 S_{t-2} + w_{t2}, \end{aligned}$$

where T_t and S_t denote trend and sinusoidal components, respectively, is sensible. Then,

$$\boldsymbol{\beta}_t = \begin{pmatrix} T_t \\ S_t \\ S_{t-1} \end{pmatrix} = \begin{pmatrix} \phi & 0 & 0 \\ 0 & \psi_1 & \psi_2 \\ 0 & 1 & 0 \end{pmatrix} \begin{pmatrix} T_{t-1} \\ S_{t-1} \\ S_{t-2} \end{pmatrix} + \begin{pmatrix} w_{t1} \\ w_{t2} \\ 0 \end{pmatrix},$$

and $Y_t = (1,1,0)\boldsymbol{\beta}_t + v_t$. Setting $V_t = \sigma_v^2$ for the variance of the independent normal sequence $\{v_t\}$ and

$$\mathbf{W}_t = \mathbf{W} = \begin{pmatrix} \sigma_{w_1}^2 & 0 & 0 \\ 0 & \sigma_{w_2}^2 & 0 \\ 0 & 0 & 0 \end{pmatrix}$$

for the covariance matrix of $\mathbf{w}_t = (w_{t1}, w_{t2}, 0)'$, the parameters that we need to estimate are $\boldsymbol{\theta} = (\phi, \psi_1, \psi_2, \sigma_v^2, \sigma_{w_1}^2, \sigma_{w_2}^2)'$, $\mathbf{b}_0$ and $\mathbf{W}_0$. Initializing the EM algorithm (see [388]) at $\mathbf{b}_0 = (3, 25, 20)', \mathbf{W}_0 = \mathbf{I}_3$, the 3×3 identity matrix, and $\boldsymbol{\theta}_0 = (0.3, -0.25, 0.25, 6, 3, 10)'$, a large number of iterations produces the following estimates, $\hat{\mathbf{b}}_0 = (8.198, 18.931, 12.448)'$, and $\hat{\boldsymbol{\theta}} = (0.415, 1.586, -1.026, 1.286, 0.326,$

$0.572)'$. We note that the estimator of the trend is less than 1, $\hat{\phi} = 0.415 < 1$, confirming that the time series exhibits a slowly decreasing trend, and that the variance of the observed process is larger than the variances of the two unobserved components.

Figure 6.2 shows the filtered estimators of the trend and sinusoidal components $T_{t|t}$ (top) and $S_{t|t}$ (bottom), both evaluated at the maximum likelihood estimates. Prediction of the monthly number of unemployed women during the next 12 months is shown in Figure 6.3. Computations were carried out using the function `kalman` available at `http://lib.stat.cmu.edu/S/`.

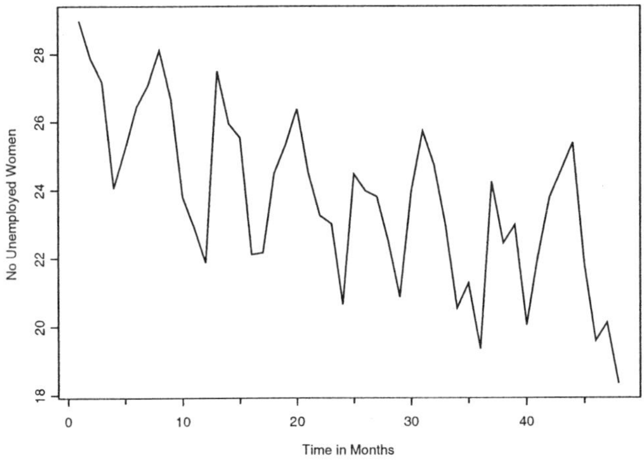

Fig. 6.1 Monthly number of unemployed women between 1997 to 2000. Data in hundreds of thousands, $N = 48$. *Source*: Bureau of Labor Statistics, Series ID LFU22001702.

6.2.3.2 Software Resources for State Space Models S–Plus functions for fitting state space models include the collection `bts.zip` which is based on algorithms described in [427] and is available at `lib.stat.cmu.edu/DOS/S/`, and the dynamic system estimation library whose description is given at the address `www.bank-banque-canada.ca/pgilbert/dse/dsedesc.htm`.

6.3 NONLINEAR AND NON-GAUSSIAN STATE SPACE MODELS

Prediction, filtering, and smoothing, can be approached more generally by using the laws of conditional probability, including Bayes theorem, and relaxing linearity and the normal assumption. In the general approach the dynamics is captured directly through the conditional densities of the observations and the states, without the formation of any particular system of equations [127], [265], [405].

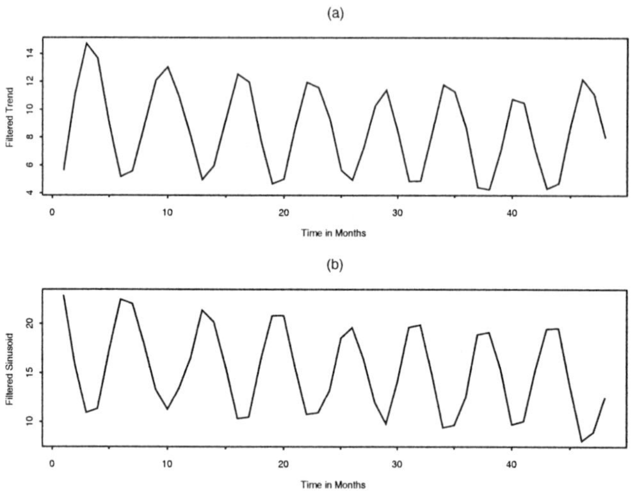

Fig. 6.2 Filtered monthly number of unemployed women between 1997 to 2000. (a) Trend component. (b) Sinusoidal component.

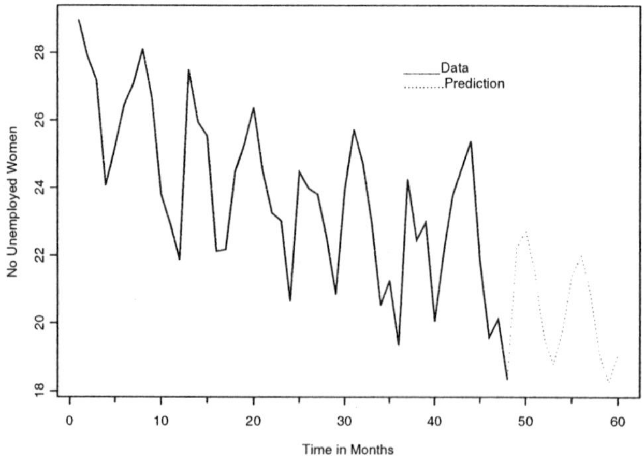

Fig. 6.3 Predicted monthly number of unemployed women for 12 months ahead.

Analogous to the linear Gaussian system represented by (6.3), (6.4), and (6.5), assume that $\{Y_t\}, t = 1, \ldots, N$ denotes the observed process and let the unobserved state process be $\{\beta_t\}, t = 0, \ldots, N$. Introducing the generic notation $f()$ for probability densities, the general state space model is formulated as follows:

$$\text{General Observation equation:} \qquad Y_t \mid \beta_t \sim f(y_t \mid \beta_t) \tag{6.20}$$

$$\text{General System equation:} \qquad \beta_t \mid \beta_{t-1} \sim f(\beta_t \mid \beta_{t-1}) \tag{6.21}$$

$$\text{Initial information:} \qquad \beta_0 \sim f(\beta_0 \mid \mathcal{F}_0) \equiv f(\beta_0) \tag{6.22}$$

with the understanding that given the states, the responses are independent, and that the sequence of unobserved states forms a Markov process. Thus, equation (6.20) means that given the state sequence $\{\beta_t\}$, the observed process $\{Y_t\}$ forms an independent sequence of random variables, and both (6.21), (6.22) mean that the sequence of unobserved states $\{\beta_t\}, t = 0, \ldots, N$ is Markov process with initial distribution $f(\beta_0)$.

It is worth pointing out that when the unobserved states assume discrete values then the definition of the general state space model is equivalent to that of a hidden Markov model; see Section 5.4. The densities in equations (6.20) and (6.21) may depend on unknown parameters referred to as *hyperparameters*. Estimation methods for hyperparameters are discussed in subsequent sections.

It is easy to verify that the linear normal state space model (6.3), (6.4), (6.5), where the corresponding conditional densities are Gaussian, is a special case of the system represented by (6.20), (6.21), and (6.22). Another special case is provided by the nonlinear and non-Gaussian state space model

$$\begin{aligned} Y_t &= h_t(\beta_t, v_t) \\ \beta_t &= \mathbf{f}_t(\beta_{t-1}, \mathbf{w}_t), \end{aligned} \tag{6.23}$$

where h_t and $\mathbf{f}_t$ are known and suitably defined functions and v_t, $\mathbf{w}_t$ are random sequences, $t = 1, \ldots, N$. An example of (6.23) is

$$\begin{aligned} Y_t &= h_t(\beta_t) + v_t \\ \beta_t &= \mathbf{f}_t(\beta_{t-1}) + \mathbf{w}_t. \end{aligned} \tag{6.24}$$

Similarly to linear state space models, the main problem is estimation of current, future and past states and their distributions given the data, that is, *filtering*, *prediction* and *smoothing*, respectively. Questions like these pose a great challenge due to nonnormality and nonlinearity as manifested by (6.20) and (6.21). In what follows we summarize several approaches to these problems before turning to Monte Carlo Markov Chain techniques.

6.3.1 General Filtering and Smoothing

Regardless of any distributional assumptions about the observation and system equations, using the definitions of conditional and marginal distributions and employing Bayes theorem, the following *density* recursions hold in general [260], [265].

Prediction: General prediction is obtained from the conditional predictive or prediction density $f(\boldsymbol{\beta}_t \mid \mathcal{F}_{t-1})$,

$$\begin{aligned} f(\boldsymbol{\beta}_t \mid \mathcal{F}_{t-1}) &= \int f(\boldsymbol{\beta}_t, \boldsymbol{\beta}_{t-1} \mid \mathcal{F}_{t-1}) d\boldsymbol{\beta}_{t-1} \\ &= \int f(\boldsymbol{\beta}_t \mid \boldsymbol{\beta}_{t-1}, \mathcal{F}_{t-1}) f(\boldsymbol{\beta}_{t-1} \mid \mathcal{F}_{t-1}) d\boldsymbol{\beta}_{t-1} \\ &= \int f(\boldsymbol{\beta}_t \mid \boldsymbol{\beta}_{t-1}) f(\boldsymbol{\beta}_{t-1} \mid \mathcal{F}_{t-1}) d\boldsymbol{\beta}_{t-1} \end{aligned} \tag{6.25}$$

for $t = 1, 2, \ldots, N$. In the linear normal case this becomes

$$f(\boldsymbol{\beta}_t \mid \mathcal{F}_{t-1}) = \int f_{\mathbf{w}}(\boldsymbol{\beta}_t - \mathbf{F}_t \boldsymbol{\beta}_{t-1}) f(\boldsymbol{\beta}_{t-1} \mid \mathcal{F}_{t-1}) d\boldsymbol{\beta}_{t-1}, \tag{6.26}$$

where $\mathbf{w}_t \sim f_{\mathbf{w}}$, and as was shown above, $f(\boldsymbol{\beta}_t \mid \mathcal{F}_{t-1})$ reduces to the density of $\mathcal{N}_p(\boldsymbol{\beta}_{t|t-1}, \mathbf{P}_{t|t-1})$.

Filtering: The filtering density $f(\boldsymbol{\beta}_t \mid \mathcal{F}_t)$ is obtained by appealing to Bayes theorem,

$$\begin{aligned} f(\boldsymbol{\beta}_t \mid \mathcal{F}_t) &= f(\boldsymbol{\beta}_t \mid \mathcal{F}_{t-1}, y_t) \\ &= \frac{f(y_t \mid \boldsymbol{\beta}_t, \mathcal{F}_{t-1}) f(\boldsymbol{\beta}_t \mid \mathcal{F}_{t-1})}{f(y_t \mid \mathcal{F}_{t-1})} \\ &= \frac{f(y_t \mid \boldsymbol{\beta}_t) f(\boldsymbol{\beta}_t \mid \mathcal{F}_{t-1})}{f(y_t \mid \mathcal{F}_{t-1})} \end{aligned} \tag{6.27}$$

for $t = 1, 2, \ldots, N$ and $f(y_t \mid \mathcal{F}_{t-1}) = \int f(y_t \mid \boldsymbol{\beta}_t) f(\boldsymbol{\beta}_t \mid \mathcal{F}_{t-1}) d\boldsymbol{\beta}_t$. In the linear normal case this simplifies to

$$f(\boldsymbol{\beta}_t \mid \mathcal{F}_t) = \frac{f_v(y_t - \mathbf{z}_t' \boldsymbol{\beta}_t) f(\boldsymbol{\beta}_t \mid \mathcal{F}_{t-1})}{f(y_t \mid \mathcal{F}_{t-1})}, \tag{6.28}$$

where $v_t \sim f_v$, and

$$f(y_t \mid \mathcal{F}_{t-1}) = \int f_v(y_t - \mathbf{z}_t' \boldsymbol{\beta}_t) f(\boldsymbol{\beta}_t \mid \mathcal{F}_{t-1}) d\boldsymbol{\beta}_t,$$

and where $f(\boldsymbol{\beta}_t \mid \mathcal{F}_t)$ is the density of $\mathcal{N}_p(\boldsymbol{\beta}_{t|t}, \mathbf{P}_{t|t})$.

Smoothing: To obtain a general smoothing density $f(\boldsymbol{\beta}_t \mid \mathcal{F}_N)$, $t < N$, consider first the joint density of $\boldsymbol{\beta}_t, \boldsymbol{\beta}_{t+1}$ given the entire history of the process $\mathcal{F}_N$,

$$\begin{aligned} f(\boldsymbol{\beta}_t, \boldsymbol{\beta}_{t+1} \mid \mathcal{F}_N) &= f(\boldsymbol{\beta}_{t+1} \mid \mathcal{F}_N) f(\boldsymbol{\beta}_t \mid \boldsymbol{\beta}_{t+1}, \mathcal{F}_N) \\ &= f(\boldsymbol{\beta}_{t+1} \mid \mathcal{F}_N) f(\boldsymbol{\beta}_t \mid \boldsymbol{\beta}_{t+1}, \mathcal{F}_t) \\ &= f(\boldsymbol{\beta}_{t+1} \mid \mathcal{F}_N) \frac{f(\boldsymbol{\beta}_t, \boldsymbol{\beta}_{t+1} \mid \mathcal{F}_t)}{f(\boldsymbol{\beta}_{t+1} \mid \mathcal{F}_t)} \\ &= f(\boldsymbol{\beta}_{t+1} \mid \mathcal{F}_N) \frac{f(\boldsymbol{\beta}_{t+1} \mid \boldsymbol{\beta}_t) f(\boldsymbol{\beta}_t \mid \mathcal{F}_t)}{f(\boldsymbol{\beta}_{t+1} \mid \mathcal{F}_t)}. \end{aligned} \tag{6.29}$$

The second equality follows from $f(\beta_t \mid \beta_{t+1}, \mathcal{F}_N) = f(\beta_t \mid \beta_{t+1}, \mathcal{F}_t)$ and the last two equations of (6.29) follow from Bayes theorem. From (6.29) we have

$$\begin{aligned} f(\beta_t \mid \mathcal{F}_N) &= \int f(\beta_t, \beta_{t+1} \mid \mathcal{F}_N) d\beta_{t+1} \\ &= f(\beta_t \mid \mathcal{F}_t) \int f(\beta_{t+1} \mid \mathcal{F}_N) \frac{f(\beta_{t+1} \mid \beta_t)}{f(\beta_{t+1} \mid \mathcal{F}_t)} d\beta_{t+1}, \quad (6.30) \end{aligned}$$

for $t = N, N-1, \ldots, 1$. In the linear normal case $f(\beta_t \mid \mathcal{F}_N)$ is the density of $\mathcal{N}_p(\beta_{t|N}, \mathbf{P}_{t|N})$.

Implementation of (6.25), (6.27), (6.30) is based on sequential integration which most of the time is complicated. To overcome the computational burden, Kitagawa [260] suggests a numerical method for piecewise linear approximation of the corresponding integrals when the dimension of the state vector is 1. Extension of this method to higher dimensions demands an additional computational cost [212]. Some other suggestions for implementation of (6.25), (6.27), (6.30) include the so called Gaussian–sum filter for approximating the integrals with mixtures of Gaussian distributions (Problem 8) and the two-filter formula for smoothing [261], [262] . More recently, a Monte Carlo method for filtering and smoothing was put forth in [263], [265], and is considered in Section 6.4.3.

6.3.2 Dynamic Generalized Linear Models

A special case of nonnormal and nonlinear state space models are the so called dynamic generalized linear models (DGLM) for time series data which broaden the class of static generalized linear models–described in Chapter 1–by allowing time varying random regression coefficients. The definition of DGLM retains the random component– recall (1.6)–while the systematic component (1.7) is augmented by a transition equation for the regression parameters. These models have been applied successfully in several diverse fields such as meteorology, epidemiology and econometrics.

Suppose that $\{Y_t\}, t = 1, \ldots, N$ denotes a response time series such that its conditional distribution given the past belongs to the exponential family of distributions

$$f(y_t \mid \theta_t) = \exp\left\{ \frac{y_t \theta_t - b(\theta_t)}{\alpha_t(\phi)} + c(y_t, \phi_t) \right\}, \quad t = 1, \ldots, N, \qquad (6.31)$$

where the parametric function $\alpha_t(\phi)$ is assumed known. Unlike the case in (1.6), the random natural parameter θ_t in (6.31) and the conditional expectation of the response $\mu_t = \mathrm{E}[Y_t \mid \theta_t]$ are dynamically linked to a changing sequence of regression or state parameters. That is, with $g(.)$ a monotone link function and a given p-dimensional covariate vector $\{\mathbf{z}_t\}$,

$$g(\mu_t) = \mathbf{z}_t' \beta_t, \qquad (6.32)$$

where $\{\beta_t\}, t = 0, \ldots, N$ is a random sequence of p–dimensional regression parameters obeying the Markov linear transition model

$$\beta_t = \mathbf{F}_t \beta_{t-1} + \mathbf{w}_t, \quad t = 1, 2, \ldots, \qquad (6.33)$$

where $\mathbf{F}_t$ is a sequence of $p \times p$ known matrices, and $\{\mathbf{w}_t\}$ is an independent sequence of random variables. It is common in practice to assume that $\mathbf{w}_t$ is normal with zero mean and covariance matrix $\mathbf{W}_t$, and that it is distributed independently of $\mathcal{F}_{t-1}$. In this case we suppose that $\boldsymbol{\beta}_0$ has a normal distribution with mean $\mathbf{b}_0$ and covariance matrix $\mathbf{W}_0$ and that is independent of $\{\mathbf{w}_t\}$ and $\mathcal{F}_{t-1}$. The state may follow other distributions depending on the context of the application.

Equations (6.31) and (6.32) where $f()$ is a member the exponential family of distributions define the *observation equation* of the model corresponding to (6.20). The evolution relation (6.33) corresponds to (6.21) where $f()$ is replaced by the normal density. This coupled with (6.31) and (6.32) conforms to the definition of dynamic generalized linear models for time series data.

Example: A State Space Model for Binary Time Series Let $\{Y_t\}$, $t = 1, \ldots, N$, be a binary time series with π_t denoting the success probability, and consider the following model which includes trend and a lagged value of the response

$$\log\left(\frac{\pi_t}{1-\pi_t}\right) = T_t + \beta_t^1 Y_{t-1}.$$

If

$$T_t = T_{t-1} + w_t^1$$

and

$$\beta_t^1 = \beta_{t-1}^1 + w_t^2$$

with $\{w_t^1\}$ and $\{w_t^2\}$ mutually uncorrelated white noise sequences, then we obtain representation (6.33) by defining $\boldsymbol{\beta}_t = \left(T_t, \beta_t^1\right)'$, $\mathbf{F}_t = \mathbf{I}_2$, the 2×2 identity matrix, and $\mathbf{w}_t = \left(w_t^1, w_t^2\right)'$. In addition set $\mathbf{z}_t = (1, y_{t-1})'$ so that (6.32) is satisfied.

The formulation of DGLM suggests several extensions suitable for nonlinear and non-Gaussian state space models as we describe in the following subsections regarding conjugate analysis and posterior mode estimation.

6.3.2.1 Conjugate Analysis and Linear Bayes Estimation Binding together conjugate analysis of prior and posterior distributions and linear Bayes estimation (see [199]), we can approach the problems of forecasting, and filtering from a Bayesian perspective [428].

Let the mean and covariance of the state vector $\boldsymbol{\beta}_t$ given $\mathcal{F}_{t-1}$ be $\boldsymbol{\beta}_{t|t-1}$ and $\mathbf{P}_{t|t-1}$, respectively. Dropping any distributional assumptions but specifying the first two moments for the error term in (6.33), and assuming independence assumptions as before, we obtain the following prediction recursions

$$\begin{aligned} \boldsymbol{\beta}_{t|t-1} &= \mathbf{F}_t\boldsymbol{\beta}_{t-1|t-1} \\ \mathbf{P}_{t|t-1} &= \mathbf{F}_t\mathbf{P}_{t-1|t-1}\mathbf{F}_t' + \mathbf{W}_t \\ \boldsymbol{\beta}_{0|0} &= \mathbf{b}_0 \\ \mathbf{P}_{0|0} &= \mathbf{W}_0. \end{aligned} \tag{6.34}$$

An alternative form of the second equation of (6.34) is given by

$$\mathbf{P}_{t|t-1} = \mathbf{B}_t\mathbf{F}_t\mathbf{P}_{t-1|t-1}\mathbf{F}_t'\mathbf{B}_t',$$

where $\mathbf{B}_t$ is a $p \times p$ diagonal matrix of the so called *discount* factors [17], [428].

Recall that for the exponential family (6.31) $b'(\theta_t) = \mu_t$, and suppose we specify a prior for $g^*(\theta_t)$ in terms of the first two moments, where $g^*(.) = (g \circ b')(.)$ such that $g^*(\theta_t) = \mathbf{z}_t'\boldsymbol{\beta}_t$. Then (6.34) implies

$$\begin{aligned} \mathrm{E}\left[g^*(\theta_t) \mid \mathcal{F}_{t-1}\right] &= \mathbf{z}_t'\boldsymbol{\beta}_{t|t-1} \equiv f_t \\ \mathrm{Var}\left[g^*(\theta_t) \mid \mathcal{F}_{t-1}\right] &= \mathbf{z}_t'\mathbf{P}_{t|t-1}\mathbf{z}_t \equiv q_t \end{aligned} \tag{6.35}$$

so that the vector $\left(g^*(\theta_t), \boldsymbol{\beta}_t'\right)'$ given $\mathcal{F}_{t-1}$ has mean vector and covariance matrix

$$\begin{pmatrix} f_t \\ \boldsymbol{\beta}_{t|t-1} \end{pmatrix} \quad \text{and} \quad \begin{pmatrix} q_t & \mathbf{z}_t'\mathbf{P}_{t|t-1} \\ \mathbf{P}_{t|t-1}\mathbf{z}_t & \mathbf{P}_{t|t-1} \end{pmatrix}, \tag{6.36}$$

respectively.

Assuming that in (6.31) $\alpha_t(\phi)$ is known for all t, the canonical parameter $\{\theta_t\}$ represents the uncertainty about the distribution of the response given the past of the process. Suppose that $\{\theta_t\}, t = 1, \ldots, N$ follows a conjugate prior of the exponential family type

$$f(\theta_t \mid \mathcal{F}_{t-1}) = \exp\left\{\gamma_t\theta_t - \delta_t b(\theta_t) + c^\star(\theta_t)\right\}. \tag{6.37}$$

Then, from (6.31), a straightforward application of Bayes theorem shows that

$$\begin{aligned} f(\theta_t \mid \mathcal{F}_t) &\propto f(y_t \mid \theta_t) f(\theta_t \mid \mathcal{F}_{t-1}) \\ &\propto \exp\left\{(\gamma_t + \frac{y_t}{\alpha_t(\phi)})\theta_t - (\delta_t + \frac{1}{\alpha_t(\phi)})b(\theta_t)\right\}, \end{aligned}$$

which shows that the posterior of θ_t given a new datum at time t belongs to the exponential family. Values for both γ_t and δ_t in (6.37) are chosen based on the fact that the canonical parameter is approximately linked to the covariates and the state parameters by $g^*(\theta_t) \approx \mathbf{z}_t'\boldsymbol{\beta}_t$ [428].

The filtering recursions are obtained by calculating $\boldsymbol{\beta}_{t|t} = \mathrm{E}\left[\boldsymbol{\beta}_t \mid \mathcal{F}_t\right]$ and $\mathbf{P}_{t|t} = \mathrm{Var}\left[\boldsymbol{\beta}_t \mid \mathcal{F}_t\right]$ using the facts (see [427, p. 528]),

$$\mathrm{E}\left[\boldsymbol{\beta}_t \mid \mathcal{F}_t\right] = \mathrm{E}\left[\mathrm{E}\left[\boldsymbol{\beta}_t \mid \theta_t, \mathcal{F}_{t-1}\right] \mid \mathcal{F}_t\right], \tag{6.38}$$

$$\mathrm{Var}\left[\boldsymbol{\beta}_t \mid \mathcal{F}_t\right] = \mathrm{Var}\left[\mathrm{E}\left[\boldsymbol{\beta}_t \mid \theta_t, \mathcal{F}_{t-1}\right] \mid \mathcal{F}_t\right] + \mathrm{E}\left[\mathrm{Var}\left[\boldsymbol{\beta}_t \mid \theta_t, \mathcal{F}_{t-1}\right] \mid \mathcal{F}_t\right]. \tag{6.39}$$

In the absence of a fully specified distribution for $\boldsymbol{\beta}_t$ given $(\theta_t, \mathcal{F}_{t-1})$, (6.36) and an appeal to the linear Bayes methodology (Problem 4) shows that the optimal linear estimator $\mathbf{d}$ of $\mathrm{E}\left[\boldsymbol{\beta}_t \mid \theta_t, \mathcal{F}_{t-1}\right]$ in the sense of minimizing the overall quadratic risk

$$r_t(\mathbf{d}) = \mathrm{tr}\left\{\mathrm{E}\left[\mathbf{A}_t(\mathbf{d}) \mid \mathcal{F}_{t-1}\right]\right\},$$

where tr denotes trace of a matrix, the expectation is taken with respect to $f(\theta_t \mid \mathcal{F}_{t-1})$ and

$$\mathbf{A}_t(\mathbf{d}) = \mathrm{E}\left[(\boldsymbol{\beta}_t - \mathbf{d})(\boldsymbol{\beta}_t - \mathbf{d})' \mid \theta_t, \mathcal{F}_{t-1}\right],$$

is given by

$$\hat{\mathrm{E}}\left[\boldsymbol{\beta}_t \mid \theta_t, \mathcal{F}_{t-1}\right] = \boldsymbol{\beta}_{t|t-1} + \mathbf{P}_{t|t-1}\mathbf{z}_t \frac{(g^*(\theta_t) - f_t)}{q_t} \tag{6.40}$$

with an associated minimum value of $r_t(\mathbf{d})$

$$\hat{\text{Var}}\left[\boldsymbol{\beta}_t \mid \theta_t, \mathcal{F}_{t-1}\right] = \mathbf{P}_{t|t-1} - \frac{\mathbf{P}_{t|t-1}\mathbf{z}_t\mathbf{z}_t'\mathbf{P}_{t|t-1}}{q_t}. \tag{6.41}$$

Both of these equations rely on (6.36) and they can be used in both (6.38) and (6.39) in the place of the *true* $\text{E}\left[\boldsymbol{\beta}_t \mid \theta_t, \mathcal{F}_{t-1}\right]$ and $\text{Var}\left[\boldsymbol{\beta}_t \mid \theta_t, \mathcal{F}_{t-1}\right]$. Then, the filtering recursions are as follows:

$$\begin{aligned} \boldsymbol{\beta}_{t|t} &= \boldsymbol{\beta}_{t|t-1} + \mathbf{P}_{t|t-1}\mathbf{z}_t \tfrac{(f_t^* - f_t)}{q_t} \\ \mathbf{P}_{t|t} &= \mathbf{P}_{t|t-1} - \mathbf{P}_{t|t-1}\mathbf{z}_t\mathbf{z}_t'\mathbf{P}_{t|t-1}\tfrac{(1-q_t^*/q_t)}{q_t}, \end{aligned} \tag{6.42}$$

where

$$\begin{aligned} \text{E}\left[g^*(\theta_t) \mid \mathcal{F}_t\right] = f_t^* \\ \text{Var}\left[g^*(\theta_t) \mid \mathcal{F}_t\right] = q_t^*. \end{aligned} \tag{6.43}$$

We see that the filtering recursions resemble those of the linear state space model with the exception of the computation of f_t^* and q_t^*.

The smoothing recursions are similar to those derived for the linear state space model and their proof is based on the application of the linear Bayes methodology [427, p. 532].

6.3.2.2 Posterior Mode Estimation Estimation of the state parameters by approximating the mode of their posterior distribution has been suggested in [136] as a generalization of the extended Kalman filter and smoother (see Problem 7). This estimation method is equivalent to the Fisher scoring algorithm.

Consider the vector of the state vector parameters up to time t, that is $\boldsymbol{\beta}_0^t = (\boldsymbol{\beta}_0, \boldsymbol{\beta}_1, \ldots, \boldsymbol{\beta}_{t-1}, \boldsymbol{\beta}_t)'$, $t = 0, \ldots, N$. Then the posterior distribution of $\boldsymbol{\beta}_0^t$ is proportional to

$$f(\boldsymbol{\beta}_0^t \mid \mathcal{F}_t) \propto \left[\prod_{s=1}^{t} f(y_s \mid \boldsymbol{\beta}_s)\right]\left[\prod_{s=1}^{t} f(\boldsymbol{\beta}_s \mid \boldsymbol{\beta}_{s-1})\right] f(\boldsymbol{\beta}_0), \tag{6.44}$$

by Bayes theorem and (6.20), (6.21). Therefore, the posterior log–likelihood is up to a constant equal to

$$\log f(\boldsymbol{\beta}_0^t \mid \mathcal{F}_t) = \sum_{s=1}^{t} \log f(y_s \mid \boldsymbol{\beta}_s) + \sum_{s=1}^{t} \log f(\boldsymbol{\beta}_s \mid \boldsymbol{\beta}_{s-1}) + \log f(\boldsymbol{\beta}_0). \tag{6.45}$$

Formula (6.45) holds quite generally and leads to

$$\begin{aligned} \log f(\boldsymbol{\beta}_0^t \mid \mathcal{F}_t) &= \sum_{s=1}^{t} l_s(\boldsymbol{\beta}_s) - \frac{1}{2}(\boldsymbol{\beta}_0 - \mathbf{b}_0)'\mathbf{W}_0^{-1}(\boldsymbol{\beta}_0 - \mathbf{b}_0) \\ &- \frac{1}{2}\sum_{s=1}^{t}(\boldsymbol{\beta}_s - \mathbf{F}_s\boldsymbol{\beta}_{s-1})'\mathbf{W}_s^{-1}(\boldsymbol{\beta}_s - \mathbf{F}_s\boldsymbol{\beta}_{s-1}) \end{aligned} \tag{6.46}$$

when both β_0 and $\{\mathbf{w}_t\}$ in (6.33) follow the Gaussian distribution. Maximizing the posterior log likelihood (6.46) yields to the following recursions [136].

Prediction:

$$\begin{aligned} \beta_{t|t-1} &= \mathbf{F}_t \beta_{t-1|t-1} \\ \beta_{0|0} &= \mathbf{b}_0 \\ \mathbf{P}_{t|t-1} &= \mathbf{F}_t \mathbf{P}_{t-1|t-1} \mathbf{F}_t' + \mathbf{W}_t \\ \mathbf{P}_{0|0} &= \mathbf{W}_0 \end{aligned} \tag{6.47}$$

for $t = 1, 2, \ldots, N$.

Filtering:

$$\begin{aligned} \beta_{t|t} &= \beta_{t|t-1} + \mathbf{P}_{t|t} \mathbf{S}_t \\ \mathbf{P}_{t|t} &= \left(\mathbf{P}_{t|t-1}^{-1} + \mathbf{G}_t\right)^{-1} \end{aligned} \tag{6.48}$$

for $t = 1, 2, \ldots, N$, where $\mathbf{S}_t = \partial \log f(Y_t \mid \beta_t, \mathcal{F}_{t-1}) / \partial \beta_t$ and $\mathbf{G}_t = -\mathrm{E}\left[\partial^2 \log f(Y_t \mid \beta_t, \mathcal{F}_{t-1}) / \partial \beta_t \partial \beta_t'\right]$, that is the score and the expected information matrix of Y_t evaluated at $\beta_{t|t-1}$. When the data follow (6.31) then an application of the matrix inversion lemma in Problem 5 leads to an equivalent representation of (6.48) (Problem 9).

Smoothing:

$$\begin{aligned} \beta_{t-1|N} &= \beta_{t-1|t-1} + \mathbf{B}_t \left(\beta_{t|N} - \beta_{t|t-1}\right) \\ \mathbf{P}_{t-1|N} &= \mathbf{P}_{t-1|t-1} + \mathbf{B}_t \left(\mathbf{P}_{t|N} - \mathbf{P}_{t|t-1}\right) \mathbf{B}_t' \end{aligned} \tag{6.49}$$

for $t = N, N-1, \ldots, 1$ and $\mathbf{B}_t = \mathbf{P}_{t-1|t-1} \mathbf{F}_t' \mathbf{P}_{t|t-1}^{-1}$. Derivations of the recursions (6.48) and (6.49) are worked out in [136], [141].

An application of posterior mode estimation to categorical time series is discussed in [137]. Some additional results regarding hyperparameters can be found in [145] and [391]. Related work in [158] considers the problem of approximating posterior moments by applying a Gauss–Hermite procedure.

6.3.2.3 Summary We have described several recursive estimation methods for nonlinear and non-normal state space models some of which are based on approximations. Several additional results are reported in Problems 7 and 8. Over the last decade important advances in computing methods and power have led to a rapid progress in inference for state space models based on simulation methods, a topic that at present is still in a development stage and hence cannot be fully described. The next section discusses some key ideas and results from this unfolding research area.

6.4 SIMULATION BASED METHODS FOR STATE SPACE MODELS

In a broad sense Monte Carlo methods refer to simulation techniques, synonymous with random experiments, for the estimation of parameters of interest. An early example is the well known Buffon's needle experiment for the estimation of π that dates back to the 18th century. Apparently, modern Monte Carlo started in earnest in the 1940s in connection with problems in mathematical physics as we learn from

Metropolis and Ulam [323] and others. Due to the rapid progress in electronic computing, at present there is a vast literature including numerous texts devoted to the study of Monte Carlo simulation methods.

Monte Carlo simulation techniques are frequently employed in the evaluation of integrals of the form

$$\int g(\mathbf{y}) f(\mathbf{y}, \boldsymbol{\theta}) d\mathbf{y},$$

where g is some integrable function and f denotes a probability density. The basic idea is to approximate the above integral by the following sum

$$\frac{1}{m} \sum_{i=1}^{m} g(\mathbf{y}_i),$$

where $\mathbf{y}_1, \mathbf{y}_2, \ldots, \mathbf{y}_m$ are independent and identically distributed realizations from the density f. Then the weak law of large numbers states that as m increases the sum converges to the desired integral [370].

There are potential difficulties with Monte Carlo based inference especially in the context of Bayesian inference where the posterior distribution might include highly correlated parameters or when the number of parameters is prohibitively large. Recent advances in computing power in connection with the introduction of what is known as Markov Chain Monte Carlo (MCMC) methods bypass these problems satisfactorily.

6.4.1 A Brief MCMC Tutorial

Introduced in [322], MCMC is applied whenever we wish to obtain samples from a distribution known up to a constant where the calculation of the constant is formidable. This is done by constructing a Markov chain whose stationary distribution is the desired distribution from which we wish to simulate data. An important application of this is in Bayesian estimation where MCMC methods are used in generating samples from the posterior distribution. In what follows, we review the useful MCMC method referred to as Gibbs sampling.

6.4.1.1 The Gibbs Sampling Algorithm The Gibbs sampling algorithm–introduced in [168]–is perhaps the most extensively used MCMC method. To describe the algorithm, suppose that $f(\boldsymbol{\theta} \mid \mathbf{y})$ denotes a posterior distribution for a p-dimensional parameter vector $\boldsymbol{\theta}$ given data $\mathbf{y}$ and consider the so called full conditionals

$$\begin{array}{l} f(\theta_1 \mid \theta_2, \theta_3, \quad \ldots, \quad \theta_p, \mathbf{y}) \\ f(\theta_2 \mid \theta_1, \theta_3, \quad \ldots, \quad \theta_p, \mathbf{y}) \\ \qquad \vdots \\ f(\theta_p \mid \theta_1, \theta_2, \quad \ldots, \quad \theta_{p-1}, \mathbf{y}). \end{array} \tag{6.50}$$

Under suitable conditions the joint density $f(\boldsymbol{\theta} \mid \mathbf{y})$ is determined by the set of full conditionals (6.50), and simulation of samples from (6.50) facilitates the approximation of the posterior density [86], [167]. More precisely, let $\boldsymbol{\theta}^0$ be a starting value of the parameter vector $\boldsymbol{\theta}$. Then the value of $\boldsymbol{\theta}^1$ is obtained by successive generation of samples from

$$\begin{array}{l} f(\theta_1^1 \mid \theta_2^0, \theta_3^0, \quad \ldots, \quad \theta_p^0, \mathbf{y}) \\ f(\theta_2^1 \mid \theta_1^1, \theta_3^0, \quad \ldots, \quad \theta_p^0, \mathbf{y}) \\ \quad \vdots \\ f(\theta_p^1 \mid \theta_1^1, \theta_2^1, \quad \ldots, \quad \theta_{p-1}^1, \mathbf{y}). \end{array} \tag{6.51}$$

That is, sample θ_1^1 from $f(\theta_1^1 \mid \theta_2^0, \theta_3^0, \ldots, \theta_p^0, \mathbf{y})$, then sample θ_2^1 from $f(\theta_2^1 \mid \theta_1^1, \theta_3^0, \ldots, \theta_p^0, \mathbf{y})$, and so on up to θ_p^1 from $f(\theta_p^1 \mid \theta_1^1, \theta_2^1, \ldots, \theta_{p-1}^1, \mathbf{y})$. This gives the first iteration. As the number of iterations $m \to \infty$, the resulting vector $\boldsymbol{\theta}^m$ is a realization from $f(\boldsymbol{\theta} \mid \mathbf{y})$ [412].

The Gibbs sampler algorithm raises several issues such as the choice of m for the number of iterations, the selection of the *burn–in* period (i.e., dropping the first few iterations due to dependence), taking into account the Markov dependence of the realizations when forming estimators of the posterior moments, choice of starting values, convergence criteria and diagnostic tools; see [162], [173] among others. Moreover, the algorithm requires knowledge of the full conditional distributions. When the full conditionals do not have a known form then the algorithm may be combined with the *rejection sampling* algorithm [370] or with the *adaptive rejection sampling* [172].

6.4.1.2 The Metropolis–Hastings Algorithm

An alternative MCMC method to the Gibbs sampling algorithm when the full conditional distributions are not available in closed from is the Metropolis–Hastings algorithm originally introduced in [322] and later generalized in [204]. The algorithm is outlined in [97].

6.4.2 MCMC Inference for State Space Models

Recall that a state space model is defined through the conditional distribution of the response given the unobserved state vector, say $f(y_t \mid \boldsymbol{\beta}_t)$, the transition density for the unobserved states $f(\boldsymbol{\beta}_t \mid \boldsymbol{\beta}_{t-1})$, and the initial distribution for $\boldsymbol{\beta}_0$, $f(\boldsymbol{\beta}_0)$. As before, set $\boldsymbol{\beta}_0^N = (\boldsymbol{\beta}_0, \boldsymbol{\beta}_1, \ldots, \boldsymbol{\beta}_N)'$. The problem of computing the posterior density $f(\boldsymbol{\beta}_0^N \mid \mathcal{F}_N)$ can be approached by appealing to the Gibbs sampling algorithm whose implementation requires the full conditional densities $f(\boldsymbol{\beta}_t \mid \boldsymbol{\beta}_{-t}, \mathcal{F}_N)$ for $t = 1, \ldots, N$, [83]. The notation $\boldsymbol{\beta}_{-t}$ denotes the set of random variables $\{\boldsymbol{\beta}_s, s \neq t\}$. According to (6.44) we obtain that

$$f(\boldsymbol{\beta}_t \mid \boldsymbol{\beta}_{-t}, \mathcal{F}_N) \propto \begin{cases} f(\boldsymbol{\beta}_{t+1} \mid \boldsymbol{\beta}_t) f(\boldsymbol{\beta}_t) & \text{if } t = 0 \\ f(\boldsymbol{\beta}_{t+1} \mid \boldsymbol{\beta}_t) f(\boldsymbol{\beta}_t \mid \boldsymbol{\beta}_{t-1}) f(y_t \mid \boldsymbol{\beta}_t) & \text{if } t = 1, \ldots, N-1 \\ f(y_t \mid \boldsymbol{\beta}_t) f(\boldsymbol{\beta}_t \mid \boldsymbol{\beta}_{t-1}) & \text{if } t = N. \end{cases} \tag{6.52}$$

In some cases it is simple to sample from the above densities. However there are instances when a rejection sampling or the Metropolis-Hastings algorithm is nested within each Gibbs sampler iteration for obtaining the full conditionals due to the fact that (6.52) is known up to a constant [83].

The following simple but motivating example illustrates the methodology. Consider the so called *random walk* model for $t = 1, \dots, N$

$$\begin{aligned} Y_t &= \beta_t + u_t \\ \beta_t &= \beta_{t-1} + w_t, \end{aligned} \tag{6.53}$$

where u_t are i.i.d. $\mathcal{N}(0, \sigma_u^2)$ and w_t are i.i.d. $\mathcal{N}(0, \sigma_w^2)$. Here both observed and unobserved states are assumed scalars. Suppose that β_0 follows a normal distribution with known mean and variance, say b_0 and σ_0^2, respectively. Suppose further that both σ_u^2 and σ_w^2 follow the inverse Gamma distribution with parameters a_u, b_u and a_w, b_w, respectively. That is,

$$f(\sigma_u^2) = \frac{1}{b_u^{a_u}\Gamma(a_u)}\left(\frac{1}{\sigma_u^2}\right)^{a_u+1} \exp\left(-\frac{1}{b_u\sigma_u^2}\right),$$

and similarly for σ_w^2. The target posterior distribution is given by $f(\boldsymbol{\beta}_0^N, \sigma^2, \sigma_w^2 \mid \mathcal{F}_N)$. Following the Gibbs sampler, it is sufficient to draw samples from the following densities:

- $f(\sigma_u^2 \mid \boldsymbol{\beta}_0^N, \sigma_w^2, \mathcal{F}_N)$
- $f(\sigma_w^2 \mid \boldsymbol{\beta}_0^N, \sigma_v^2, \mathcal{F}_N)$
- $f(\beta_t \mid \boldsymbol{\beta}_{-t}, \sigma_u^2, \sigma_w^2, \mathcal{F}_N)$.

The first two densities are easily sampled since the inverse Gamma distribution is conjugate to the normal with respect to the precision parameter (1/variance). Indeed,

$$\begin{aligned} f(\sigma_u^2 \mid \boldsymbol{\beta}_0^N, \sigma_w^2, \mathcal{F}_N) &\propto f(\sigma_u^2)\left[\prod_{t=1}^{N} f(y_t \mid \beta_t, \sigma_u^2)\right] \\ &\propto \left(\frac{1}{\sigma_u^2}\right)^{a_u+1} \exp\left(-\frac{1}{b_u\sigma_u^2}\right) \\ &\times \left(\frac{1}{\sigma_u^2}\right)^{N/2} \exp\left(-\frac{1}{2\sigma_u^2}\sum_{t=1}^{N}(y_t - \beta_t)^2\right). \end{aligned}$$

The first two factors are due to the functional form of the inverse Gamma while the other two appear as a consequence of the observation model from (6.53). This calculation shows that the full conditional of σ_u^2 is inverse Gamma with parameters $a_u + N/2$ and $\left(1/b_u + \sum_{t=1}^{n}(y_t - \beta_t)^2/2\right)^{-1}$. Similarly the full conditional of σ_w^2 is inverse Gamma with parameters $a_w + N/2$ and $\left(1/b_w + \sum_{t=1}^{n}(\beta_t - \beta_{t-1})^2/2\right)^{-1}$. Turning now to the full conditional of the state parameter given the rest, (6.52) shows that:

- For $t = 0$

$$f(\beta_0 \mid \boldsymbol{\beta}_{-0}, \sigma_u^2, \sigma_w, \mathcal{F}_N) \propto f(\beta_1 \mid \beta_0) f(\beta_0)$$

$$\propto \exp\left[-\frac{1}{2\sigma_w^2}(\beta_1-\beta_0)^2 - \frac{1}{2\sigma_0^2}(\beta_0-b_0)^2\right]$$
$$\propto \exp\left\{-\frac{1}{2}\left[\left(\frac{1}{\sigma_w^2}+\frac{1}{\sigma_0^2}\right)\beta_0^2 - 2\left(\frac{\beta_1}{\sigma_w^2}+\frac{b_0}{\sigma_0^2}\right)\beta_0\right]\right\},$$

which shows that this is a normal density with mean $(\beta_1/\sigma_w^2 + b_0/\sigma_0^2)(1/\sigma_w^2 + 1/\sigma_0^2)^{-1}$ and variance $(1/\sigma_w^2 + 1/\sigma_0^2)^{-1}$.

- For $t = 1, \ldots, N-1$

$$f(\beta_t \mid \boldsymbol{\beta}_{-t}, \sigma_u^2, \sigma_w, \mathcal{F}_N)$$
$$\propto f(\beta_{t+1} \mid \beta_t, \sigma_w^2) f(\beta_t \mid \beta_{t-1}, \sigma_w^2) f(y_t \mid \beta_t, \sigma_u^2)$$
$$\propto \exp\left[-\frac{(\beta_{t+1}-\beta_t)}{2\sigma_w^2} - \frac{(\beta_t-\beta_{t-1})}{2\sigma_w^2} - \frac{(y_t-\beta_t)}{2\sigma_u^2}\right]$$
$$\propto \exp\left\{-\frac{1}{2}\left[\left(\frac{2}{\sigma_w^2}+\frac{1}{\sigma_u^2}\right)\beta_t^2 - 2\left(\frac{(\beta_{t+1}+\beta_{t-1})}{\sigma_w^2}+\frac{y_t}{\sigma_u^2}\right)\beta_t\right]\right\},$$

which is a normal distribution with mean $((\beta_{t+1}+\beta_{t-1})/\sigma_w^2 + y_t/\sigma_u^2)(2/\sigma_w^2 + 1/\sigma_u^2)^{-1}$ and variance $(2/\sigma_w^2 + 1/\sigma_u^2)^{-1}$.

- For $t = N$, it can be shown in the same way that the conditional distribution of β_N given the rest of the parameters is normal with mean $(\beta_{N-1}/\sigma_w^2 + y_N/\sigma_u^2)(1/\sigma_w^2 + 1/\sigma_u^2)^{-1}$ and variance $(1/\sigma_w^2 + 1/\sigma_u^2)^{-1}$.

It follows that a after a large number of iterations the output of the Gibbs sampler is a random draw from the posterior distribution $f(\boldsymbol{\beta}_0^N, \sigma_u^2, \sigma_w^2 \mid \mathcal{F}_N)$. Running the algorithm only K times leads to $k = 1, \ldots, K$ i.i.d. vectors $(\boldsymbol{\beta}_{0,k}^N, \sigma_{u,k}^2, \sigma_{w,k}^2)'$ from the posterior distribution. Hence, an estimate of the smoothing density is given by

$$\hat{f}(\beta_t \mid \sigma_u^2, \sigma_w^2, \mathcal{F}_N) = \frac{1}{K}\sum_{k=1}^{K} f(\beta_t \mid \boldsymbol{\beta}_{-t,k}, \sigma_{u,k}^2, \sigma_{w,k}^2, \mathcal{F}_N).$$

For the linear state space model considered above the full conditionals are given in a closed form, a fact which does not hold in general. If closed from conditionals are not available, then the resulting *output* is still useful for the approximation of the posterior means and covariances provided that these quantities exists. This is so since the available sample is drawn from $f(\boldsymbol{\beta}_t \mid \mathcal{F}_N)$ and therefore standard Monte Carlo integrations shows that the average

$$\frac{1}{K}\sum_{k=1}^{K} \boldsymbol{\beta}_{t,k}$$

approximates the mean $\mathrm{E}[\boldsymbol{\beta}_t \mid \mathcal{F}_N]$. A similar calculation holds for the covariance matrix.

Although the example is about linear models with random hyperparameters, it also shows that this approach covers a broad class of nonlinear models. However, the above method generates the states in a *single* update, that is the states are generated one at a time which results in slow convergence. Slow convergence of single update methods is a typical problem in the application of this methodology (see [84]). In addition, the method does not allow for sequential updating which means that when a new observation becomes available the algorithm needs to restart from the beginning. Because of these issues alternative approaches have been sought to the problem of simulating samples from the posterior density.

The use of *multiple* update algorithms, where the generation of the states proceeds simultaneously by using the time ordering of the state space model and sampling from $f(\beta_0^N \mid \mathcal{F}_N)$ instead of sampling one at a time from (6.52), has been suggested to address the problem of slow convergence [84], [85], [159] [382]. At least empirically, the convergence of multiple update algorithms is faster when compared to single update algorithms. In particular, efficient MCMC inference for the class of normal dynamic linear models is based on the idea of sampling the entire set of state vectors β_0^N which can be accomplished by employing the Markovian structure of the model

$$f(\beta_0^N \mid \mathcal{F}_N) = f(\beta_N \mid \mathcal{F}_N) \prod_{t=1}^{N-1} f(\beta_t \mid \beta_{t+1}, \mathcal{F}_{t-1}). \tag{6.54}$$

Equation (6.54) indicates that β_0^N can be sampled efficiently by generating β_t, $t = 0, \ldots, N$ backwards. Furthermore, if the Gaussian linear dynamic model holds, then all the densities in (6.54) are Gaussian so it is sufficient to compute their means and variances. However for $t = N-1, N-2, \ldots, 1, 0$ (see Problem 11)

$$\begin{aligned} \mathrm{E}[\beta_t \mid \beta_{t+1}, \mathcal{F}_{t-1}] &= \beta_{t|t} + \mathbf{K}_t(\beta_{t+1} - \beta_{t+1|t}) \\ \mathrm{Var}[\beta_t \mid \beta_{t+1}, \mathcal{F}_{t-1}] &= \mathbf{P}_{t|t} - \mathbf{K}_t \mathbf{P}_{t+1|t} \mathbf{K}_t', \end{aligned} \tag{6.55}$$

where $\mathbf{K}_t$ is the Kalman gain defined in (6.12). Thus the algorithm for linear dynamic models runs as follows:

1. Run the standard Kalman filter for $t = 0, \ldots, N$ to obtain prediction and filtering estimates.

2. Sample β_N form the normal distribution with mean $\beta_{N|N}$ and variance $\mathbf{P}_{N|N}$.

3. Sample β_t for $t = N-1, N-2, , \ldots, 1, 0$ from the normal distribution with mean and variance given by (6.55), respectively.

This algorithm is termed forward–filtering (step 1) backward–sampling (steps 2 and 3).

An intermediate approach between single and multiple update algorithms divides the state vector into several blocks and is called *block move* algorithm [384]. Another alternative to the problem of slow convergence reparametrizes the model in terms of independent system disturbances and is applicable to time series that follow generalized linear models [163] (see also [116]). Both of these methods require Fisher scoring steps which can be avoided according to a block move algorithm proposed in [270]. More recently, Durbin and Koopman [130] building on an earlier work in [129] consider the analysis of non-Gaussian state space models using importance sampling and antithetic variables without resorting to any MCMC methods.

The introduction of MCMC techniques had a profound impact on the analysis of state space models over the last decade. An indication of this we see from the following diverse applications. Single move Gibbs sampler applied to non-Gaussian observations [138], consideration of importance sampling algorithms and adaptive density estimators for inference in dynamic nonlinear models [426], implementation of MCMC in modeling categorical time series [82], application of Gibbs sampling to random level–shift models, additive outliers and missing values in autoregression [314], use of MCMC inference in the analysis of stochastic volatility models [226], and an application of MCMC in the study of autoregressive time series subject to regime switching [12]. Other examples include the application of Bayesian mixture models in nonlinear analysis of autoregressive processes [332], and certain extensions of Gibbs sampling [84], [95], [96]. More recently, Cargnoni et al. [81] applied MCMC inference to multinomial observations, Gerlach et al. [170] proposed an efficient MCMC algorithm for the estimation of linear Gaussian state space models generalizing results in [85], and Frühwirth-Schnatter [160] introduced permutation sampling in switching and mixture models.

6.4.3 Sequential Monte Carlo Sampling Methods

The preceding discussion advocates the use of multiple update algorithms instead of single update methods as far as convergence is concerned. However multiple update algorithms are not recursive as they do not allow sequential processing of the data. To amend this, Gordon et al. [178] and Kitagawa [263], among others, suggested the so called *particle filter* method whose inception can be traced back to Handschin et al. [195] and Handschin [194]. The recent review article in [127] and the recent collections in [126] and [266] manifest the importance of this methodology and list additional references.

Following [127] and recalling that $\beta_0^t = (\beta_0, \ldots, \beta_t)'$, $\mathcal{F}_t = \{Y_1, \ldots, Y_t\}$ and (6.20) (6.21), the following calculation leads to a recursive formula for $f(\beta_0^{t+1} \mid \mathcal{F}_{t+1})$:

$$
\begin{aligned}
f(\beta_0^{t+1} \mid \mathcal{F}_{t+1}) &= \frac{f(\mathcal{F}_{t+1} \mid \beta_0^{t+1}) f(\beta_0^{t+1})}{f(\mathcal{F}_{t+1})} \\
&= \frac{f(\mathcal{F}_t \mid \beta_0^t) f(y_{t+1} \mid \beta_{t+1}) f(\beta_0^t) f(\beta_{t+1} \mid \beta_t)}{f(y_{t+1} \mid \mathcal{F}_t) f(\mathcal{F}_t)} \\
&= f(\beta_0^t \mid \mathcal{F}_t) \frac{f(y_{t+1} \mid \beta_{t+1}) f(\beta_{t+1} \mid \beta_t)}{f(y_{t+1} \mid \mathcal{F}_t)}. \qquad (6.56)
\end{aligned}
$$

The first equation is based on Bayes theorem and the second on model assumptions (6.20) and (6.21). The denominator of (6.56) cannot be calculated in closed form but an appeal to Bayesian importance sampling (Problem 12) which consists of obtaining a sample from an importance density $\pi()$ with the property

$$\pi(\beta_0^t \mid \mathcal{F}_t) = \pi(\beta_0) \prod_{s=1}^{t} \pi(\beta_s \mid \beta_0^{s-1}, \mathcal{F}_s) \tag{6.57}$$

yields the following algorithm [127].

Sequential Importance Sampling (SIS):

Let $t = 0, 1, 2, \ldots\ldots, N$.

- For $i = 1, \ldots, n$ draw a sample $\beta_{t,i}$ from $\pi(\beta_t \mid \beta_{0,i}^{t-1}, \mathcal{F}_t)$ and put $\beta_{0,i}^t = (\beta_{0,i}^{t-1}, \beta_{t,i})$.
- For $i = 1, \ldots, n$ calculate the importance weights up to a normalizing constant

$$\hat{w}_{t,i} = \hat{w}_{t-1,i} \frac{f(y_t \mid \beta_{t,i}) f(\beta_{t,i} \mid \beta_{t-1,i})}{\pi(\beta_{t,i} \mid \beta_{0,i}^{t-1}, \mathcal{F}_t)}. \tag{6.58}$$

- For $i = 1, \ldots, n$, normalize the importance weights by

$$w_{t,i} = \frac{\hat{w}_{t,i}}{\sum_{l=1}^{n} \hat{w}_{t,l}}. \tag{6.59}$$

It is instructive to consider the above recursions in some detail. Suppose that at time $t = 0$, a sample of size n is available from $\pi(\beta_0)$. Then (6.58) implies that the importance weights are given up to a normalizing constant by

$$\hat{w}_{0,i} = \frac{f(\beta_{0,i})}{\pi(\beta_{0,i})}$$

and therefore the normalizing weights are

$$w_{0,i} = \frac{\hat{w}_{0,i}}{\sum_{l=1}^{n} \hat{w}_{0,l}}.$$

That is the standard importance sampling scheme to obtain a random draw from $f(\beta_0)$ since it is the desired target density at $t = 0$. At time $t = 1$, draw a random sample of size n from the importance density $\pi(\beta_1 \mid \beta_{0,i}, \mathcal{F}_1)$ and set

$$\hat{w}_{1,i} = \hat{w}_{0,i} \frac{f(y_1 \mid \beta_{1,i}) f(\beta_{1,i} \mid \beta_{0,i})}{\pi(\beta_{1,i} \mid \beta_{0,i}, \mathcal{F}_1)}.$$

The recursion is a consequence of (6.56) which states that the joint posterior density of (β_0, β_1) is proportional to the product of $f(\beta_0)$ and $f(y_1 \mid \beta_1) f(\beta_1 \mid \beta_0)$. Since the denominator of (6.56) cannot be calculated explicitly, the importance weights need to be calculated up to a normalizing constant first and then be normalized. Hence (6.58) and (6.59) follow and therefore $(\beta_{0,i}, \beta_{1,i})$ is a random sample from $\pi(\beta_0^1 \mid \mathcal{F}_1)$. The normalized weights at $t = 1$ are given by

$$w_{1,i} = \frac{\hat{w}_{1,i}}{\sum_{l=1}^{n} \hat{w}_{1,l}}.$$

Having available the normalized weights $\{w_{1,i}\}$ and the random sample $(\beta_{0,i}, \beta_{1,i})$, $i = 1, \ldots, n$, estimation of the following integral

$$I = \int \mathbf{h}(\beta_0, \beta_1) f(\beta_0^1 \mid \mathcal{F}_1) d\beta_0^1$$

is accomplished by the sum

$$\sum_{i=1}^{n} \mathbf{h}(\beta_{0,i}, \beta_{1,i}) w_{1,i},$$

provided that the integral exists.

The iteration process continues until time $t = N$, leading to a sample which is used in the estimation of expectations of $f(\beta_0^N \mid \mathcal{F}_N)$. The above discussion points to the advantage of sequential importance sampling when compared to MCMC methods. Namely, sequential importance sampling is an *on–line* estimation procedure with the advantage that when a new observation becomes available the process need not start from the beginning.

The choice of the importance function is crucial and it is sensible to choose an importance function so that the variance of the importance weights (6.58) conditional on $\beta_{0,i}^{t-1}$ and $\mathcal{F}_t$ is minimized. Otherwise the algorithm is degenerate in the sense that all but one of the normalized weights (6.59) approach zero after a few iterations. It can be proved that $\pi(\beta_t \mid \beta_{0,i}^{t-1}, \mathcal{F}_t) = f(\beta_t \mid \beta_{0,i}^{t-1}, \mathcal{F}_t)$ is the *optimal* importance function (Problem 13). This choice of importance function has been considered by various authors including [94]. However, there are limitations on the choice of this importance function and several alternative strategies have been considered [44], [127], [263], [304], [410].

To avoid degeneracy of the sequential importance sampling algorithm several authors have considered resampling methods where the key idea is to resample from the generated sample under a predetermined condition [373]. See also [127], [211], [216], [271], [304], and [349] among others.

The simulated sample derived either by sequential importance sampling or by any other resampling method is used for the purpose of prediction, filtering, and smoothing. Consider the problem of filtering, that is, estimation of the density $f(\beta_t \mid \mathcal{F}_t)$. Using (6.59), an estimate of $f(\beta_0^t \mid \mathcal{F}_t)$ is given by

$$\hat{f}(\beta_0^t) = \sum_{i=1}^{n} w_{t,i} \delta_{\beta_{0,i}^t}(\beta_0^t),$$

for *any* $t = 0, \ldots, N$, where $\delta_{\beta_{0,i}^t}(\beta_0^t)$ denotes a point mass at the vector point. Consequently an estimator of the filtering density $f(\beta_t \mid \mathcal{F}_t)$ is obtained by keeping only the corresponding simulated component $\boldsymbol{\beta}_{t,i}$,

$$\hat{f}(\boldsymbol{\beta}_t) = \sum_{i=1}^{n} w_{t,i}\delta_{\boldsymbol{\beta}_{t,i}}(\boldsymbol{\beta}_t).$$

Similarly the prediction density $f(\boldsymbol{\beta}_t \mid \mathcal{F}_{t-1})$ is approximated by

$$\hat{f}(\boldsymbol{\beta}_t \mid \mathcal{F}_{t-1}) = \sum_{i=1}^{n} w_{t-1,i}\delta_{\boldsymbol{\beta}_{t,i}}(\boldsymbol{\beta}_t),$$

where $\boldsymbol{\beta}_{t,i}$ has been drawn from $f(\boldsymbol{\beta}_t \mid \boldsymbol{\beta}_{t-1,i})$. Similar recursions hold for smoothing [127].

6.4.4 Likelihood Inference

In applications both observation and transition densities may depend on unknown parameters as illustrated by the example in Section 6.4.2. Thus, suppose that the observation density $f(y_t; \boldsymbol{\theta}_1 \mid \boldsymbol{\beta}_t, \mathcal{F}_{t-1})$ depends on $\boldsymbol{\theta}_1$, and the transition density $f(\boldsymbol{\beta}_t, \boldsymbol{\theta}_2 \mid \boldsymbol{\beta}_{t-1})$ depends on $\boldsymbol{\theta}_2$. Regarding both $\boldsymbol{\theta}_1$ and $\boldsymbol{\theta}_2$ as random variables with independent prior distributions $f(\boldsymbol{\theta}_1)$ and $f(\boldsymbol{\theta}_2)$, respectively, then the full conditionals

$$f(\boldsymbol{\theta}_1 \mid \boldsymbol{\beta}_0^N, \mathcal{F}_N, \boldsymbol{\theta}_2) \propto f(\boldsymbol{\theta}_1) \left[\prod_{t=1}^{N} f(y_t; \boldsymbol{\theta}_1 \mid \boldsymbol{\beta}_t)\right]$$

and

$$f(\boldsymbol{\theta}_2 \mid \boldsymbol{\beta}_0^N, \mathcal{F}_N, \boldsymbol{\theta}_2) \propto f(\boldsymbol{\theta}_2) \left[f(\boldsymbol{\beta}_0, \boldsymbol{\theta}_2) \prod_{t=1}^{N} f(\boldsymbol{\beta}_t; \boldsymbol{\theta}_2 \mid \boldsymbol{\beta}_{t-1})\right]$$

can be used in inference about $\boldsymbol{\theta}_1$ and $\boldsymbol{\theta}_2$, respectively. This approach was already discussed in Section 6.4.2 where the MCMC methodology was applied to a linear state space model. When in addition the response belongs to the exponential family then a good choice of the conjugate prior eases the computation of the posterior density.

Regarding the general likelihood

$$\begin{aligned} f(y_1, \ldots, y_N) &= \prod_{t=1}^{N} f(y_t \mid \mathcal{F}_{t-1}) \\ &= \prod_{t=1}^{N} \int f(y_t \mid \boldsymbol{\beta}_t) f(\boldsymbol{\beta}_t \mid \mathcal{F}_{t-1}) d\boldsymbol{\beta}_t, \end{aligned} \tag{6.60}$$

MCMC inference can be avoided as noted by several authors who consider approximating the likelihood by sequential importance sampling as described in Section 6.4.3. See [127], [216], and [264]. The asymptotic behavior of the maximum likelihood estimator for state space models has been discussed in [228].

6.4.5 Longitudinal Data

State space modeling is also useful in the analysis of longitudinal data. Jones [233] shows that the linear mixed model has a state space representation and develops its inference under normality. More recently, multivariate state space models for mixed responses–both continuous and discrete responses–using exponential dispersion models have been studied in [235], and in [236] the authors consider a state space model for *multivariate* count data (Problem 14). The recent volume [121] includes further results on generalized linear models for longitudinal data from a Bayesian perspective.

6.5 KALMAN FILTERING IN SPACE-TIME DATA

Recently there has been a growing interest in the application and extension of dynamic models to space–time data such as environmental data. Non-Bayesian models for space-time data have been considered by Huang and Cressie [215] who analyze snow–water equivalent data using a vector autoregressive process with spatially independent innovations, and by Mardia et al. [308] who combine kriging and state space models. Wikle and Cressie [432] (see also [73]) propose a model suitable for a large number of sites whereby the observations are generated as a sum of an unobserved process which incorporates space–time dependence and an unobserved process which is spatially and temporally uncorrelated. Kalman recursions for space-time models within the Bayesian framework are discussed in [142], [375], [406], and [431] among others.

6.6 PROBLEMS AND COMPLEMENTS

1. Verify by substitution that the ARMA(1, 1) process

$$Y_t = \phi Y_{t-1} + \theta v_{t-1} + v_t, \quad t = 0, \pm 1, \pm 2, ...,$$

has the state space representation

$$\begin{aligned} Y_t &= \beta_t + v_t \\ \beta_t &= \phi\beta_{t-1} + (\theta + \phi) v_{t-1}. \end{aligned}$$

2. Consider a slight reinterpretation of the general state space model (6.21), (6.22) as in

General Observation equation:	$Y_t \mid \boldsymbol{\beta}_t \sim f(y_t \mid \boldsymbol{\beta}_t, \mathcal{F}_{t-1})$
General System equation:	$\boldsymbol{\beta}_t \mid \boldsymbol{\beta}_{t-1} \sim f(\boldsymbol{\beta}_t \mid \boldsymbol{\beta}_{t-1})$
Initial information:	$\boldsymbol{\beta}_0 \sim f(\boldsymbol{\beta}_0 \mid \mathcal{F}_0) \equiv f(\boldsymbol{\beta}_0).$

Prove the following recursions in k:

(a) Prediction densities for the states

$$f(\boldsymbol{\beta}_{t+k} \mid \mathcal{F}_t) = \int f(\boldsymbol{\beta}_{t+k} \mid \boldsymbol{\beta}_{t+k-1}) f(\boldsymbol{\beta}_{t+k-1} \mid \mathcal{F}_t) d\boldsymbol{\beta}_{t+k-1}.$$

(b) Prediction densities for the observations

$$f(y_{t+k} \mid \mathcal{F}_t) = \int f(y_{t+k} \mid \boldsymbol{\beta}_{t+k}) f(\boldsymbol{\beta}_{t+k} \mid \mathcal{F}_t) d\boldsymbol{\beta}_{t+k}.$$

Further results and alternative computational methods for filtering and smoothing densities are given in [280].

3. Suppose that the vector $\mathbf{Y}_t = (Y_{t1}, \ldots, Y_{tk})'$ follows the multinomial distribution

$$f(\mathbf{y}_t \mid \boldsymbol{\pi}_t) = \frac{n_t!}{\prod_{j=1}^k y_{tj}!} \prod_{j=1}^k \pi_{tj}^{y_{tj}},$$

for $t = 1, \ldots, N$, where $n_t = \sum_{j=1}^k Y_{tj}$ and $\boldsymbol{\pi}_t = (\pi_{t1}, \ldots, \pi_{tk})'$. Suppose that the distribution of $\boldsymbol{\pi}_{t-1}$ given $\mathcal{F}_{t-1}$ is Dirichlet, that is,

$$f(\boldsymbol{\pi}_{t-1} = \boldsymbol{\pi} \mid \mathcal{F}_{t-1}) = C(\mathbf{r}_t) \prod_{j=1}^k \pi^{r_{tj}},$$

where $\mathbf{r}_t = (r_{t1}, \ldots, r_{tk})'$ and $\boldsymbol{\pi}$ belongs to k–dimensional simplex.

(a) Calculate the distribution of $\mathbf{Y}_t$ given $\mathcal{F}_{t-1}$.

(b) Calculate the distribution of $\boldsymbol{\pi}_t$ given $\mathcal{F}_t$.

4. *Linear Bayes Methodology* [199], [427, Ch. 4.9.2]. Assume that the joint distribution of data $\mathbf{Y}$ and a parameter $\boldsymbol{\theta}$ is partially specified in terms of the first and second moments such that

$$\mathrm{E}\{(\boldsymbol{\theta}, \mathbf{Y})'\} = (\boldsymbol{\mu}_{\boldsymbol{\theta}}, \boldsymbol{\mu}_{\mathbf{Y}})',$$

and

$$\mathrm{Var}\{(\boldsymbol{\theta}, \mathbf{Y})'\} = \begin{pmatrix} \boldsymbol{\Sigma}_\theta & \mathbf{A}\boldsymbol{\Sigma}_{\mathbf{Y}} \\ \boldsymbol{\Sigma}_{\mathbf{Y}}\mathbf{A}' & \boldsymbol{\Sigma}_{\mathbf{Y}} \end{pmatrix}.$$

Since the posterior risk cannot be calculated, consider the overall risk

$$r(\mathbf{d}) = \mathrm{trace}\mathrm{E}\left[(\boldsymbol{\theta} - \mathbf{d})(\boldsymbol{\theta} - \mathbf{d})'\right],$$

where the expectation is taken *unconditional* on $\mathbf{Y}$. Show that if $\mathbf{d}$ is a linear function of the data such that $\mathbf{d} = \mathbf{h} + \mathbf{H}\mathbf{Y}$, then a linear Bayes estimate of $\boldsymbol{\theta}$ in the sense of minimizing $r(\mathbf{d})$ is given by

$$\mathbf{m} = \boldsymbol{\mu}_{\boldsymbol{\theta}} + \mathbf{A}(\mathbf{Y} - \boldsymbol{\mu}_{\mathbf{Y}})$$

with associated risk matrix

$$\mathbf{C} = \mathbf{\Sigma}_\theta - \mathbf{A}\mathbf{\Sigma}_\mathbf{Y}\mathbf{A}'.$$

Thus the minimum risk is equal to the trace of $\mathbf{C}$.

5. *Matrix inversion lemma.* Let $\mathbf{P}$, $\mathbf{U}$, $\mathbf{R}$ be $p \times p$, $p \times p$ and $k \times k$ symmetric matrices, respectively, and $\mathbf{H}$ an arbitrary $k \times p$ matrix. Assuming that the indicated inverses exist, prove the following matrix relations [265, p. 85], [366, p. 33].

   (a) $\left(\mathbf{P}^{-1} + \mathbf{H}'\mathbf{R}^{-1}\mathbf{H}\right)^{-1} = \mathbf{P} - \mathbf{P}\mathbf{H}'\left(\mathbf{H}\mathbf{P}\mathbf{H}' + \mathbf{R}\right)^{-1}\mathbf{H}\mathbf{P}$

   (b) $\left(\mathbf{P}^{-1} + \mathbf{H}'\mathbf{R}^{-1}\mathbf{H}\right)^{-1}\mathbf{H}'\mathbf{R}^{-1} = \mathbf{P}\mathbf{H}'\left(\mathbf{H}\mathbf{P}\mathbf{H}' + \mathbf{R}\right)^{-1}$

   (c) $\left(\mathbf{P}^{-1} - \mathbf{U}^{-1}\right)^{-1} = \mathbf{P}\left(\mathbf{U} - \mathbf{P}\right)^{-1}\mathbf{U}$

   (d) $\mathbf{U} + \left(\mathbf{P}^{-1} - \mathbf{U}^{-1}\right)^{-1} = \mathbf{U}\left(\mathbf{U} - \mathbf{P}\right)^{-1}\mathbf{U}$

6. *Lag-One Covariance Smoother* [388, p. 320]. With the Kalman gain $\mathbf{K}_t$ (6.12), $\mathbf{B}_t$ in (6.15), and

$$\mathbf{P}_{t_1,t_2|s} = \mathrm{E}[(\boldsymbol{\beta}_{t_1} - \boldsymbol{\beta}_{t_1|s})(\boldsymbol{\beta}_{t_2} - \boldsymbol{\beta}_{t_2|s})'],$$

   verify that

$$\mathbf{P}_{n,n-1|n} = (\mathbf{I} - \mathbf{K}_n\mathbf{z}_n')\mathbf{F}_t\mathbf{P}_{n-1|n-1}$$

   and show that for $t = n, n-1, \ldots, 2$,

$$\mathbf{P}_{t-1,t-2|n} = \mathbf{P}_{t-1|t-1}\mathbf{B}_{t-1}' + \mathbf{B}_t[\mathbf{P}_{t,t-1|n} - \mathbf{F}_t\mathbf{P}_{t-1|t-1}]\mathbf{B}_{t-1}'.$$

7. *Extended Kalman Filter* [18, Ch. 8.2], [427, Ch. 13.2]. Consider the following *non-linear* state space model

$$\begin{aligned} Y_t &= \mathbf{h}_t(\boldsymbol{\beta}_t) + v_t \\ \boldsymbol{\beta}_t &= \mathbf{f}_t(\boldsymbol{\beta}_{t-1}) + \mathbf{w}_t, \end{aligned} \tag{6.61}$$

   for $t = 1, \ldots, N$ and known functions $\mathbf{h}_t$, $\mathbf{f}_t$. Assume that $\{v_t\}$ and $\{\mathbf{w}_t\}$ are independent and both follow Gaussian distributions $\mathcal{N}(0, \sigma_t^2)$ and $\mathcal{N}_p(\mathbf{0}, \mathbf{W}_t)$, respectively. Also, suppose that $\boldsymbol{\beta}_0$ follows the normal distribution with mean $\mathbf{0}$ and covariance matrix $\mathbf{W}_0$ and is independent of $\{v_t\}$ and $\{\mathbf{w}_t\}$. By assuming differentiability of $\mathbf{h}_t$ and $\mathbf{f}_t$ and using Taylor expansion show that model (6.61) is equivalent to

$$\begin{aligned} Y_t &= \mathbf{H}_t'\boldsymbol{\beta}_{t-1} + \left(\mathbf{h}_t(\boldsymbol{\beta}_{t|t-1}) - \mathbf{H}_t'\boldsymbol{\beta}_{t|t-1}\right) + v_t, \\ \boldsymbol{\beta}_t &= \mathbf{F}_t\boldsymbol{\beta}_{t-1} + \left(\mathbf{f}_t(\boldsymbol{\beta}_{t-1|t-1}) - \mathbf{F}_t\boldsymbol{\beta}_{t-1|t-1}\right) + \mathbf{w}_t, \end{aligned} \tag{6.62}$$

   where

$$\mathbf{F}_t = \frac{\partial \mathbf{f}_t(\boldsymbol{\beta})}{\partial \boldsymbol{\beta}}\Big|_{\boldsymbol{\beta}_{t-1|t-1}} \quad \text{and} \quad \mathbf{H}_t' = \frac{\partial \mathbf{h}_t(\boldsymbol{\beta})}{\partial \boldsymbol{\beta}}\Big|_{\boldsymbol{\beta}_{t|t-1}}.$$

Thus conclude that the so called extended Kalman filter is given by the following recursions.

$$\begin{aligned}
\boldsymbol{\beta}_{t|t-1} &= \mathbf{f}_t(\boldsymbol{\beta}_{t-1|t-1}) \\
\mathbf{P}_{t|t-1} &= \mathbf{F}_t \mathbf{P}_{t-1|t-1} \mathbf{F}_t' + \mathbf{W}_t \\
\boldsymbol{\beta}_{t|t} &= \boldsymbol{\beta}_{t|t-1} + \mathbf{B}_t \left(Y_t - \mathbf{h}_t(\boldsymbol{\beta}_{t|t-1}) \right) \\
\mathbf{P}_{t|t} &= \left(\mathbf{I} - \mathbf{B}_t \mathbf{H}_t' \right) \mathbf{P}_{t|t-1},
\end{aligned}$$

where the Kalman gain is given by

$$\mathbf{B}_t = \mathbf{P}_{t|t-1} \mathbf{H}_t \left(\mathbf{H}_t' \mathbf{P}_{t|t-1} \mathbf{H}_t + \sigma_t^2 \right)^{-1}.$$

8. *Gaussian sum approximation* [14], [18, Ch. 8.4], [427, Ch. 12]. The rationale of the Gaussian sum approximation is the fact that any probability density on R^n is approximated by a Gaussian mixture in $L_1(R^n)$ distance. Thus, let

$$f(\mathbf{x}) = \sum_{i=1}^{m} \alpha_i \phi\left(\mathbf{x}; \boldsymbol{\mu}_i, \boldsymbol{\Sigma}_i\right),$$

where $\phi\left(\mathbf{x}; \boldsymbol{\mu}_i, \boldsymbol{\Sigma}_i\right)$ is the density of an n–dimensional normal random variable with mean $\boldsymbol{\mu}_i$ and positive definite covariance matrix $\boldsymbol{\Sigma}_i$, $\alpha_i > 0$ such that $\sum_i \alpha_i = 1$, and $f(.)$ is a density function. Show that if $\mathbf{X}$ is a random vector with pdf $f(\mathbf{x})$ then

$$\mathrm{E}[\mathbf{X}] = \sum_{i=1}^{m} \alpha_i \boldsymbol{\mu}_i \tag{6.63}$$

$$\mathrm{Var}[\mathbf{X}] = \sum_{i=1}^{m} \alpha_i \left(\boldsymbol{\Sigma}_i + (\mathrm{E}[\mathbf{X}] - \boldsymbol{\mu}_i)^2 \right). \tag{6.64}$$

The above equations facilitate calculation of posterior means and covariance matrices due to the next two theorems. Indeed, show that:

(a) With state space model (6.61) and with

$$f(\boldsymbol{\beta}_t \mid \mathcal{F}_{t-1}) = \sum_{i=1}^{m} \alpha_{i(t-1)} \phi(\boldsymbol{\beta}_t, \bar{\boldsymbol{\mu}}_{it}, \bar{\boldsymbol{\Sigma}}_{it})$$

the updated density $f(\boldsymbol{\beta}_t \mid \mathcal{F}_t)$ approaches the Gaussian mixture

$$\sum_{i=1}^{m} \alpha_{it} \phi(\boldsymbol{\beta}_t, \boldsymbol{\mu}_{it}, \boldsymbol{\Sigma}_{it})$$

uniformly in $\boldsymbol{\beta}_t$ and Y_t as $\bar{\boldsymbol{\Sigma}}_{it} \to 0$ for $i = 1, 2, \ldots, m$, where

$$\begin{aligned}
\boldsymbol{\mu}_{it} &= \bar{\boldsymbol{\mu}}_{it} + \mathbf{K}_{it}\left(Y_t - \mathbf{h}_t(\bar{\boldsymbol{\mu}}_{it})\right) \\
\boldsymbol{\Sigma}_{it} &= \bar{\boldsymbol{\Sigma}}_{it} - \bar{\boldsymbol{\Sigma}}_{it}\mathbf{H}_{it}\left(\mathbf{H}'_{it}\bar{\boldsymbol{\Sigma}}_{it}\mathbf{H}_{it} + \sigma_t^2\right)^{-1}\mathbf{H}'_{it}\bar{\boldsymbol{\Sigma}}_{it} \\
\mathbf{K}_{it} &= \bar{\boldsymbol{\Sigma}}_{it}\mathbf{H}_{it}\left(\mathbf{H}'_{it}\bar{\boldsymbol{\Sigma}}_{it}\mathbf{H}_{it} + \mathbf{W}_t\right)^{-1} \\
\alpha_{it} &= \frac{\alpha_{i(t-1)}\phi\left(\mathbf{h}_t(\bar{\boldsymbol{\mu}}_{it}), \mathbf{H}'_{it}\bar{\boldsymbol{\Sigma}}_{it}\mathbf{H}_{it} + \mathbf{W}_t\right)}{\sum_{i=1}^{m}\alpha_{i(t-1)}\phi\left(\mathbf{h}_t(\bar{\boldsymbol{\mu}}_{it}), \mathbf{H}'_{it}\bar{\boldsymbol{\Sigma}}_{it}\mathbf{H}_{it} + \mathbf{W}_t\right)} \\
\mathbf{H}'_{it} &= \frac{\partial \mathbf{h}_t(\boldsymbol{\beta})}{\partial \boldsymbol{\beta}}|_{\boldsymbol{\beta}=\bar{\boldsymbol{\mu}}_{ti}}.
\end{aligned} \tag{6.65}$$

(b) With state space model (6.61) and with

$$f(\boldsymbol{\beta}_t \mid \mathcal{F}_t) = \sum_{i=1}^{m} \alpha_{it}\phi(\boldsymbol{\beta}_t, \bar{\boldsymbol{\mu}}_{it}, \bar{\boldsymbol{\Sigma}}_{it})$$

the one–step ahead prediction density $f(\boldsymbol{\beta}_{t+1} \mid \mathcal{F}_t)$ approaches the Gaussian mixture

$$\sum_{i=1}^{m} \alpha_{it}\phi(\boldsymbol{\beta}_t, \boldsymbol{\mu}_{i(t+1)}, \boldsymbol{\Sigma}_{i(t+1)})$$

uniformly in $\boldsymbol{\beta}_t$ as $\bar{\boldsymbol{\Sigma}}_{it} \to 0$ for $i = 1, 2, \ldots, m$, where

$$\begin{aligned}
\boldsymbol{\mu}_{i(t+1)} &= \mathbf{f}_t(\bar{\boldsymbol{\mu}}_{it}) \\
\boldsymbol{\Sigma}_{i(t+1)} &= \mathbf{F}_{it}\bar{\boldsymbol{\Sigma}}_{it}\mathbf{F}'_{it} + \mathbf{W}_t \\
\mathbf{F}_{it} &= \frac{\partial \mathbf{f}_t(\boldsymbol{\beta})}{\partial \boldsymbol{\beta}}|_{\boldsymbol{\beta}=\bar{\boldsymbol{\mu}}_{ti}}.
\end{aligned} \tag{6.66}$$

Some early references are [198] and [396]. Further properties of mixtures in state space modeling are studied in [427, Ch. 12].

9. An alternative form of the smoothing recursions (6.48) can be derived with the aid of the matrix inversion lemma discussed in Problem 5 and the properties of the exponential family. More specifically show that when the data follow the exponential family of distributions then (6.48) are modified as

$$\begin{aligned}
\boldsymbol{\beta}_{t|t} &= \boldsymbol{\beta}_{t|t-1} + \mathbf{B}_t\left(Y_t - \mu_t\right) \\
\mathbf{P}_{t|t} &= \left(\mathbf{I} - \mathbf{B}_t\frac{\partial \mu_t}{\partial \eta_t}\mathbf{z}'_t\right)\mathbf{P}_{t|t-1},
\end{aligned}$$

where μ_t is the conditional expectation of the response and $\eta_t = \mathbf{z}'_t\boldsymbol{\beta}_{t|t-1}$. The matrix $\mathbf{B}_t$ is referred as *Kalman gain* and is equal to

$$\mathbf{P}_{t|t-1}\mathbf{z}_t\frac{\partial \mu_t}{\partial \eta_t}\left(\mathbf{z}'_t\left(\frac{\partial \mu_t}{\partial \eta_t}\right)^2\mathbf{P}_{t|t-1}\mathbf{z}_t + \sigma_t^2\right)^{-1},$$

where σ_t^2 is the conditional variance of Y_t. Notice that a similar recursion holds for multivariate data.

10. Consider a count time series $\{Y_t\}, t = 1, \ldots, N$ such that

$$f(y_t \mid \beta_t) = \frac{\beta_t^{y_t} exp(-\beta_t)}{y_t!},$$

where β_t $t = 1, \ldots, N$ is a sequence of unobserved states. Assume further that the distribution of β_{t-1} is Gamma with parameters $a = a_{t-1|t-1}$ and $b = b_{t-1|t-1}$ such that

$$f(\beta_{t-1} \mid \mathcal{F}_{t-1}) = \frac{\exp(-b\beta_{t-1})\beta_{t-1}^{a-1}}{\Gamma(a)b^a}.$$

Suppose that β_t given $\mathcal{F}_{t-1}$ has also a Gamma distribution with parameters $a_{t|t-1} = \omega a_{t-1|t-1}$ and $b_{t|t-1} = \omega b_{t-1|t-1}$ for some $\omega \in (0, 1]$.

   (a) Calculate $\mathrm{E}[\beta_t \mid \mathcal{F}_{t-1}]$ and $\mathrm{Var}[\beta_t \mid \mathcal{F}_{t-1}]$ and compare your answer with the standard dynamic linear model recursions.

   (b) Compute $f(\beta_t \mid \mathcal{F}_t)$.

   (c) Calculate the predictive p.d.f. $f(y_t \mid \mathcal{F}_{t-1})$ and find its mean and variance.

Dynamic models for count time series with conjugate priors have been considered by Harvey and Fernandes [201] who also develop parallel methodology for binomial, multinomial and negative binomial responses.

11. Under the Gaussian linear state space model show that the means and variances of the conditional densities $f(\boldsymbol{\beta}_t \mid \boldsymbol{\beta}_{t+1}, \mathcal{F}_{t-1})$ for $t = N-1, \ldots, 0$ are given by (6.55).

12. *Importance Sampling* [171]. Consider the problem of approximating the following expectation:

$$I(\theta) = \int g(x;\theta) f(x) dx$$

with respect to f, assuming that it exists. When it is not possible to sample from $f()$ a sensible approach is to sample from another density, say $h()$, which approximates $f()$ and then use the sampled valued $x_1, \ldots, x_n$ to form the estimator

$$\hat{I}(\theta) = \frac{\sum_{i=1}^n w_i g(x_i, \theta)}{\sum_{i=1}^n w_i},$$

where $w_i = f(x_i)/h(x_i)$. Show that if the support of $h(x)$ includes the support of $f(x)$ then the estimator $\hat{I}(\theta)$ is strongly consistent for $I(\theta)$. Obtain an estimator for the Monte Carlo standard error of $\hat{I}(\theta)$.

13. *Optimal Importance Function*. Show that $\pi(\boldsymbol{\beta}_t \mid \boldsymbol{\beta}_{0,i}^{t-1}, \mathcal{F}_t) = f(\boldsymbol{\beta}_t \mid \boldsymbol{\beta}_{0,i}^{t-1}, \mathcal{F}_t)$ is the importance function which minimizes the variance of the importance weights (6.58) conditional on $\boldsymbol{\beta}_{0,i}^{t-1}$ and $\mathcal{F}_t$. How do the weights (6.58) transform for this choice of importance function?

14. *A state space model for multivariate longitudinal count data* [236]. Consider a multivariate time series of counts $\mathbf{Y}_t = (Y_{t1}, \ldots, Y_{td})'$, $t = 1, \ldots, N$ and suppose that the conditional distribution of Y_{it} given an unobserved process θ_t is Poisson with parameters $a_{it}\theta_t$ with $a_{it} = \exp(\mathbf{x}_t'\boldsymbol{\alpha}_i)$. Here $\mathbf{x}_t$ denotes short–term covariates and $\boldsymbol{\alpha}_i$ are k–dimensional regression parameters, $i = 1, \ldots, d$. Assume that $\theta_0 = 1$ and suppose that the conditional distribution of θ_t given θ_{t-1} is Gamma with mean $b_t\theta_{t-1}$ and squared coefficient of variation equal to σ^2/θ_{t-1}. Here σ^2 denotes a dispersion parameter and b_t depends on the so called long–term covariates $\mathbf{z}_t$ through the model $b_t = \exp(\Delta\mathbf{z}_t'\boldsymbol{\beta})$, where $\Delta\mathbf{z}_t = \mathbf{z}_t - \mathbf{z}_{t-1}$ and $\mathbf{z}_0 = 0$.

(a) Calculate the conditional expectation and variance of θ_t given $\theta_0, \ldots, \theta_{t-1}$. What do you observe?

(b) Show that

$$\mathrm{E}[\theta_t] = b_1 \cdots b_t,$$

to conclude

$$\log \mathrm{E}[\theta_t] = \mathbf{z}_t'\boldsymbol{\beta}.$$

In addition, show that

$$\mathrm{Var}[\theta_t] = \phi_t \mathrm{E}[\theta_t]\sigma^2$$

and

$$\mathrm{Cov}(\theta_t, \theta_{t+k}) = \phi_t \mathrm{E}[\theta_{t+k}]\sigma^2,$$

where $\phi_t = b_t + b_t b_{t-1} + b_t b_{t-1} \ldots b_1$.

(c) Turning to the moment structure of the observed process, define $\mathbf{a}_t = (a_{1t}, \ldots, a_{kt})'$ and $\boldsymbol{\Lambda}_t = \mathrm{diag}(a_{1t}, \ldots, a_{kt})$ to prove that

$$\mathrm{E}[\mathbf{Y}_t] = \mathbf{a}_t \mathrm{E}[\theta_t]$$

and

$$\mathrm{Var}[\mathbf{Y}_t] = \boldsymbol{\Lambda}_t \mathrm{E}[\theta_t] + \mathbf{a}_t\mathbf{a}_t'\phi_t\sigma^2\mathrm{E}[\theta_t].$$

The last expression shows that the variance of $\mathbf{Y}_t$ consists of two components; the first term is Poisson variance and the second term represents overdispersion. Conclude that $\log \mathrm{E}[Y_{it}] = \mathbf{x}_t'\boldsymbol{\alpha}_i + \mathbf{z}_t'\boldsymbol{\beta}$, a fact that follows from the log–link.

(d) Discuss Kalman prediction and filtering for this model.

7

Prediction and Interpolation

7.1 INTRODUCTION

Prediction in the context of time series analysis usually means prediction of future values of a primary series of interest from past observations of the primary series itself as well as, when feasible, from past observations of covariate series. We saw already an example of this in time series that follow generalized linear models where the prediction is facilitated by an assumed model. Another example is prediction in autoregressive moving average (ARMA) series where the nature of statistical dependence within and between series enables successful forecasting from past observations, perhaps after some differencing or some other linear filtering [63]. Nonlinear alternatives include neural network prediction models where the input series and the primary output series are connected by a sequence of layers each containing several nodes, and a Volterra type expansion relating the input and output by nonlinear filters [358], [425]. Closely related is prediction in state space models where the objective is the estimation of a future state value from past observations.

The problem, however, is not prediction per se, for experience shows that, at least for the short term, different prediction methods give very similar forecasts. Even a simple time averaging or exponential smoothing ([307]) can perform adequately, let alone some common sense and good judgment. The problem is really the construction of reliable prediction intervals and the assessment of the goodness or quality of prediction.

In many cases a time series may not follow a generalized linear model or any ARMA model, may not conform to state space formalism, may be observed irregularly with gaps of missing data, and may be far too short for fitting complex models such as neural networks or Volterra type models. In addition, the data may be governed by a skewed distribution far from Gaussian, which is often the case with economic and geophysical data as illustrated by Figure 7.1. An interesting example concerning distribution skewness is also provided by a time series of area *averaged* rain rate as discussed in [253] and elsewhere. As we shall see, these problems, namely, very short time series, data gaps, irregularly observed time series, non-Gaussian data, and incorporation of covariate data can be alleviated surprisingly well by a Bayesian approach centered about the estimation of the predictive density: the conditional density of the value to be predicted given the observed data.

We would like to approach the prediction problem more broadly from an angle that takes into account to some extent all these points, and specifically also addresses the problem of interpolation. In this chapter we treat the more general problem of spatial prediction (really interpolation) of which the prediction problem of time series is a special case. In recent years several very useful and sensible methods of spatial prediction have emerged, including different versions of kriging and Bayesian procedures whose use is widespread particularly in geophysical and environmental studies. These methods, originally motivated by spatial considerations, are quite general and are also useful for time series. We turn to Bayesian spatial prediction, the main topic of this chapter, after a brief yet helpful introduction to stationary random fields and ordinary kriging.

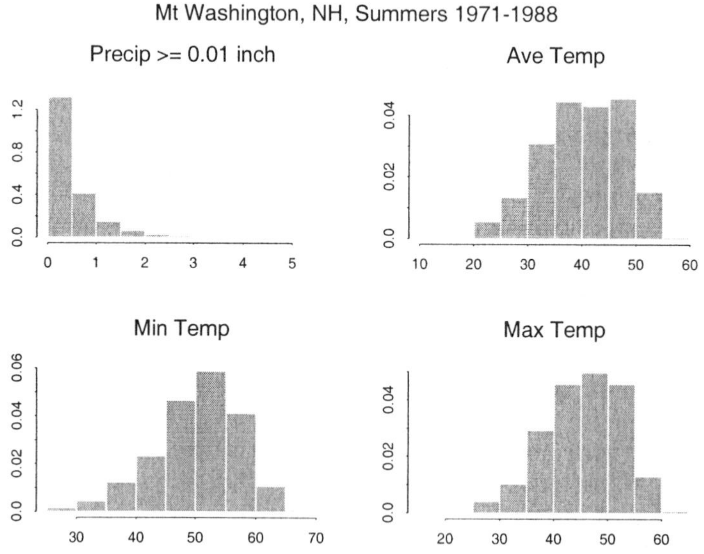

Fig. 7.1 Histograms of daily precipitation at Mt. Washington, NH, in inches and hundredths ($\geq$ 0.01 inch) and corresponding temperature in degrees Fahrenheit. Period: all summers (6/1 to 9/15) from 1971 to 1988.

7.1.1 Elements of Stationary Random Fields

We consider very briefly stationary random fields in order to introduce some basic ideas and terminology to be used later in spatial prediction. Stationarity is not absolutely needed and it is not a precondition for our development, but it eases the exposition. For authoritative treatments of the rapidly growing area of spatial processes and their applications we refer the reader to [98], [109], [399], [441].

Let $\{Z(\mathrm{s})\}$, $\mathrm{s} \in D$, be a spatial process or a random field. By this we mean that D is a subset of R^d, $d \geq 1$, and $Z(\mathrm{s})$ is a random variable for each $\mathrm{s} \in D$. A random field $\{Z(\mathrm{s})\}$ is Gaussian if for all $\mathrm{s}_1, ..., \mathrm{s}_n \in D$, the joint distribution of $(Z(\mathrm{s}_1), ..., Z(\mathrm{s}_n))$ is multivariate normal.

We say that $\{Z(\mathrm{s})\}$ is (second-order) stationary, when

$$\mathrm{E}(Z(\mathrm{s})) = \mu, \quad \mathrm{s} \in D$$

and

$$\mathrm{Cov}(Z(\mathrm{s}+\mathbf{h}), Z(\mathrm{s})) \equiv C(\mathbf{h}), \quad \mathrm{s}+\mathbf{h}, \mathrm{s} \in D.$$

The function $C(\cdot)$ is called the covariogram or covariance function. We shall further assume that $C(\mathbf{h})$ depends only on the distance $||\mathbf{h}||$ between the locations $\mathrm{s}+\mathbf{h}$ and s but not on the direction of $\mathbf{h}$. In this case the covariance function as well as the process are called *isotropic*. The corresponding isotropic correlation function is given by $K(l) = C(l)/C(0)$, where $l = ||\mathbf{h}||$ is the distance between points.

There are several useful parametric correlation functions depending on a parameter vector $\boldsymbol{\theta}$ as follows.

(a) Matérn correlation:

$$K_{\boldsymbol{\theta}}(l) = \begin{cases} \frac{1}{2^{\theta_2-1}\Gamma(\theta_2)} \left(\frac{l}{\theta_1}\right)^{\theta_2} \kappa_{\theta_2}\left(\frac{l}{\theta_1}\right) & \text{if } l \neq 0 \\ 1 & \text{if } l = 0, \end{cases}$$

where $\theta_1 > 0, \theta_2 > 0$, and κ_{θ_2} is a modified Bessel function of the third kind of order θ_2. The parameter θ_2 is considered the more critical parameter controlling the mean square differentiability of $\{Z(\mathrm{s})\}$ [399].

(b) Spherical correlation:

$$K_{\boldsymbol{\theta}}(l) = \begin{cases} 1 - \frac{3}{2}\left(\frac{l}{\theta}\right) + \frac{1}{2}\left(\frac{l}{\theta}\right)^3 & \text{if } l \leq \theta \\ 0 & \text{if } l > \theta, \end{cases}$$

where $\theta > 0$ controls the correlation range.

(c) Exponential correlation:

$$K_{\boldsymbol{\theta}}(l) = \exp(l^{\theta_2} \log \theta_1),$$

where $\theta_1 \in (0, 1)$ controls the correlation range and $\theta_2 \in (0, 2]$ controls the smoothness of $\{Z(\mathrm{s})\}$.

(d) Rational quadratic:

$$K_{\boldsymbol{\theta}}(l) = \left(1 + \frac{l^2}{\theta_1^2}\right)^{-\theta_2},$$

where $\theta_1 > 0$ controls the correlation range and $\theta_2 > 0$ the correlation attenuation.

By varying the correlation function and its parameters, many different types of realizations are obtained as illustrated in Figures 7.2, and 7.3 in terms of three-dimensional Gaussian fields with mean zero and variance 1. Figure 7.4 conveys similar but coarser information in terms of two-dimensional clipped or quantized isotropic realizations at three levels such that each of the corresponding "four colors" covers on average 25% of the clipped discrete image. Recall that for threshold levels $-\infty = \rho_0 < \rho_1 < \ldots < \rho_k = \infty$, a clipped field $\{X(\mathbf{s})\}$ is defined by thresholding a primary field $\{Z(\mathbf{s})\}$: $X(\mathbf{s}) = m \quad \text{iff} \quad \rho_m \leq Z(\mathbf{s}) < \rho_{m+1}, m = 0, \ldots, k-1$.

The realizations and their clipped versions were obtained by using the web program at `http://www.math.umd.edu/~bnk/bak/generate.cgi?4}` created by Kozintsev [276]. Here "4" is the default value of the number of colors k corresponding to $k - 1$ threshold or clipping levels; k admits integer values from 1 to 14 [276], [277].

7.1.2 Ordinary Kriging

To introduce some basic ideas and terms and also point out some inherent problems in spatial prediction, we describe next a simple yet very useful and widely used method called *kriging*, synonymous with best linear unbiased prediction. The method has several variants of which ordinary kriging to be dealt with next is the simplest [98], [109], [399]. The method is attractive as it requires second-order moments only.

Consider a stationary real valued field $\{Z(\mathbf{s})\}$, $\mathbf{s} \in D \subset R^d$, with an unknown mean $\mathrm{E}(Z(\mathbf{s})) = \mu$, $\mu \in R$, admitting the representation

$$Z(\mathbf{s}) = \mu + \delta(\mathbf{s}), \quad \mathbf{s} \in D, \tag{7.1}$$

where $\delta(\mathbf{s})$ is a zero-mean, second-order stationary process with covariogram $C(\mathbf{h}), \mathbf{h} \in R^d$.

Given the data

$$\mathbf{Z} \equiv (Z(\mathbf{s}_1), \ldots, Z(\mathbf{s}_n))'$$

observed at locations $\{\mathbf{s}_1, \ldots, \mathbf{s}_n\}$ in D, the problem is to predict (or estimate) $Z(\mathbf{s}_0)$ at location $\mathbf{s}_0$ using the best linear unbiased predictor (BLUP) obtained by minimizing

$$\mathrm{E}\left(Z(\mathbf{s}_0) - \sum_{i=1}^{n} \lambda_i Z(\mathbf{s}_i)\right)^2 \quad \text{subject to} \quad \sum_{i=1}^{n} \lambda_i = 1. \tag{7.2}$$

Thus we minimize

$$\mathrm{E}\left(Z(\mathbf{s}_0) - \sum_{i=1}^{n} \lambda_i Z(\mathbf{s}_i)\right)^2 - 2m\left(\sum_{i=1}^{n} \lambda_i - 1\right) \tag{7.3}$$

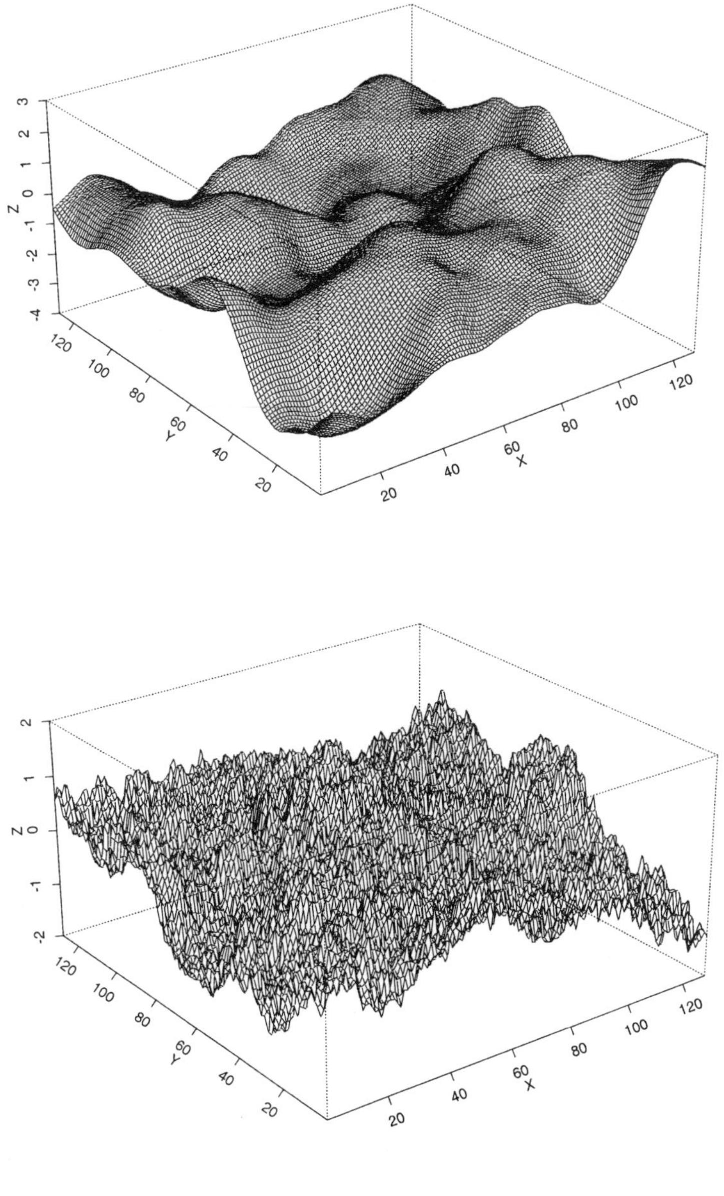

Fig. 7.2 Two 128 × 128 realizations from isotropic Gaussian random fields with different correlation functions. Top: Matérn ($\theta_1 = 8, \theta_2 = 3$). Bottom: spherical ($\theta = 120$).

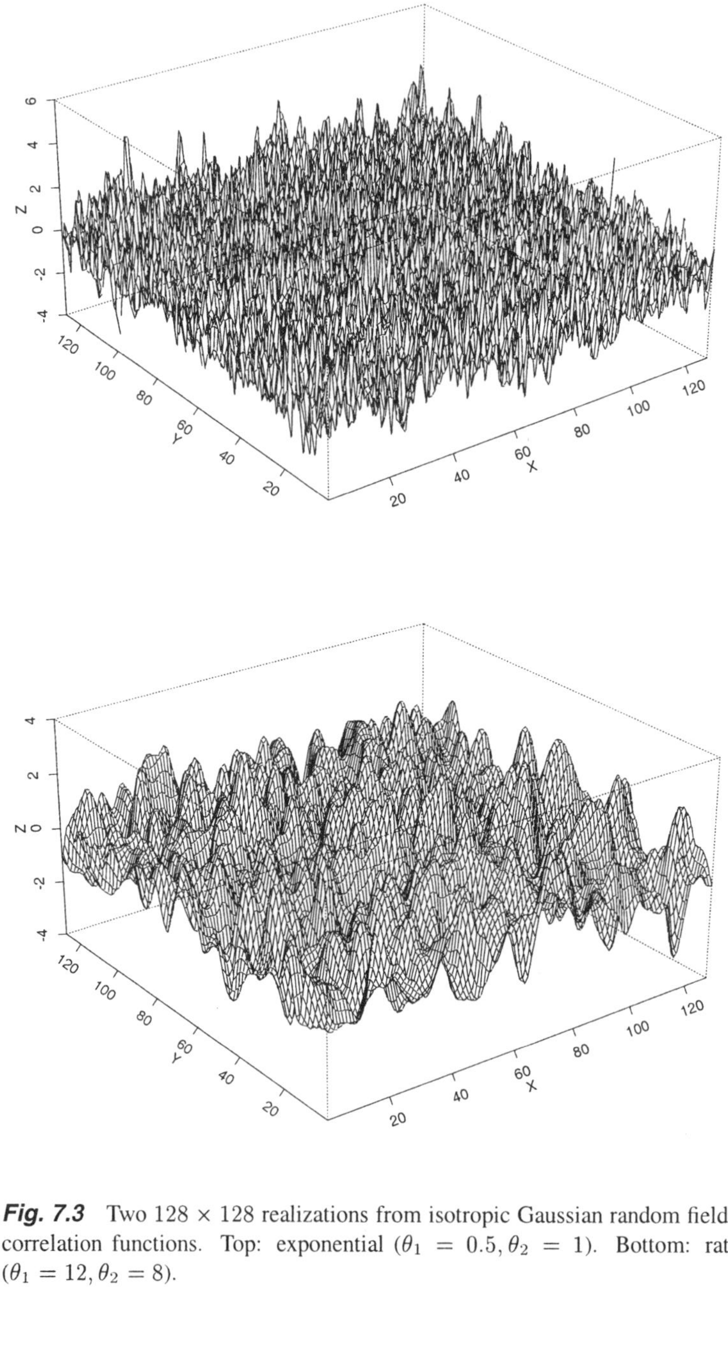

Fig. 7.3 Two 128 × 128 realizations from isotropic Gaussian random fields with different correlation functions. Top: exponential ($\theta_1 = 0.5, \theta_2 = 1$). Bottom: rational quadratic ($\theta_1 = 12, \theta_2 = 8$).

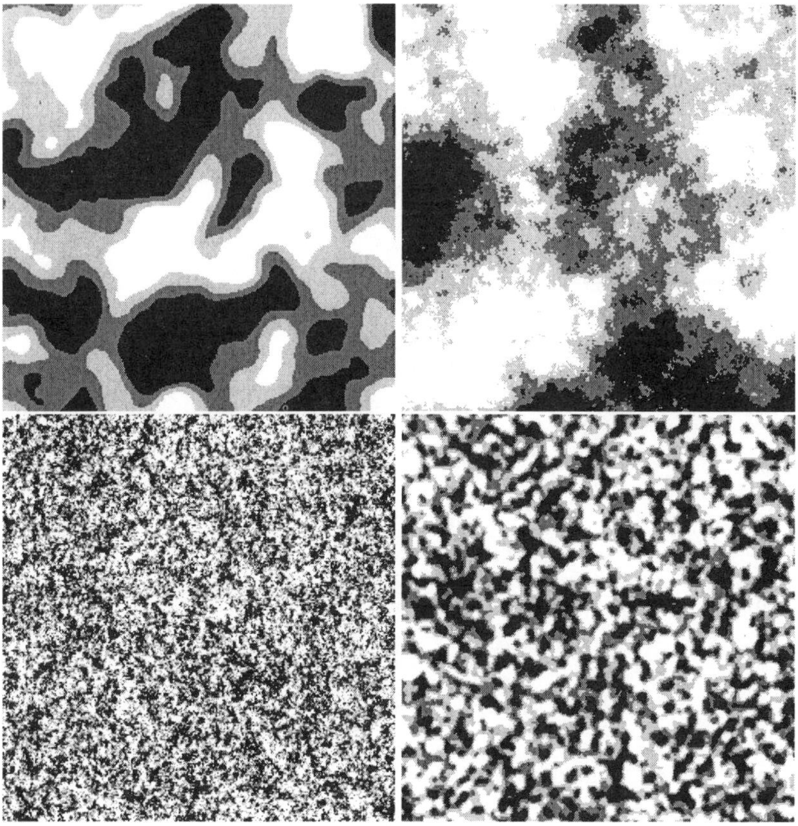

Fig. 7.4 Clipped, at 3 levels, 256×256 realizations from isotropic Gaussian random fields with different correlation functions. Top left: Matérn ($\theta_1 = 8, \theta_2 = 3$). Top right: spherical ($\theta = 120$). Bottom left: exponential ($\theta_1 = 0.5, \theta_2 = 1$). Bottom right: rational quadratic ($\theta_1 = 12, \theta_2 = 8$).

with respect to $\lambda_1, \ldots, \lambda_n$ and the Lagrange multiplier m. This is done by appealing to the second-order properties of the random field.

Clearly, the condition that the λ_i sum to 1 guarantees an unbiased predictor,

$$\mathrm{E}\left(\sum_{i=1}^{n} \lambda_i Z(\mathbf{s}_i)\right) = \mathrm{E}(Z(\mathbf{s}_0)) \sum_{i=1}^{n} \lambda_i = \mu.$$

Using the notation

$$\begin{aligned} \mathbf{1} &= (1, 1, \ldots, 1)', \quad 1 \times n \text{ vector} \\ \mathbf{c} &= \big(C(\mathbf{s}_0 - \mathbf{s}_1), \ldots, C(\mathbf{s}_0 - \mathbf{s}_n)\big)' \\ \mathbf{C} &= \big(C(\mathbf{s}_i - \mathbf{s}_j)\big), \quad i, j = 1, \ldots, n \\ \boldsymbol{\lambda} &= (\lambda_1, \lambda_2, \ldots, \lambda_n)' \\ \mathbf{m} &= m\mathbf{1} \end{aligned}$$

the minimization leads to the system

$$\begin{cases} \mathbf{C}\boldsymbol{\lambda} = \mathbf{c} + \mathbf{m} \\ \sum_{i=1}^{n} \lambda_i = 1. \end{cases} \tag{7.4}$$

Therefore

$$1 = \mathbf{1}'\boldsymbol{\lambda} = \mathbf{1}'\mathbf{C}^{-1}(\mathbf{c} + \mathbf{m}) = \mathbf{1}'\mathbf{C}^{-1}\mathbf{c} + \mathbf{1}'\mathbf{C}^{-1}m\mathbf{1}$$

or

$$m = \frac{1 - \mathbf{1}'\mathbf{C}^{-1}\mathbf{c}}{\mathbf{1}'\mathbf{C}^{-1}\mathbf{1}} \tag{7.5}$$

and

$$\hat{\boldsymbol{\lambda}} = \mathbf{C}^{-1}\left(\mathbf{c} + \frac{1 - \mathbf{1}'\mathbf{C}^{-1}\mathbf{c}}{\mathbf{1}'\mathbf{C}^{-1}\mathbf{1}}\mathbf{1}\right). \tag{7.6}$$

The ordinary kriging predictor is then

$$\hat{Z}(\mathbf{s}_0) = \hat{\boldsymbol{\lambda}}'\mathbf{Z}.$$

The minimized mean-square prediction error, denoted by $\sigma_k^2(\mathbf{s}_0)$, is called the *kriging variance* and is given by

$$\sigma_k^2(\mathbf{s}_0) = \mathrm{E}\big(Z(\mathbf{s}_0) - \hat{Z}(\mathbf{s}_0)\big)^2 = C(\mathbf{0}) - \hat{\boldsymbol{\lambda}}'\mathbf{c} + m. \tag{7.7}$$

It follows that when $Z(\mathbf{s})$ is Gaussian,

$$\hat{Z}(\mathbf{s}_0) \pm 1.96\sigma_k(\mathbf{s}_0) \tag{7.8}$$

is a 95% prediction interval for $Z(\mathbf{s}_0)$. For non-Gaussian fields this may not hold, as we shall soon see.

To summarize, the ordinary kriging algorithm requires the values $Z(\mathbf{s}_1), \ldots, Z(\mathbf{s}_n)$ at points $\mathbf{s}_1, \ldots, \mathbf{s}_n$ in space, the covariogram $C(\cdot)$, and the prediction location of interest $\mathbf{s}_0$, and it runs as follows. Define:

- $\mathbf{Z} = (Z(\mathbf{s}_1), \ldots, Z(\mathbf{s}_n))'$
- $\mathbf{c} = \big(C(\mathbf{s}_0 - \mathbf{s}_1), \ldots, C(\mathbf{s}_0 - \mathbf{s}_n)\big)'$
- $\mathbf{C} = (C_{ij}) = (C(\mathbf{s}_i - \mathbf{s}_j)), i, j = 1, \ldots n$
- $m = \frac{1 - \mathbf{1}'\mathbf{C}^{-1}\mathbf{c}}{\mathbf{1}'\mathbf{C}^{-1}\mathbf{1}}$
- $\hat{\boldsymbol{\lambda}} = \mathbf{C}^{-1}(\mathbf{c} + \mathbf{m})$, where $\mathbf{m} = (m, \ldots, m)'$ is a $1 \times n$ vector
- $\sigma_k^2(\mathbf{s}_0) = C(\mathbf{0}) - \hat{\boldsymbol{\lambda}}'\mathbf{c} + m.$

Then:

- The ordinary kriging predictor of $Z(\mathbf{s}_0)$ is $\hat{Z}(\mathbf{s}_0) = \hat{\boldsymbol{\lambda}}'\mathbf{Z}$.
- A 95% prediction interval in the Gaussian case is

$$\{\hat{Z}(\mathbf{s}_0) - 1.96\sigma_k(\mathbf{s}_0), \hat{Z}(\mathbf{s}_0) + 1.96\sigma_k(\mathbf{s}_0)\}.$$

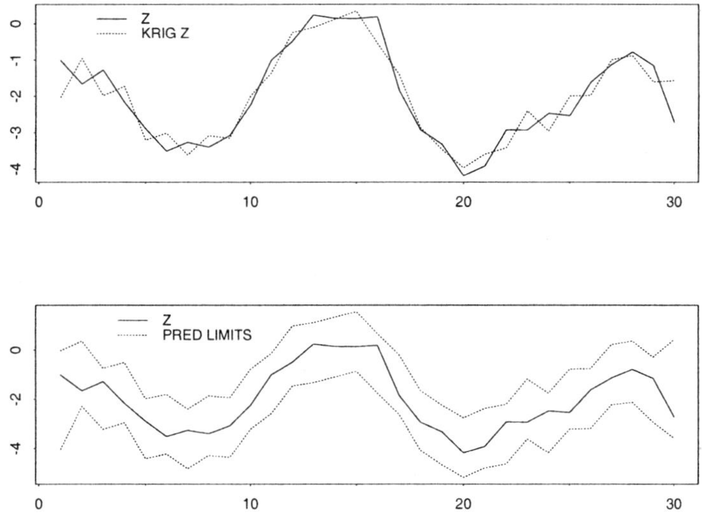

Fig. 7.5 Kriging interpolation and prediction limits in a Gaussian ARMA(1,1) time series.

Obviously, the kriging algorithm is applicable to stationary time series as well. As an example we consider kriging interpolation of the kth value Z_k from

$$Z_1, ..., Z_{k-1}, Z_{k+1}, ..., Z_N,$$

$k = 1, ..., N, N = 30$, in a stationary Gaussian ARMA(1,1) time series with AR and MA parameter values 0.6 and -0.5, respectively (see Problem 2). Figure 7.5 shows the interpolation results and also the 95% prediction intervals for each k. Here kriging is doing a fine job. The results, however, may deteriorate for time series with highly skewed probability distributions as seen from Figure 7.6. To rectify the problem a log transformation is used, but then the prediction is done not at the original scale.

The last example points to the usefulness of transformations in temporal and spatial prediction, an important special case of which is the so called *trans-Gaussian kriging* [109]. Trans-Gaussian kriging assumes that $Z(\mathrm{s})$ is obtained from a Gaussian random field $Y(\mathrm{s})$ by the nonlinear transformation,

$$Z(\mathrm{s}) = \phi(Y(\mathrm{s})) \quad \mathrm{s} \in D, \tag{7.9}$$

where ϕ is a one-to-one twice-differentiable function. Prediction is done in two steps, first $Y(\mathrm{s}_0)$ is predicted from the Y data at the Y scale, and then the result is transformed to the Z scale, corrected for bias, to obtain a predictor for $Z(\mathrm{s}_0)$ and an approximate prediction interval by the δ–method.

In general, suitable normalizing transformations are very desirable as Gaussian data enable likelihood based inference, linear prediction from related covariates, and the construction of reliable prediction and confidence intervals. In this connection two problems arise. First, how to get the normalizing transformation, and second, how to overcome the problematic scale change. In answering these questions, we describe next a Bayesian approach where the transformation is known to belong to a certain parametric class for some unknown parameter, and the prediction is carried out at the original Z scale.

7.2 BAYESIAN SPATIAL PREDICTION

So far we have introduced several general key issues and terms as well as problems akin to spatial prediction and interpolation. Against this backdrop we present now the main idea of the chapter, namely, a certain Bayesian approach to spatial prediction. By fixing one coordinate and letting the other represent time, the method can be applied also in time series prediction and interpolation. Recently there has been an upsurge in popularity of Bayesian methods in spatial analysis [40].

7.2.1 An Auxiliary Gaussian Process

The Gaussian assumption offers a very convenient and attractive basis for statistical inference in general, and it is difficult to exaggerate its importance concerning multivariate data in particular. At the same time, temporal and spatial data governed

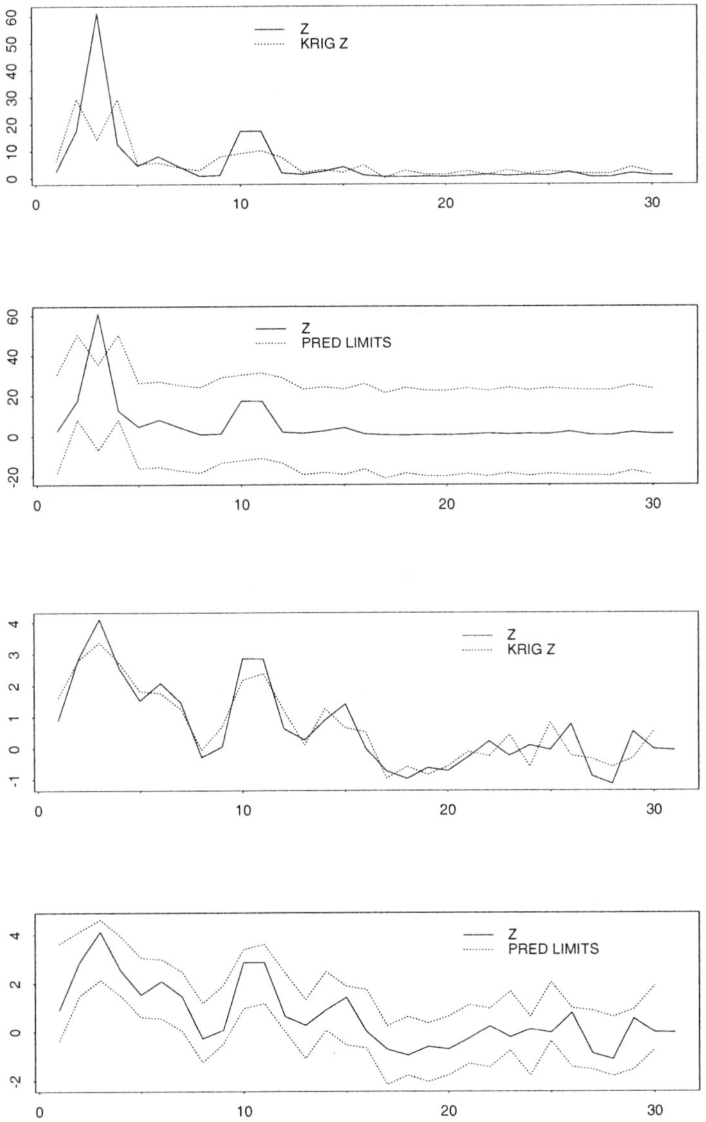

Fig. 7.6 Kriging interpolation and prediction limits. Top two: lognormal time series. Bottom two: The corresponding log-transformed time series.

by non-Gaussian skewed distributions are ubiquitous in nature. To reconcile these two facts, we shall assume, much like in trans-Gaussian kriging, that the data were obtained from an underlying latent Gaussian field by some monotone transformation known to belong to a certain parametric family. When such an assumption is credible, the ensuing Gaussian based inference is straightforward.

What differentiates this approach from that of trans-Gaussian kriging is that it is a Bayesian approach that leads to prediction at the original scale from the so called *predictive density* function,

$$\begin{aligned} p(\mathbf{z}_o|\mathbf{z}) &= \int_\Omega p(\mathbf{z}_o, \boldsymbol{\eta}|\mathbf{z})d\boldsymbol{\eta} \\ &= \int_\Omega p(\mathbf{z}_o|\boldsymbol{\eta}, \mathbf{z})p(\boldsymbol{\eta}|\mathbf{z})d\boldsymbol{\eta}, \end{aligned} \tag{7.10}$$

where $\mathbf{z}_o = (z_{o1}, \ldots, z_{ok})'$ is the value to be predicted, and $\boldsymbol{\eta}$ is the vector of model parameters taking values in Ω. The predictive density is both an analytical as well as pictorial device whose dispersion magnitude conveys fast the degree of goodness and quality of prediction. A very dispersed predictive density points to questionable prediction, whereas a concentrated density indicates a more conclusive prediction. We use the median of the predictive density as the predictor, instead of the mean, as the later may not exist for some commonly used transformations. The method is referred to as *Bayesian Transformed Gaussian* (BTG), a term coined by De Oliveira [117], [118].

This section follows the development in [117], [118], extending the Bayesian methodology presented in [192], [193] for prediction in Gaussian random fields. The Bayesian prediction paradigm applied to Gaussian random fields has been considered previously by several authors including [74], [164], [192], [268], [267], [339].

7.2.2 The Likelihood

As before, $\{Z(\mathbf{s}),\ \mathbf{s} \in D\}$, $D \subset R^2$, is a random field observed at n locations in D, $\mathbf{Z} = (Z(\mathbf{s}_1), \ldots, Z(\mathbf{s}_n))'$, and we wish to predict the unobserved random vector $\mathbf{Z}_0 = (Z(\mathbf{s}_{01}), \ldots, Z(\mathbf{s}_{0k}))'$, at known distinct locations $\mathbf{s}_{01}, \ldots, \mathbf{s}_{0k}$ in D using (7.10). The extra generality of predicting a vector rather than a scalar is obtained essentially at no additional cost.

Let $G = \{g_\lambda(\cdot) : \lambda \in \Lambda\}$ be a parametric family of nonlinear monotone transformations where each $g_\lambda(\cdot) \in G$ has a continuous derivative $g'_\lambda(x)$. Assume that for some unknown $\lambda \in \Lambda$ the process

$$\{Y(\mathbf{s}) = g_\lambda(Z(\mathbf{s})),\ \mathbf{s} \in D\}$$

is a Gaussian random field. In analogy with a log-Gaussian random field, the original field $\{Z(\mathbf{s}), \mathbf{s} \in D\}$ is referred to as g_λ-Gaussian random field.

We shall make the following moment assumptions.

Mean in terms of covariates:

$$\mathrm{E}(Y(\mathbf{s})) = \sum_{j=1}^{p} \beta_j f_j(\mathbf{s}) = \beta' \underline{f}(\mathbf{s}),\ \mathbf{s} \in D,$$

where $\beta = (\beta_1, \ldots, \beta_p)' \in R^p$ are unknown regression parameters, and $\underline{f}(\mathbf{s}) = (f_1(\mathbf{s}), \ldots, f_p(\mathbf{s}))'$ is a set of known location-dependent covariates.

Variance in terms of precision:

$$\mathrm{Var}(Y(\mathbf{s})) = \tau^{-1},$$

where τ is the precision of the random field.

Parametric Correlation:

$$\mathrm{Cov}(Y(\mathbf{s}), Y(\mathbf{u})) = \frac{1}{\tau} K_{\boldsymbol{\theta}}(||\mathbf{s} - \mathbf{u}||), \quad \mathbf{s}, \mathbf{u} \in D.$$

The parameter $\boldsymbol{\theta} = (\theta_1, \ldots, \theta_q)' \in \Theta \subset R^q$ is a structural parameter controlling the range of correlation and/or the smoothness of the random field, and for every $\boldsymbol{\theta} \in \Theta$, $K_{\boldsymbol{\theta}}(\cdot)$ is an isotropic correlation function. Here isotropy is not essential, and any parametric $K_{\boldsymbol{\theta}}(\cdot)$ will do.

For any vector $\mathbf{a} = (a_1, \ldots, a_n)$ define $\underline{g}_\lambda(\mathbf{a}) \equiv (g_\lambda(a_1), \ldots, g_\lambda(a_n))$. Then the Gaussian assumption about $Y(\mathbf{s})$ implies,

$$(\underline{g}_\lambda(\mathbf{Z}_0), \underline{g}_\lambda(\mathbf{Z}) | \beta, \tau, \boldsymbol{\theta}, \lambda)' \sim \mathcal{N}_{k+n}\left(\begin{pmatrix} \mathbf{X}_0\beta \\ \mathbf{X}\beta \end{pmatrix}, \frac{1}{\tau} \begin{pmatrix} \mathbf{E}_{\boldsymbol{\theta}} & \mathbf{B}_{\boldsymbol{\theta}} \\ \mathbf{B}'_{\boldsymbol{\theta}} & \boldsymbol{\Sigma}_{\boldsymbol{\theta}} \end{pmatrix} \right) \tag{7.11}$$

for some $\lambda \in \Lambda$ and $(\beta, \tau, \boldsymbol{\theta})' \in R^p \times (0, \infty) \times \Theta$. The matrices $\mathbf{X}$ and $\mathbf{X}_0$ are known $n \times p$ and $k \times p$ design matrices, respectively, defined by $X_{ij} = f_j(\mathbf{s}_i)$, $X_{0,ij} = f_j(\mathbf{s}_{0i})$, and $\mathbf{E}_{\boldsymbol{\theta}}$, $\mathbf{B}_{\boldsymbol{\theta}}$ and $\boldsymbol{\Sigma}_{\boldsymbol{\theta}}$ are respectively $k \times k$, $k \times n$, and $n \times n$, correlation matrices defined as

$$\begin{aligned}
\mathbf{E}_{\boldsymbol{\theta},ij} &= K_{\boldsymbol{\theta}}(||\mathbf{s}_{0i} - \mathbf{s}_{0j}||) \\
\mathbf{B}_{\boldsymbol{\theta},ij} &= K_{\boldsymbol{\theta}}(||\mathbf{s}_{0i} - \mathbf{s}_j||) \\
\boldsymbol{\Sigma}_{\boldsymbol{\theta},ij} &= K_{\boldsymbol{\theta}}(||\mathbf{s}_i - \mathbf{s}_j||).
\end{aligned}$$

It is assumed that $\mathbf{X}$ has full rank and that the matrix $\boldsymbol{\Sigma}_{\boldsymbol{\theta}}$, $\boldsymbol{\theta} \in \Theta$, is nonsingular.

From (7.11) the likelihood $L(\boldsymbol{\eta}; \mathbf{z}) \equiv p(\mathbf{z}|\boldsymbol{\eta})$ of the model parameters $\boldsymbol{\eta} = (\beta, \tau, \boldsymbol{\theta}, \lambda)'$ based on the *original data* $\mathbf{z} = (z_1, \ldots, z_n)'$ is given by

$$\begin{aligned}
L(\boldsymbol{\eta}; \mathbf{z}) = {} & \left(\frac{\tau}{2\pi}\right)^{n/2} |\boldsymbol{\Sigma}_{\boldsymbol{\theta}}|^{-1/2} \\
& \times \exp\left\{ -\frac{\tau}{2} (\underline{g}_\lambda(\mathbf{z}) - \mathbf{X}\beta)' \boldsymbol{\Sigma}_{\boldsymbol{\theta}}^{-1} (\underline{g}_\lambda(\mathbf{z}) - \mathbf{X}\beta) \right\} J_\lambda
\end{aligned} \tag{7.12}$$

for $z_i \in g_\lambda^{-1}(R)$, and is 0 otherwise, where $J_\lambda = \prod_{i=1}^{n} |g'_\lambda(z_i)|$ is the Jacobian of the transformation.

7.2.3 Prior and Posterior of Model Parameters

In general, the choice of a prior distribution for model parameters reflects the statistician's prior belief and experience, and at times inexperience, and hence it cannot be objective. The Bayesian paradigm calls for the update of the prior distribution by means of Bayes theorem taking into account the likelihood of the observed data. The update, called the posterior distribution, is a function of the prior and the likelihood, and forms the basis for inference.

Insightful arguments in [60], [409], and our own experience leads us to the prior

$$p(\beta, \tau, \boldsymbol{\theta}, \lambda) \propto \frac{p(\boldsymbol{\theta})p(\lambda)}{\tau J_\lambda^{p/n}}, \tag{7.13}$$

where $p(\boldsymbol{\theta})$ and $p(\lambda)$ are the prior marginals of $\boldsymbol{\theta}$ and λ, respectively. Observe that (7.13) is an unusual prior in the sense that it depends on the data. Alternative priors in connection with transformed models are discussed in [118], [281], [347], [409]. In general, noninformative priors for spatial analysis may produce improper posterior distributions [40], [118]. However, it has been shown in [40] that certain noninformative "reference" prior distributions do yield proper posterior distributions and also have some desirable properties.

The joint posterior distribution of the model parameters given the data is obtained from its factors,

$$p(\eta|\mathbf{z}) = p(\beta, \tau, \boldsymbol{\theta}, \lambda|\mathbf{z}) = p(\beta, \tau|\boldsymbol{\theta}, \lambda, \mathbf{z})p(\boldsymbol{\theta}, \lambda|\mathbf{z}).$$

A compact $\Theta \times \Lambda$ is a sufficient condition for this posterior to be proper.

From the fact that conditional on $\boldsymbol{\theta}$ and λ, $\underline{g}_\lambda(\mathbf{z})$ is a linear model in terms of the transformed data, we obtain (Problem 6),

$$(\beta|\tau, \boldsymbol{\theta}, \lambda, \mathbf{z}) \sim \mathcal{N}_p\left(\hat{\beta}_{\boldsymbol{\theta},\lambda}, \frac{1}{\tau}(\mathbf{X}'\boldsymbol{\Sigma}_{\boldsymbol{\theta}}^{-1}\mathbf{X})^{-1}\right)$$

$$(\tau|\boldsymbol{\theta}, \lambda, \mathbf{z}) \sim Ga\left(\frac{n-p}{2}, \frac{2}{\tilde{q}_{\boldsymbol{\theta},\lambda}}\right), \tag{7.14}$$

where

$$\hat{\beta}_{\boldsymbol{\theta},\lambda} = (\mathbf{X}'\boldsymbol{\Sigma}_{\boldsymbol{\theta}}^{-1}\mathbf{X})^{-1}\mathbf{X}'\boldsymbol{\Sigma}_{\boldsymbol{\theta}}^{-1}\underline{g}_\lambda(\mathbf{z})$$

is the weighted least squares estimate of β based on the transformed data, and

$$\tilde{q}_{\boldsymbol{\theta},\lambda} = (\underline{g}_\lambda(\mathbf{z}) - \mathbf{X}\hat{\beta}_{\boldsymbol{\theta},\lambda})'\boldsymbol{\Sigma}_{\boldsymbol{\theta}}^{-1}(\underline{g}_\lambda(\mathbf{z}) - \mathbf{X}\hat{\beta}_{\boldsymbol{\theta},\lambda}).$$

It follows that the conditional posterior

$$p(\beta, \tau|\boldsymbol{\theta}, \lambda, \mathbf{z}) = p(\beta|\tau, \boldsymbol{\theta}, \lambda, \mathbf{z})p(\tau|\boldsymbol{\theta}, \lambda, \mathbf{z})$$

is Normal-Gamma.

To compute the second factor in the joint posterior note that

$$p(\boldsymbol{\theta}, \lambda|\mathbf{z}) = p(\beta, \tau, \boldsymbol{\theta}, \lambda|\mathbf{z})/p(\beta, \tau|\boldsymbol{\theta}, \lambda, \mathbf{z}),$$

and by applying Bayes theorem to the numerator we obtain (Problem 7),

$$p(\boldsymbol{\theta}, \lambda|\mathbf{z}) \propto |\boldsymbol{\Sigma}_{\boldsymbol{\theta}}|^{-1/2}|\mathbf{X}'\boldsymbol{\Sigma}_{\boldsymbol{\theta}}^{-1}\mathbf{X}|^{-1/2}\tilde{q}_{\boldsymbol{\theta},\lambda}^{-\frac{n-p}{2}} J_{\lambda}^{1-\frac{p}{n}} p(\boldsymbol{\theta})p(\lambda) \tag{7.15}$$

7.2.4 Prediction of Z_0

In addition to the joint posterior distribution $p(\boldsymbol{\eta}|\mathbf{z})$ derived above, the predictive density $p(\mathbf{z}_o|\mathbf{z})$ in (7.10) requires $p(\mathbf{z}_o|\boldsymbol{\eta}, \mathbf{z})$. The latter can be derived as follows. From (7.11),

$$(\underline{g}_{\lambda}(\mathbf{Z}_0)|\beta, \tau, \boldsymbol{\theta}, \lambda, \mathbf{z}) \sim \mathcal{N}_k\left(\mathbf{M}_{\beta,\boldsymbol{\theta},\lambda}, \frac{1}{\tau}\mathbf{D}_{\boldsymbol{\theta}}\right),$$

where

$$\begin{aligned}
\mathbf{M}_{\beta,\boldsymbol{\theta},\lambda} &= \mathbf{B}_{\boldsymbol{\theta}}\boldsymbol{\Sigma}_{\boldsymbol{\theta}}^{-1}\underline{g}_{\lambda}(\mathbf{z}) + \mathbf{H}_{\boldsymbol{\theta}}\beta \\
\mathbf{H}_{\boldsymbol{\theta}} &= \mathbf{X}_0 - \mathbf{B}_{\boldsymbol{\theta}}\boldsymbol{\Sigma}_{\boldsymbol{\theta}}^{-1}\mathbf{X} \\
\mathbf{D}_{\boldsymbol{\theta}} &= \mathbf{E}_{\boldsymbol{\theta}} - \mathbf{B}_{\boldsymbol{\theta}}\boldsymbol{\Sigma}_{\boldsymbol{\theta}}^{-1}\mathbf{B}_{\boldsymbol{\theta}}'.
\end{aligned}$$

Therefore,

$$\begin{aligned}
p(\mathbf{z}_o|\boldsymbol{\eta}, \mathbf{z}) &= \left(\frac{\tau}{2\pi}\right)^{k/2} |\mathbf{D}_{\boldsymbol{\theta}}|^{-1/2} \prod_{j=1}^{k} |g_{\lambda}'(z_{oj})| \\
&\times \exp\left\{-\frac{\tau}{2}(\underline{g}_{\lambda}(\mathbf{z}_o) - \mathbf{M}_{\beta,\boldsymbol{\theta},\lambda})'\mathbf{D}_{\boldsymbol{\theta}}^{-1}(\underline{g}_{\lambda}(\mathbf{z}) - \mathbf{M}_{\beta,\boldsymbol{\theta},\lambda})\right\}.
\end{aligned} \tag{7.16}$$

We now have the integrand $p(\mathbf{z}_o|\boldsymbol{\eta}, \mathbf{z})p(\boldsymbol{\eta}|\mathbf{z})$ needed for $p(\mathbf{z}_o|\mathbf{z})$. By integrating out β and τ in (7.10) we obtain the simplified form

$$\begin{aligned}
p(\mathbf{z}_o|\mathbf{z}) &= \int_{\Lambda}\int_{\Theta} p(\mathbf{z}_o|\boldsymbol{\theta}, \lambda, \mathbf{z})p(\boldsymbol{\theta}, \lambda|\mathbf{z})d\boldsymbol{\theta}d\lambda \\
&= \frac{\int_{\Lambda}\int_{\Theta} p(\mathbf{z}_o|\boldsymbol{\theta}, \lambda, \mathbf{z})p(\mathbf{z}|\boldsymbol{\theta}, \lambda)p(\boldsymbol{\theta})p(\lambda)d\boldsymbol{\theta}d\lambda}{\int_{\Lambda}\int_{\Theta} p(\mathbf{z}|\boldsymbol{\theta}, \lambda)p(\boldsymbol{\theta})p(\lambda)d\boldsymbol{\theta}d\lambda}.
\end{aligned} \tag{7.17}$$

Now,

$$(\underline{g}_{\lambda}(\mathbf{Z}_0)|\boldsymbol{\theta}, \lambda, \mathbf{z}) \sim T_k(n-p, \mathbf{m}_{\boldsymbol{\theta},\lambda}, (\tilde{q}_{\boldsymbol{\theta},\lambda}\mathbf{C}_{\boldsymbol{\theta}})^{-1})$$

is a k-variate Student t-distribution with $n-p$ degrees of freedom, location parameter vector

$$\mathbf{m}_{\boldsymbol{\theta},\lambda} = \mathbf{B}_{\boldsymbol{\theta}}\boldsymbol{\Sigma}_{\boldsymbol{\theta}}^{-1}\underline{g}_{\lambda}(\mathbf{z}) + \mathbf{H}_{\boldsymbol{\theta}}\hat{\beta}_{\boldsymbol{\theta},\lambda}$$

and scale matrix $\tilde{q}_{\boldsymbol{\theta},\lambda}\mathbf{C}_{\boldsymbol{\theta}}$, where

$$\mathbf{C}_{\boldsymbol{\theta}} = \mathbf{D}_{\boldsymbol{\theta}} + \mathbf{H}_{\boldsymbol{\theta}}(\mathbf{X}'\boldsymbol{\Sigma}_{\boldsymbol{\theta}}^{-1}\mathbf{X})^{-1}\mathbf{H}'_{\boldsymbol{\theta}}$$

and $\tilde{q}_{\boldsymbol{\theta},\lambda}$ is as defined earlier. Therefore,

$$\begin{aligned} p(\mathbf{z}_o|\boldsymbol{\theta},\lambda,\mathbf{z}) &= \frac{\Gamma(\frac{n-p+k}{2})\prod_{j=1}^{k}|g'_\lambda(z_{oj})|}{\Gamma(\frac{n-p}{2})\pi^{k/2}|\tilde{q}_{\boldsymbol{\theta},\lambda}\mathbf{C}_{\boldsymbol{\theta}}|^{1/2}} \\ &\times \ [1+(\underline{g}_\lambda(\mathbf{z}_o)-\mathbf{m}_{\boldsymbol{\theta},\lambda})'(\tilde{q}_{\boldsymbol{\theta},\lambda}\mathbf{C}_{\boldsymbol{\theta}})^{-1} \\ &\times \ (\underline{g}_\lambda(\mathbf{z}_o)-\mathbf{m}_{\boldsymbol{\theta},\lambda})]^{-\frac{n-p+k}{2}}. \end{aligned} \tag{7.18}$$

From (7.15) and Bayes theorem we have

$$p(\mathbf{z}|\boldsymbol{\theta},\lambda) \propto |\boldsymbol{\Sigma}_{\boldsymbol{\theta}}|^{-1/2}|\mathbf{X}'\boldsymbol{\Sigma}_{\boldsymbol{\theta}}^{-1}\mathbf{X}|^{-1/2}\tilde{q}_{\boldsymbol{\theta},\lambda}^{-\frac{n-p}{2}}J_\lambda^{1-\frac{p}{n}}, \tag{7.19}$$

where the proportionality constant is independent of $\boldsymbol{\theta}$ and λ so that its value cancels out in the computation of $p(\mathbf{z}_o|\mathbf{z})$ in (7.17). We see that the predictive density function in (7.17) is a mixture of transformed t-distribution with mixing distribution $p(\boldsymbol{\theta},\lambda|\mathbf{z})$. Observe that by integrating out λ the resulting $p(\mathbf{z}_o|\mathbf{z})$ is free of λ.

In practice, the integration in (7.17) is approximated numerically as done in the next section for the case of $k = 1$, prediction at a single location.

7.2.5 Numerical Algorithm for the Case $\mathrm{k} = 1$

Consider $p(\mathbf{z}_o|\mathbf{z})$ in (7.17) for the case $k = 1$. A Monte Carlo algorithm for the approximation of $p(z_o|\mathbf{z})$ based on (7.17) and assuming that $p(\boldsymbol{\theta})$ and $p(\lambda)$ are proper densities is as follows [117], [118], [164].

BTG Algorithm: Predictive Density Approximation

1. Let $S = \{z_o^{(j)} : j = 1, \ldots, r\}$ be the set of values obtained by discretizing the effective range of Z_0.
2. Generate independently $\boldsymbol{\theta}_1, \ldots, \boldsymbol{\theta}_m$ i.i.d. $\sim p(\boldsymbol{\theta})$ and $\lambda_1, \ldots, \lambda_m$ i.i.d. $\sim p(\lambda)$.
3. For $z_o \in S$, the approximation to $p(z_o|\mathbf{z})$ is given by

$$\hat{p}_m(z_o|\mathbf{z}) = \frac{\sum_{i=1}^{m} p(z_o|\boldsymbol{\theta}_i,\lambda_i,\mathbf{z})p(\mathbf{z}|\boldsymbol{\theta}_i,\lambda_i)}{\sum_{i=1}^{m} p(\mathbf{z}|\boldsymbol{\theta}_i,\lambda_i)}, \tag{7.20}$$

where $p(z_o|\boldsymbol{\theta},\lambda,\mathbf{z})$ and $p(\mathbf{z}|\boldsymbol{\theta},\lambda)$ were respectively given in (7.18) and (7.19).

It can be shown under mild regularity conditions that $\hat{p}_m(z_o|\mathbf{z})$ is consistent and asymptotically normal as m increases [171].

An actual point predictor $\hat{Z}_0$ for $Z_0 = Z(\mathrm{s}_0)$ is a functional of $p(z_o|\mathbf{z})$ that can be obtained in more than one way. The first thing that comes to mind is to use the conditional expectation $\mathrm{E}[Z_0|\mathbf{Z}]$, the mean of the predictive density, as the predictor for Z_0. This predictor is optimal under squared error loss but it may not exist for some important transformations $g_\lambda(\cdot)$ [118], [448], a case in point is the Box-Cox family of transformations to be defined and used below. To circumvent this difficulty, we use as our predictor for Z_0 the median of the predictive density,

$$\hat{Z}_0 = \text{Median of } (Z_0|\mathbf{Z}).$$

The median is the optimal predictor corresponding to the absolute error loss function [5]. As a measure of prediction uncertainty we use a $100(1-\alpha)\%$ prediction interval symmetric about $\hat{Z}_0$, which is readily approximated from $\hat{p}_m(z_o|\mathbf{z})$.

7.2.6 Normalizing Transformations

Rendering the data "normal" by a suitable transformation is a sensible widespread practice as many statistical procedures, such as the classical analysis of variance, do require normally distributed data, or simply because the normal assumption leads to mathematical simplification and tractability. A case in point is the assumption of normal errors in classical regression. A useful parametric family of transformations that is often used in applications to "normalize" positive data $(x > 0)$ is the Box-Cox family of power transformations [60],

$$g_\lambda(x) = \begin{cases} \frac{x^\lambda - 1}{\lambda} & \text{if } \lambda \neq 0 \\ \log(x) & \text{if } \lambda = 0. \end{cases} \tag{7.21}$$

The probability distributions resulting from the inverse of (7.21) include in particular the Gaussian $(\lambda = 1)$ and the log-Gaussian $(\lambda = 0)$ distributions. For more on the Box-Cox family of transformations and related work see [47], [118].

Another useful family of transformations is the Aranda-Ordaz [22] parametric family used in symmetrizing the distribution of proportions x, $0 < x < 1$,

$$g_\lambda(x) = \begin{cases} \log\left(\frac{(1-x)^{-\lambda} - 1}{\lambda}\right) & \text{if } \lambda \neq 0 \\ \log(-\log(1-x)) & \text{if } \lambda = 0. \end{cases} \tag{7.22}$$

Throughout the rest of this chapter we shall use only the Box-Cox family in conjunction with our spatial prediction approach.

7.2.7 Software for BTG Implementation

The BTG algorithm is fairly straightforward and there exists software for its implementation, the `btg` program, developed by David S. Bindel, taking as input positive data and using the Box-Cox family of transformations. The code is a hybrid of C++, Tcl/Tk, and FORTRAN 77 routines. See [52] for the web address from which the software as well as an accompanying user manual can be downloaded. The user manual also contains information about the program organization, interface code, support code, and error messages.

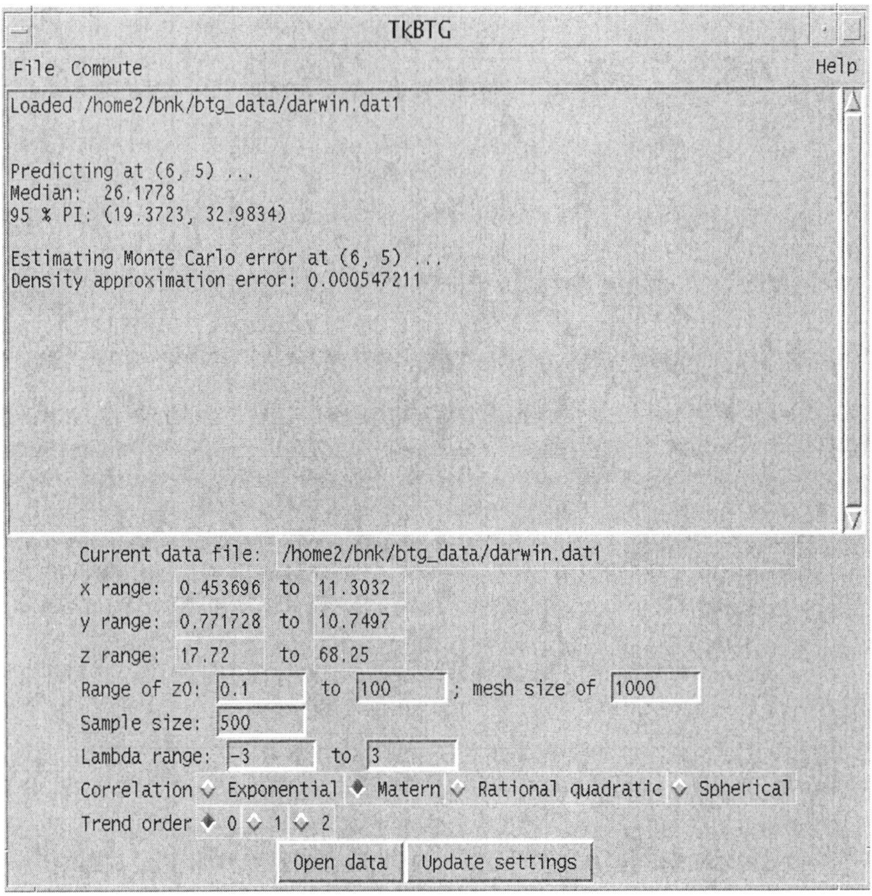

Fig. 7.7 The `tkbtg` interface layout.

The program assigns the correlation parameters noninformative priors as follows. In the exponential case, θ_1 and θ_2 are assumed independent random variables uniformly distributed as Unif(0,1) and Unif(0,2], respectively. In the Matérn, spherical, and rational quadratic correlation functions, the transformed parameters

$$\exp(-\theta_1), \quad \exp(-\theta_2) \quad .$$

are assumed independent random variables uniformly distributed as Unif(0,1). In all cases the default distribution of λ is Unif$(-3, 3)$.

For a given location, correlation function (chosen from Matérn, exponential, spherical, rational quadratic), and trend of degree $n, n = 0, 1, 2$, in the location coordinates, the `btg` program provides in particular the estimated predictive density and the corresponding point prediction and 95% prediction interval. The program also provides the estimated prediction map over the region of interest, and cross-validation at all data points. Except for the trend option in the location coordinates, at present the software does not allow the incorporation of covariates. However, users with C++ programming background can use the existing computational engine to extend the present program. The prediction map and estimated predictive density outputs are written to ASCII text files in directory `btg_data` which also contains the data files.

The `tkbtg` program is an X window graphical interface to the `btg` program. With it users can interactively control and access the applications of the `btg` engine.

The `tkbtg` console, shown in Figure 7.7, has three major parts. The menu bar at the top offers computational tasks and load/save script (from previous/current settings) operations. Program output and status messages are displayed on the middle part of the window. The bottom part, console panel, displays information about the current data set, and allows users to update program settings. It is important to note that the user must supply the range of z_o and then click "Update settings" to effect the changes.

In most practical cases, positive data as required by the `btg` application can be induced by simply adding to every data point a constant which later can be removed from the predicted values. It should be noted that since the algorithm requires the evaluation of the inverse and determinant of the $n \times n$ correlation matrix $\boldsymbol{\Sigma}_{\boldsymbol{\theta}}$, in its present form the algorithm is suitable for small and at best moderate sample sizes, for example $20 < n < 150$.

7.3 APPLICATIONS OF BTG

7.3.1 Spatial Rainfall Prediction

The BTG algorithm is applied here in rainfall prediction using the `btg` program with the Box-Cox transformation family (7.21), $\lambda \sim \text{Unif}(-3, 3)$, $m = 500$, and two different parametric correlation functions, Matérn and exponential. In the absence of additional covariate information we assume a model with constant mean, that is, $p = 1$ and $\text{E}[Y(\mathbf{s})] = \beta_1$.

Table 7.1 and Figure 7.8 give the weekly rainfall totals in mm and the corresponding locations of measurement from the 76th to the 82th day of 1991 in Darwin, Australia, during the rainy season there [259]. The rainfall was measured using tipping buckets in $n = 24$ stations scattered throughout a region, called the D-scale, of about 12×12 km^2 as described in Figure 7.8. The region is part of the coastal plain of the Adelaide River, where the terrain is flat with no orographic patterns. A histogram of the 24 observations in Figure 7.9 suggests that the rainfall data follow a skewed distribution.

The prediction surfaces over the region $(0, 11.875) \times (0, 11.875)$ with mesh size 0.125 and the corresponding contour plots are shown in Figures 7.10 (Matérn) and 7.11 (exponential) and are seen to be very similar. Assuming the Matérn correlation, the predictive density obtained from a cross-validation study, where a chosen observation is predicted from the remaining 23 observations, is shown in four cases in Figure 7.12. We see that the predictive density also provides useful information about the prediction quality, being less informative for the peripheral point 63.1 but much more informative for the other three cases. In all four cases the true observation is contained in its 95% prediction interval. The cross-validation results including 95% prediction intervals regarding all the observations are given in Table 7.1. In all the cases the true observation is contained in its 95% prediction interval. Notice the differences between Table 7.1 and Figure 7.12 regarding the predictors and prediction intervals in the four cases reported in the figure. These are due to the fact that the results in the table and the figure were obtained from different runs of the algorithm and the predictor as well as the prediction interval are random.

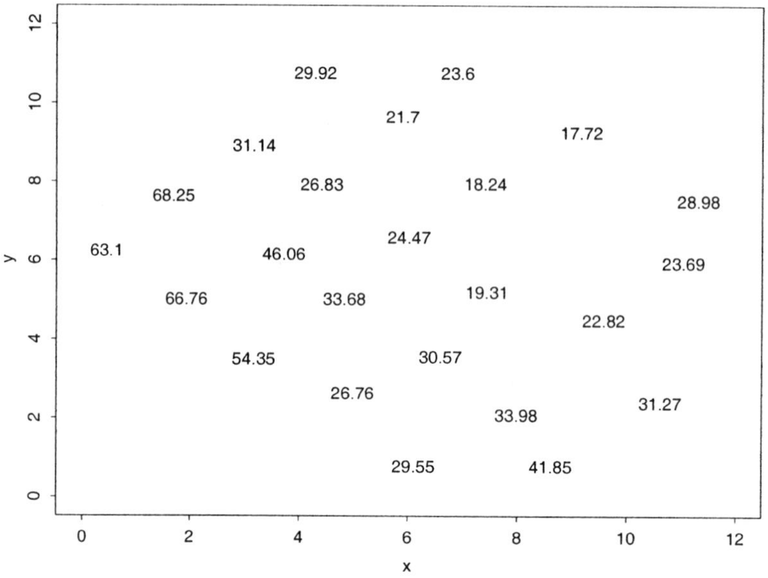

Fig. 7.8 Rain gauge positions and weekly rainfall totals (in mm) corresponding to Table 7.1.

Table 7.1 Weekly rainfall totals (in mm) from 24 rain gauges in Darwin, Australia, at locations $\mathbf{s} = (x, y)$. The predictors and 95% prediction intervals are the result of cross-validation using the Matérn correlation. *Source*: Australian Bureau of Meteorology Research Center (BMRC). Processing the data to the present useful form is courtesy of Goddard Space Flight Center, National Aeronautics and Space Administration (NASA/GSFC).

No.	x	y	z	$\hat{z}$	95% PI
1	6.104768	0.771728	29.55	33.74	(10.70, 56.78)
2	8.613696	0.771728	41.85	34.32	(15.29, 53.35)
3	4.986304	2.625432	26.76	36.71	(20.93, 52.49)
4	7.978304	2.069432	33.98	33.39	(18.49, 48.29)
5	10.608000	2.378568	31.27	32.99	(9.53, 56.45)
6	3.173696	3.490568	54.35	39.13	(16.46, 61.79)
7	6.588928	3.551728	30.57	24.45	(15.40, 33.59)
8	9.580928	4.479136	22.82	23.09	(11.52, 34.66)
9	1.934464	5.004000	66.76	64.12	(28.25, 100)
10	4.835072	5.004000	33.68	35.16	(18.01, 52.30)
11	7.434304	5.189704	19.31	24.51	(15.62, 33.40)
12	11.031232	5.930296	23.69	26.45	(15.35, 37.54)
13	0.453696	6.239432	63.10	72.07	(44.14, 100)
14	3.717696	6.147136	46.06	40.60	(18.58, 62.63)
15	6.014464	6.579704	24.47	22.32	(14.00, 30.63)
16	11.303232	7.506000	28.98	21.62	(13.43, 29.81)
17	1.692928	7.629432	68.25	46.54	(19.25, 73.84)
18	4.412928	7.907432	26.83	29.52	(16.57, 42.46)
19	7.404928	7.938568	18.24	19.00	(10.79, 27.21)
20	3.173696	8.896000	31.14	37.36	(20.33, 54.39)
21	5.893696	9.637704	21.70	22.97	(14.71, 31.22)
22	9.157696	9.236272	17.72	22.69	(11.64, 33.74)
23	4.291072	10.749704	29.92	26.56	(12.21, 40.91)
24	6.890304	10.749704	23.60	21.83	(10.85, 32.81)

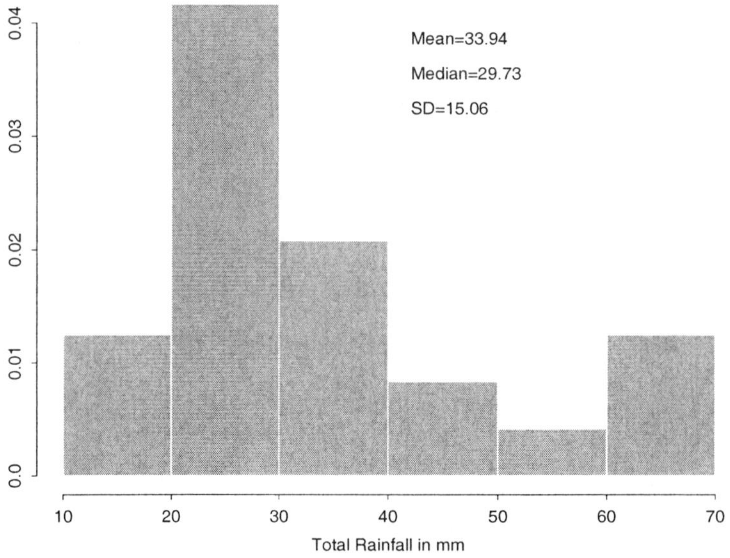

Fig. 7.9 Histogram of weekly rainfall totals (in mm) corresponding to Table 7.1.

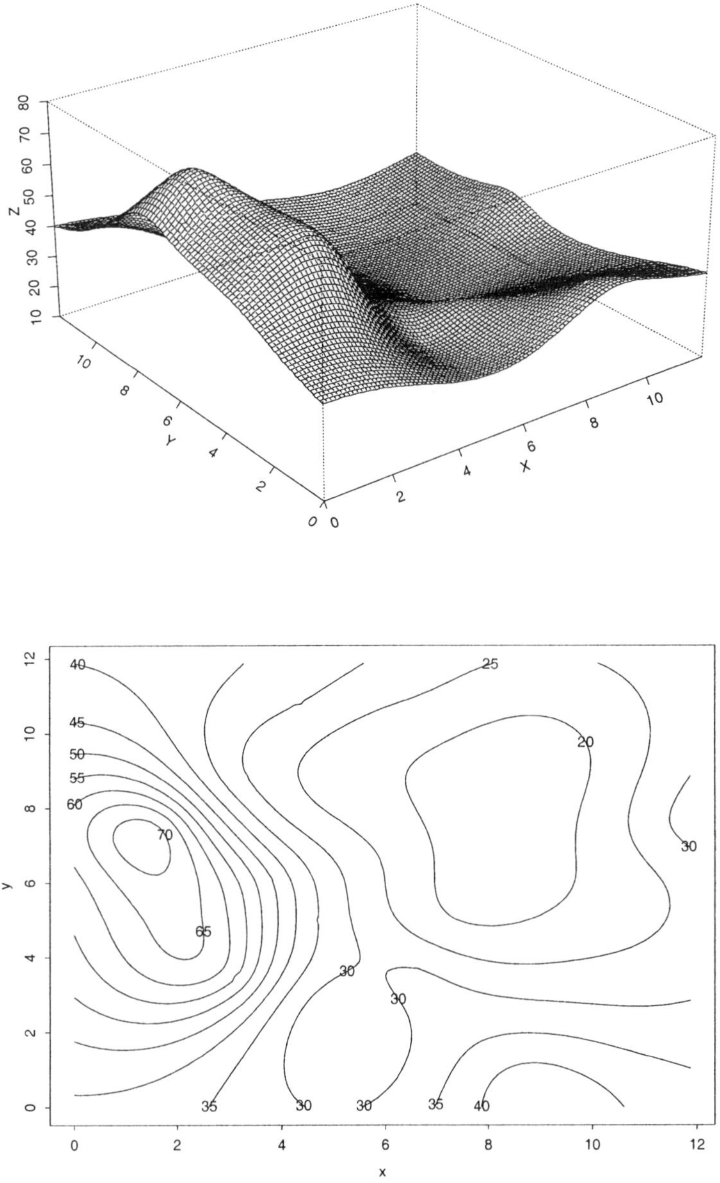

Fig. 7.10 Spatial prediction and contour maps from the Darwin data using Matérn correlation.

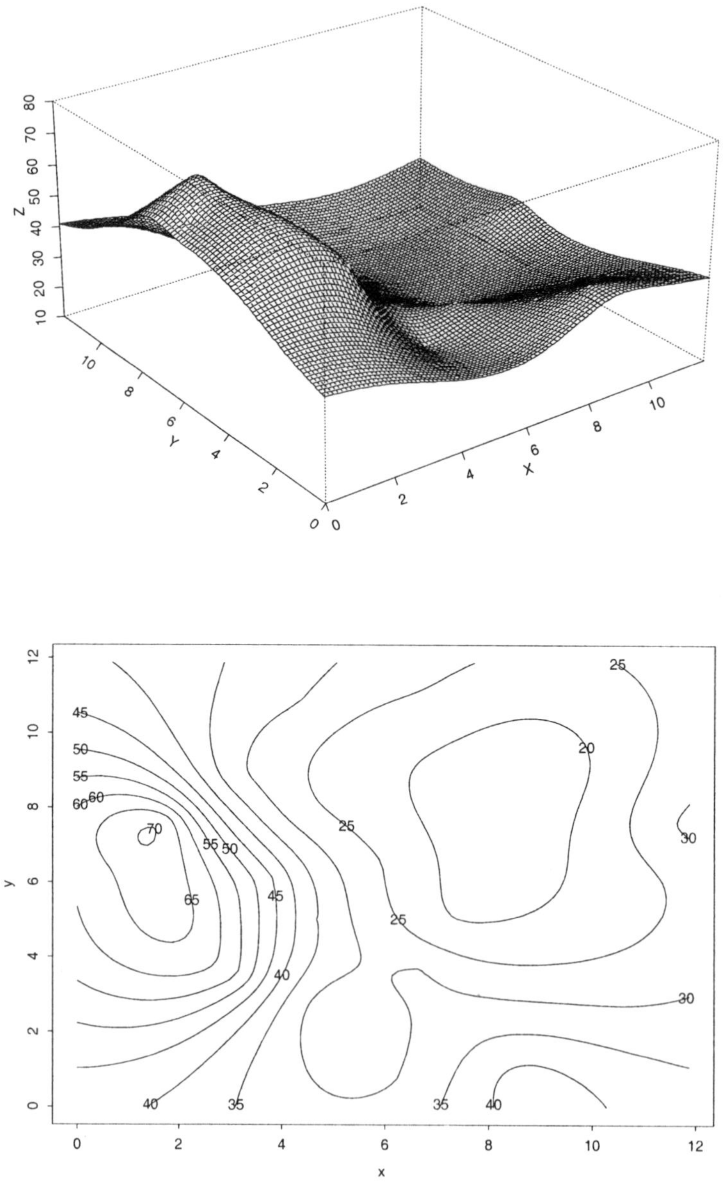

Fig. 7.11 Spatial prediction and contour maps from the Darwin data using exponential correlation.

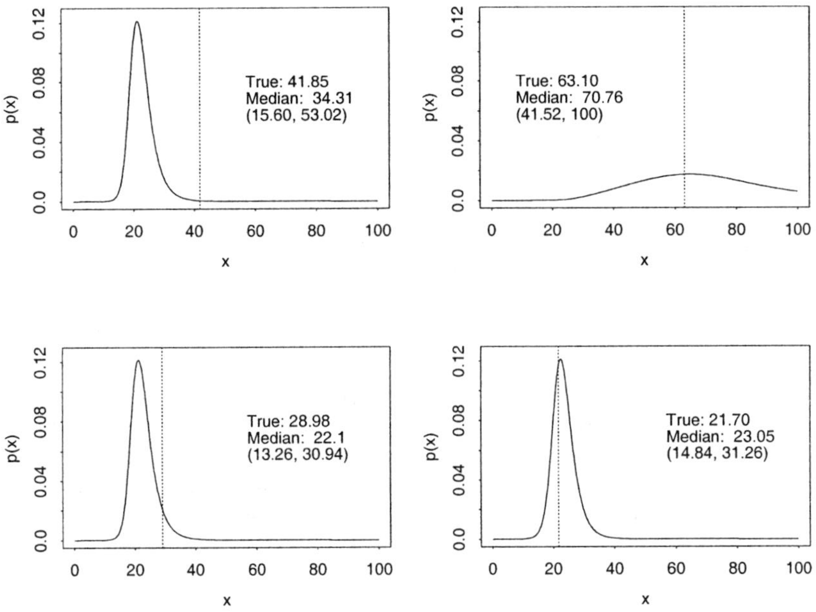

Fig. 7.12 Predictive densities, 95% prediction intervals, and cross-validation: Predicting a true value from the remaining 23 observations using Matérn correlation. The vertical line marks the location of the true value.

7.3.2 Comparison with Kriging

It is instructive to compare the BTG Bayesian method with kriging and trans-Gaussian kriging using artificial data. In the following comparison the prediction from kriging and trans-Gaussian kriging was obtained with a known Box-Cox transformation parameter λ and a known correlation parameter $\boldsymbol{\theta}$, while in the BTG prediction these parameters were not known in accordance with our Bayesian approach. In practice, the parameters must be estimated for all three methods. In this example there were no auxiliary covariates.

A stationary Gaussian random field Y, with a given correlation function, mean 5, and variance 1, was generated on a 50×50 grid. Then for a chosen λ an inverse Box-Cox transformation was applied to Y to produce a random field of interest Z. The data consist of a sample of size 50 points taken from the 2500 Z-values. The results of cross-validation, predicting each sample value from the remaining 49 observations, are summarized in Table 7.2. The entries in the table consist of the mean square prediction error (MSE), average length of the 95% prediction interval (AvePI), and the percentage of observations not contained in their respective prediction intervals (% out).

From Table 7.2 we see that for $\lambda = 0, 0.5$, the kriging prediction intervals (7.8), based on the incorrect Gaussian assumption, are unrealistically narrow and fail to contain most of the true Z-values. This is remedied dramatically for normal data when $\lambda = 1$. Both trans-Gaussian kriging and BTG offer an appreciable improvement in the reliability of the prediction intervals in the non-Gaussian cases, with both methods performing roughly the same despite the unknown parameters disadvantage in the BTG case. In general, BTG tends to give wider albeit realistic intervals as seen in particular from the log-normal data where $\lambda = 0$. The three methods performed similarly for normal data where $\lambda = 1$.

The same comparison was carried out with 50×50 radar reflectivity snapshots from ground radars in two locations. Interestingly, the three methods, all with estimated parameters, gave similar results [278].

7.3.3 Time Series Prediction

The next example is an application of BTG prediction, with the Matérn correlation, constant mean mean function (a mean function linear in t may be more appropriate), to a short time series with 48 observations. For this purpose, the y coordinate was set to 0 at all "locations", $s = (t, y) = (t, 0)$, where the first coordinate represents time. The BTG method treats forward and backward prediction as well as interpolation of missing data in exactly the same way.

Table 7.3 displays four years of monthly time series, not seasonally adjusted, of unemployed women 20 years of age and older from 1997 to 2000. Recall that we have already examined this series in the previous chapter where its plot is given in Figure 6.1. One and two step forward and backward prediction is presented in Figure 7.13. In the forward prediction the 48th observation is predicted from the preceding 47 (one step) and 46 (two step) observations, and similarly in the backward prediction

Table 7.2 Cross-validation comparison between kriging (KRG), trans-Gaussian kriging (TGK), and Bayesian transformed Gaussian (BTG) using exponential, Matérn, spherical, and rational quadratic correlation functions with the indicated parameters. *Source*: [278].

	Exponential($e^{-0.03}$, 1)			Matérn(1,10)		
λ	0	0.5	1	0	0.5	1
KRG MSE	12212.32	1.83	0.13	68397.48	7.15	0.58
TGK MSE	11974.73	1.84	0.13	55260.90	7.08	0.58
BTG MSE	12520.70	1.89	0.14	64134.30	7.31	0.56
KRG AvePI	1.45	1.43	1.45	2.42	2.51	2.42
TGK AvePI	267.92	5.24	1.45	291.80	8.21	2.42
BTG AvePI	466.69	6.10	1.63	330.68	10.23	2.87
KRG % out	98%	64%	2%	100%	48%	6%
TGK % out	20%	4%	2%	18%	8%	6%
BTG % out	6%	2%	2%	12%	6%	6%
	Spherical(50)			Rat. Quad.(0.9,1)		
λ	0	0.5	1	0	0.5	1
KRG MSE	3024.65	1.74	0.16	28423.78	10.35	0.91
TGK MSE	2886.04	1.73	0.16	28362.65	10.35	0.91
BTG MSE	2788.95	1.68	0.15	31120.60	10.42	0.92
KRG AvePI	1.42	1.42	1.42	3.91	3.91	3.91
TGK AvePI	151.64	4.74	1.42	552.24	13.59	3.91
BTG AvePI	150.37	5.58	1.64	305.31	14.12	4.34
KRG % out	100%	56%	6%	98%	56%	2%
TGK % out	12%	6%	6%	4%	0%	2%
BTG % out	10%	4%	4%	20%	2%	4%

of the first observation from the following 47 and 46 observations, respectively. In the four cases considered the true values are captured within the corresponding 95% prediction intervals. As expected, in the two step forward and backward prediction the intervals are longer and the predictive densities are more dispersed, exhibiting a greater variability.

Figure 7.14 shows the results of cross-validation for the unemployed women series where a *missing* observation at time t, $t = 1, ..., 48$, is predicted from the remaining 47 observations. Out of the 48 observations, those at $t = 12, 13, 36$ are outside their 95% prediction intervals so that the coverage rate is 94% approximately. This example shows that the BTG procedure also gives reasonable prediction/interpolation when applied to integer-valued time series.

Table 7.3 Monthly data in thousands, not seasonally adjusted, of unemployed women over 20. *Source*: Bureau of Labor Statistics, Current Population Survey, Series ID LFU22001702.

Month	1997	1998	1999	2000
Jan.	2898	2750	2447	2425
Feb.	2788	2595	2398	2245
Mar.	2718	2554	2381	2298
Apr.	2406	2213	2250	2005
May	2520	2218	2086	2208
June	2645	2449	2397	2379
July	2708	2532	2573	2459
Aug.	2811	2639	2475	2539
Sep.	2666	2449	2299	2182
Oct.	2380	2326	2054	1959
Nov.	2292	2302	2127	2012
Dec.	2187	2065	1935	1834
Avg.	2585	2424	2285	2212

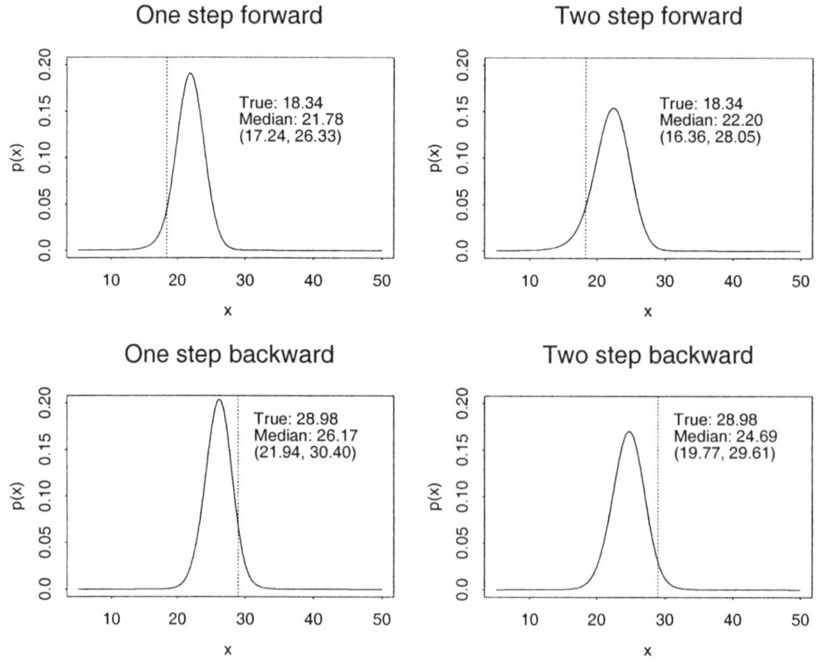

Fig. 7.13 Forward and backward one- and two-step prediction in the unemployed women series. Data in hundreds of thousands.

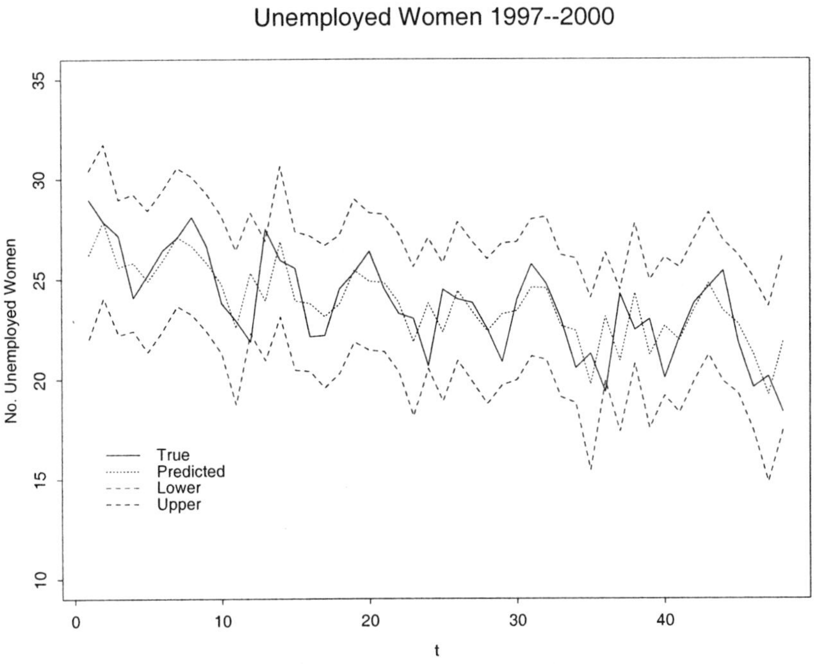

Fig. 7.14 Cross-validation and 95% prediction intervals in the unemployed women series. Observations at times $t = 12, 13, 36$ are outside their 95%. Data in hundreds of thousands.

7.3.4 Seasonal Time Series

The next example is an application of BTG prediction with the Matérn correlation and a constant mean to a short seasonal time series extracted from a longer series. Again, the y coordinate is set to 0 at all "locations", $s = (t, y) = (t, 0)$, where the first coordinate represents time.

Table 7.5 and Figure 7.15 provide a monthly time series of the number (in thousands) of international airline passengers from January 1949 to December 1960. The series is markedly seasonal and evolves about a monotone increasing trend. It is interesting to see the results of BTG prediction applied to this series. We applied the BTG algorithm to a much shorter section of the series of length $N = 36$ from March 53 to February 56 or $t = 51, ..., 86$. The results of cross-validation are plotted in Figure 7.16. Notice that performing cross-validation for the last observation at time $t = 86$ amounts to prediction there from the previous 35 observations. Except for times $t = 62, 63$, all the other true values are within their respective 95% prediction intervals, giving a success rate, that is, an empirical coverage, of 94.4%. The individual predictive densities for times $t = 62$ (miss) and $t = 86$ (success) are shown in Figure 7.17.

The one, two, three, and four step ahead predictors at times $t = 87, 88, 89, 90$ from the data at times $t = 51, ..., 86$ are given in Table 7.4. The first three prediction intervals contain the true observation.

We can see that the BTG algorithm with the Matérn correlation captured the seasonality very well and gave very reasonable prediction and interpolation from a short time series.

Table 7.4 Prediction of the monthly number of airline passengers at times $t = 87, 88, 89, 90$ from data at $t = 51, ..., 86$. The last prediction interval corresponding to the four step ahead prediction does not contain the true observation.

t	z	$\hat{z}$	95% PI
87	317	269.64	(217.92, 321.35)
88	313	263.65	(186.54, 340.76)
89	318	259.75	(168.35, 351.15)
90	374	257.40	(159.92, 354.89)

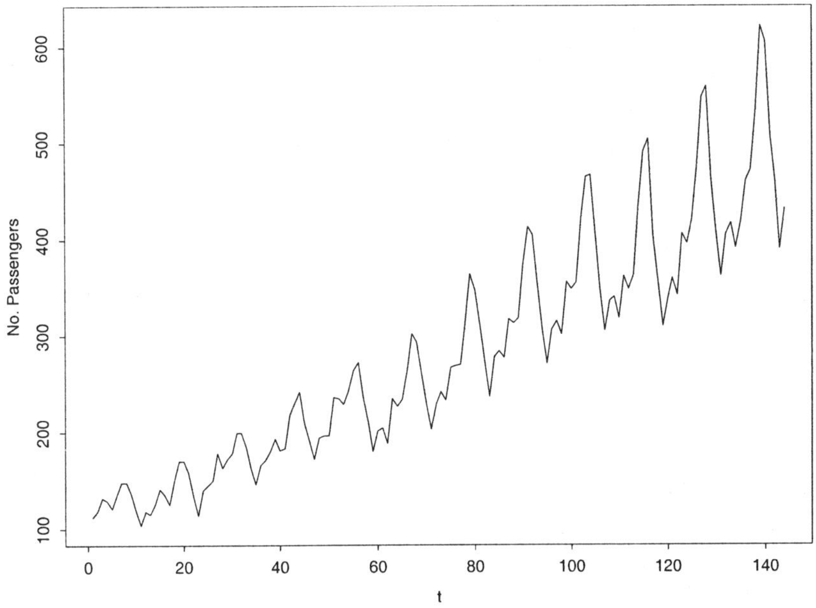

Fig. 7.15 Time series of monthly international airline passengers in thousands, January 1949–December 1960. $N = 144$. *Source*: [63].

Table 7.5 Monthly data in thousands of international airline passengers, January 1949–December 1960. $N = 144$. *Source*: [63]. Reprinted with permission from Pearson Education, Upper Saddle River, NJ.

Year	Jan.	Feb.	Mar.	Apr.	May	June	July	Aug.	Sep.	Oct.	Nov.	Dec.
1949	112	118	132	129	121	135	148	148	136	119	104	118
1950	115	126	141	135	125	149	170	170	158	133	114	140
1951	145	150	178	163	172	178	199	199	184	162	146	166
1952	171	180	193	181	183	218	230	242	209	191	172	194
1953	196	196	236	235	229	243	264	272	237	211	180	201
1954	204	188	235	227	234	264	302	293	259	229	203	229
1955	242	233	267	269	270	315	364	347	312	274	237	278
1956	284	277	317	313	318	374	413	405	355	306	271	306
1957	315	301	356	348	355	422	465	467	404	347	305	336
1958	340	318	362	348	363	435	491	505	404	359	310	337
1959	360	342	406	396	420	472	548	559	463	407	362	405
1960	417	391	419	461	472	535	622	606	508	461	390	432

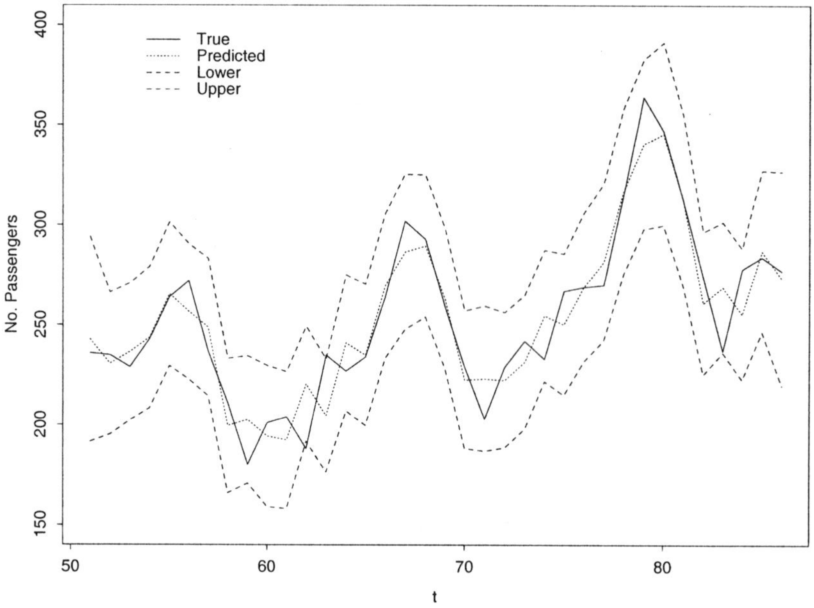

Fig. 7.16 BTG cross-validation and prediction intervals for the monthly airline passengers series, $t = 51, ..., 86$, using Matérn correlation. Observations at $t = 62, 63$ are outside the PI.

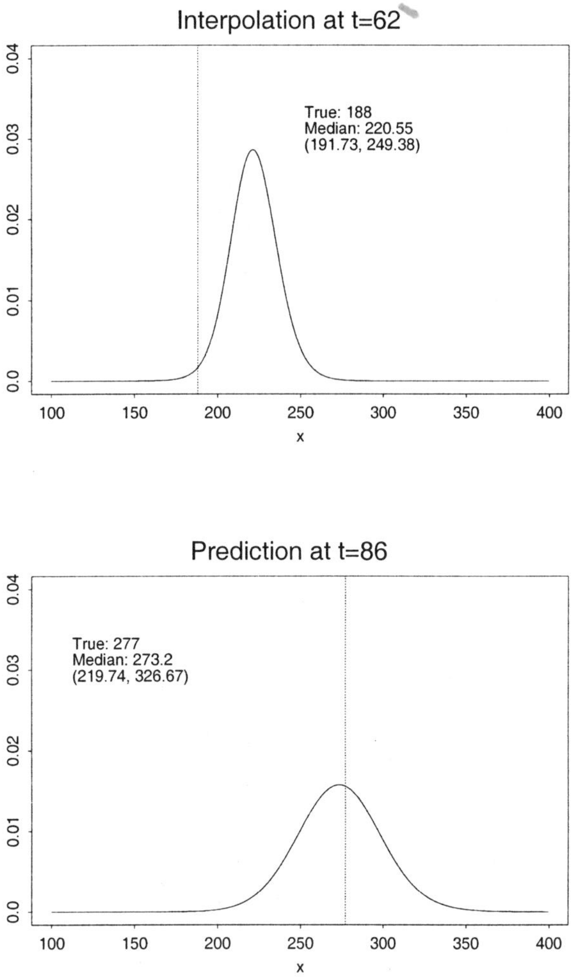

Fig. 7.17 Predictive densities and 95% prediction intervals for $t = 62$ (interpolation) and $t = 86$ (prediction), using Matérn correlation. The vertical line marks the location of the true observation.

7.4 PROBLEMS AND COMPLEMENTS

1. Use the unemployment data in Table 7.3 in the BTG prediction of unemployed women 20 years of age and older in the first six months of the year 2001. Compare the results for various correlation functions, and in each case obtain 95% prediction limits. The true monthly observations in thousands are given in Table 7.6. *Source*: Bureau of Labor Statistics, Current Population Survey, Series ID LFU22001702.

Table 7.6 Monthly data of unemployed women from 01/2001 to 06/2001.

Jan.	2404
Feb.	2329
Mar.	2285
Apr.	2175
May	2245
June	2492

2. Let $\{Z_t\}$ be a weakly stationary mixed first order autoregressive moving average process, ARMA(1,1), with mean 0 and autocovariance $C(h), h = 0, \pm 1, \pm 2,$

$$Z_t - \phi Z_{t-1} = u_t - \theta u_{t-1},$$

where $\{u_t\}$ is a sequence of uncorrelated random variables with mean 0 and variance σ_u^2. Show that:

$$\begin{aligned} C(0) &= \frac{1 + \theta^2 - 2\phi\theta}{1 - \phi^2}\sigma_u^2 \\ C(1) &= \frac{(1 - \phi\theta)(\phi - \theta)}{1 - \phi^2}\sigma_u^2 \\ C(h) &= \phi C(h-1), \; h \geq 2. \end{aligned}$$

3. A die has two states of nature, balanced and unbalanced, with respective prior probabilities. The die is tossed k times to produce a sequence of k 5's. It happens. Use Bayes theorem to show that asymptotically in k the posterior probability that the die is unbalanced given the data goes to either 0 or 1, depending on the likelihood of k 5's given the state of nature.

4. Given p, $X_1, ..., X_n$ are independent Bernoulli(p) random variables. Let $Y = \sum_{i=1}^n X_i$, and assume that the prior of p is Beta(α, β).

   (a) Obtain $\mathrm{E}(p|Y)$, the Bayes estimator of p with respect to quadratic loss.

   (b) Show that $\mathrm{E}(p|Y)$ is a convex combination of the sample mean and the prior mean.

   (c) Show that $X_1, ..., X_n$ are marginally dependent.

(d) Show that the marginal distribution of Y depends on the prior parameters. What do you make of this?

(e) Find the mean and variance of the marginal distribution of Y.

(f) Show that marginally, Y/n converges to p with probability one.

5. *Predicting successes from previous successes in Bernoulli populations* [59]. Consider an infinite Bernoulli population with probability of success θ, and take T random samples of sizes $m_1, m_2, \ldots, m_T$. Let Y_i be the number of successes in sample i, and let X be the number of successes in a subsequent sample of size n. If θ has a uniform prior, show that the posterior is Beta$(\sum_{i=1}^{T} y_i + 1, \sum_{i=1}^{T} m_i - \sum_{i=1}^{T} y_i + 1)$, and the predictive distribution of X given $Y_1, Y_2, \ldots, Y_T$ is beta-binomial,

$$\Pr(X = x \mid y_1, y_2, \ldots, y_T) = \frac{\binom{x+\sum_{i=1}^{T} y_i}{x}\binom{n-x+\sum_{i=1}^{T} m_i - \sum_{i=1}^{T} y_i}{n-x}}{\binom{n+\sum_{i=1}^{T} m_i + 1}{n}}.$$

6. [268], [449]. Derive the distributions of $(\boldsymbol{\beta}|\tau, \boldsymbol{\theta}, \lambda, \mathbf{z})$ and $(\tau|\boldsymbol{\theta}, \lambda, \mathbf{z})$ in (7.14).

7. Derive the distribution of $p(\boldsymbol{\theta}, \lambda|\mathbf{z})$ as given in equation (7.15).

8. *An estimation problem in random fields: Regressing the area average on the fractional area* [256], [386]. In certain situations in remote sensing it is difficult to obtain precise measurements due to the great distance between the measuring instrument and the quantity to be observed. In many cases however, the instrument can still be used qualitatively in deciding when a measurement exceeds a given threshold. This gives rise to the so called *threshold method* described here in terms of rain intensity (speed).

   By rain rate we mean *instantaneous* rain intensity measured in mm/hr at points in space. By a *snapshot* we mean the map of rain rate over a given region at a fixed instant. Such maps can be obtained by a precipitation radar. The *area* refers to the area of a given large region. Consider a snapshot taken at an instant of time, and delineate the fraction of the snapshot where the rate exceeds a fixed threshold τ. This is the instantaneous *fractional area*. It has been observed that the area average and fractional area are highly correlated, where for the optimal threshold τ the sample correlation can exceed in some cases 98%. This suggests that the area average can be inferred from the fractional area by simple linear regression. Once the regression line is estimated, all that is required of an instrument measuring rain rate is the ability to distinguish correctly between rain rate values that are either above or below a given threshold, and not precise point measurements.

   Suppose rain rate is observed over a given region A and throughout a specific period $[0, T]$. Then, each space–time point $\omega \equiv (\mathbf{a}, t)$, $\mathbf{a} \in A, t \in [0, T]$ is endowed with an instantaneous rain rate measurement. The set of all these endowed ω's constitutes the sample space $\Omega \equiv A \times [0, T]$ of interest. Let $X(\omega)$

be the random variable that gives the value of rain rate associated with ω, so that $X : A \times [0, T] \longrightarrow [0, \infty)$. Then, X has a *mixed* distribution because it admits the value 0 (no rain) with positive probability, say, $1 - p$. That is, $P(X = 0) = 1 - p$. But otherwise, that is conditional on rain, X admits values in a continuum and therefore has a continuous distribution. Let $G(x)$ denote the distribution function of X.

a. Argue that $G(x)$ is of a *mixed type*,

$$G(x) \quad = \quad (1 - p)H(x) + pF(x), \tag{7.23}$$

where $H(x)$ is a step function defined by

$$H(x) = \begin{cases} 0 & \text{if } x < 0 \\ 1 & \text{if } x \geq 0 \end{cases}$$

and $F(x)$ is a continuous distribution function with positive density $f(x) = F'(x),\ x > 0$.

b. Argue that under an assumption on $f(x)$, for each snapshot at time t the following linear relationship holds approximately

$$< X_t > \simeq \beta(\tau) < I[X_t > \tau] >, \tag{7.24}$$

where $\beta(\tau)$ is a constant for each given τ, $< X_t >$ is the area average, and $< I[X_t > \tau] >$ is the fractional area.

c. Assume that $f(x)$ is a lognormal density

$$f(x; \mu, \sigma) = \frac{1}{\sqrt{2\pi}\sigma x} \exp[\frac{-(\log(x) - \mu)^2}{2\sigma^2}],\ x > 0,$$

where $-\infty < \mu < \infty, \sigma > 0$. Obtain an approximation to the correlation between $< X_t >$ and $< I[X_t > \tau] >$ as a function of τ, and suggest a way to derive an optimal threshold.

9. Consider a stationary and isotropic random field $\{Z(\mathbf{s})\}$, $\mathbf{s} = (s_1, s_2)' \in R^2$ with covariance function

$$C(\mathbf{h}) = \rho(\sqrt{h_1^2 + h_2^2}),$$

where $\rho(.)$ is some function and $\mathbf{h} = (h_1, h_2)'$. Show that the spectral density function of the field, assuming that it exists, is given in [441]

$$f(\omega_1, \omega_2) = \frac{1}{2\pi} \int_0^\infty \kappa_0(r\eta)\rho(r)r dr,$$

where $\eta = \sqrt{\omega_1^2 + \omega_2^2}$ and κ_0 is the Bessel function of the first kind of zero order. If c_1, c_2 are two positive constants and $\rho(r) = c_1 \exp(-c_2 r)$, what is the form of the spectral density?

APPENDIX
Elements of Stationary Processes

The theory of stationary processes is presented here briefly in its most basic level concerning mainly discrete time processes. All the results carry over to the continuous time case with the appropriate modifications. We highlight the basic notions of autocovariance and autocorrelation and their relation to the spectrum, and discuss briefly methods for autocovariance/autocorrelation estimation.

There are many excellent books on stationary processes and their applications in various fields. References that emphasize spectral and correlation analyses and related statistical procedures and applications are [20], [69], [72], [184], [272], [357], [372] and [441], while more theoretical treatments can be found in [125], and [385]. Model building and statistical inference based on autoregressive moving average processes and certain extensions thereof are discussed in [63].

The Notion of Stationarity

A stochastic or random process $\{Y_t\}$ is a collection of random variables, real or complex-valued, indexed by t where t takes values in some index set T. Examples of T are $[0, 1]$, $[0, \infty)$, $\{\ldots, -2, -1, 0, 1, 2, \ldots\}$, and $\{1, 2, 3, \ldots\}$. For our purposes it is convenient to think of t as time. When T is an interval, such as $(-\infty, \infty)$, the process is called a continuous time process, and it is called a discrete time process when T is a discrete set such as $\{\ldots, -2, -1, 0, 1, 2, \ldots\}$. In the sequel we confine ourselves almost exclusively to discrete time processes. But clearly, regardless of the type of the time parameter t, the random variables Y_t may be continuous (e.g. monthly average temperature), discrete (e.g. daily number of customers), or of mixed type (e.g. hourly rainfall amount).

For any collection of n time points $t_1 < t_2 < \ldots < t_n$, the probability distribution of the random vector $(Y_{t_1}, Y_{t_2}, \ldots, Y_{t_n})$,

$$P(Y_{t_1} \leq y_1, Y_{t_2} \leq y_2, \ldots, Y_{t_n} \leq y_n)$$

is called a *finite dimensional (probability) distribution*. In principle, we must specify the finite dimensional distributions in order to describe the stochastic process probabilistically. In practice, however, we are quite content with much less and instead of distributions provide some moment conditions to describe the process.

From the finite dimensional distributions we can obtain probabilities and, when they exist, moments. Suppose $f_{t_1}(y)$ is the probability density function of Y_{t_1}. Then the mean (first moment) of Y_{t_1} is given by

$$\text{E}[Y_{t_1}] = \int_{-\infty}^{\infty} y f_{t_1}(y) dy.$$

In general, the mean may be different at different time points, and if we define $m(t) \equiv \text{E}[Y_t]$, the result is a function of time called the *mean function*. The mean function may assume any form, but $\text{E}[Y_t - m(t)] = 0$ always.

Suppose $f_{t_1,t_2}(y_1, y_2)$ is the joint probability density function of (Y_{t_1}, Y_{t_2}). Then a second-order moment is given by the expectation

$$\text{E}[Y_{t_1} Y_{t_2}] = \int_{-\infty}^{\infty} \int_{-\infty}^{\infty} y_1 y_2 f_{t_1,t_2}(y_1, y_2) dy_1 dy_2.$$

Recall that the covariance $\text{Cov}[Y_{t_1}, Y_{t_2}]$ of Y_{t_1} and Y_{t_2} is given by

$$\text{Cov}[Y_{t_1}, Y_{t_2}] = \text{E}\{[Y_{t_1} - \text{E}(Y_{t_1})][Y_{t_2} - \text{E}(Y_{t_2})]\} = \text{E}[Y_{t_1} Y_{t_2}] - \text{E}[Y_{t_1}]\text{E}[Y_{t_2}].$$

Clearly, $\text{Cov}[Y_{t_1}, Y_{t_1}] = \text{Var}[Y_{t_1}]$. The function $c(s,t) \equiv \text{Cov}[Y_s, Y_t]$ is called the *covariance* or *autocovariance function*.

The *correlation* or *autocorrelation function*, $\rho(s,t)$, is defined from the normalized covariance function

$$\rho(s,t) \equiv \frac{c(s,t)}{\sqrt{c(s,s)}\sqrt{c(t,t)}}$$

and satisfies $\rho(s,t) = \rho(t,s)$, and $|\rho(s,t)| \leq \rho(t,t) = 1$.

An important example of a stochastic process is a *Gaussian* or *normal* process. A real-valued process $\{Y_t\}$, $t \in T$, is called Gaussian process if for all $t_1, t_2, \ldots, t_n \in T$, the joint distribution of $(Y_{t_1}, Y_{t_2}, \ldots, Y_{t_n})$ is multivariate normal.

Since a multivariate normal is completely specified by the vector of means and the covariance matrix, the finite dimensional distributions are completely determined from the mean function $m(t) = \mathrm{E}[Y_t]$ and the covariance function $c(s, t) = \mathrm{Cov}[Y_s, Y_t]$. Thus, for a Gaussian process we need to supply only first- and second-order moment information in order to obtain the finite dimensional distributions, a fact which makes Gaussian processes easier to handle than many other types of processes.

A stochastic process $\{Y_t\}$ is said to be a *strictly stationary* process if the joint distribution of $(Y_{t_1}, Y_{t_2}, \ldots, Y_{t_n})$ is the same as the joint distribution of the shifted vector $(Y_{t_1+\tau}, Y_{t_2+\tau}, \ldots, Y_{t_n+\tau})$ for all $t_1, t_2, \ldots, t_n$, n, and τ. In other words, as long as the relative distances between the time points are fixed, the joint distribution does not change.

Strict stationarity implies that Y_t and $Y_{t+\tau}$ have the same distribution for all t and τ. Hence, if first-order moments exist, put $\tau = -t$ to see that

$$\mathrm{E}[Y_t] = \mathrm{E}[Y_{t+\tau}] = \mathrm{E}[Y_0] = m, \tag{A.1}$$

where m is a constant. Likewise, stationarity implies that (Y_t, Y_s) and $(Y_{t+\tau}, Y_{s+\tau})$ have the same distribution for all t, s, and τ, and in particular for $\tau = -s$. Therefore, if second-order moments exist,

$$\mathrm{E}[Y_t Y_s] = \mathrm{E}[Y_{t+\tau} Y_{s+\tau}] = \mathrm{E}[Y_{t-s} Y_0]$$

or

$$c(t, s) = \mathrm{Cov}[Y_t, Y_s] = c(t - s),$$

where we use the same symbol c in an obvious way. Thus, the autocovariance function of a real valued stationary process is a function of the time *lag* τ only,

$$c(\tau) = \mathrm{Cov}[Y_{t+\tau}, Y_t] = \mathrm{Cov}[Y_\tau, Y_0]. \tag{A.2}$$

Clearly, $\mathrm{Var}[Y_t] = c(0)$, and $c(\tau) = c(-\tau)$. Likewise, the autocorrelation is a function of τ only,

$$\rho(\tau) = \frac{c(\tau)}{c(0)}, \tag{A.3}$$

assuming that $c(0) \neq 0$. The autocorrelation $\rho(\tau)$ measures the correlation between $Y_{t+\tau}$ and Y_t as a function of the indices difference independently of t. Obviously, $\rho(\tau) = \rho(-\tau)$, and $|\rho(\tau)| \leq \rho(0) = 1$.

A great deal of the theory of stationary processes only requires the fulfillment of the conditions (A.1) and (A.2). In general, a process that satisfies (A.1) and (A.2) is called *weakly stationary* or *stationary in the wide sense* or sometimes is said to be *second-order stationary*. A strictly stationary process need not be weakly stationary unless second order moments exist, and conversely, since weak stationarity is less restrictive it does not imply strict stationarity.

Complex-Valued Stationary Processes

So far we have spoken of real-valued processes; however, there is nothing special about real processes, and the theory can easily be extended to cover complex-valued processes as well.

A complex random variable X has the general form $X = U + iV$ where U, V are real random variables. We have, the complex conjugate $\overline{X} = U - iV$, $|X|^2 = X\overline{X} = U^2 + V^2$, $\mathrm{E}[X] \equiv \mathrm{E}[U] + i\mathrm{E}[V]$, $\mathrm{E}[|X|^2] = \mathrm{E}[U^2] + \mathrm{E}[V^2]$. For two complex-valued random variables X, Y, the covariance is defined by

$$\mathrm{Cov}(X, Y) \equiv E\{[X - \mathrm{E}(X)]\overline{[Y - \mathrm{E}(Y)]}\} = \mathrm{E}[X\overline{Y}] - \mathrm{E}[X]\mathrm{E}[\overline{Y}]$$

and $\mathrm{Var}[X] = \mathrm{Cov}[X, X] = \mathrm{E}[|X - \mathrm{E}(X)|^2]$. When $\mathrm{E}(X) = \mathrm{E}(Y) = 0$, then $\mathrm{Cov}(X, Y) = \mathrm{E}[X\overline{Y}]$, and when $\mathrm{Cov}(X, Y) = 0$, we say that X and Y are uncorrelated.

Similarly, we define a complex-valued stochastic process $\{Y_t\}$ from two real-valued processes $\{U_t\}$,$\{V_t\}$,

$$Y_t = U_t + iV_t.$$

The complex process is described probabilistically from the finite dimensional distributions of the vectors, $(U_{t_1}, U_{t_2}, \ldots, U_{t_n}, V_{t_1}, V_{t_2}, \ldots, V_{t_n})$, for all possible time values $t_1, t_2, \ldots, t_n$, and n, and is said to be strictly stationary if the finite dimensional distributions remain unaltered under time shifts, just as in the real case. When $\{Y_t\}$ is strictly stationary, so are the real-valued processes $\{U_t\}$ and $\{V_t\}$.

Wide or weak sense stationarity is defined as in the real case with the appropriate modifications. Thus we say that $\{Y_t\}$ is weakly stationary if the mean is a (complex) constant m, say,

$$\mathrm{E}[Y_t] = \mathrm{E}[U_t] + i\mathrm{E}[V(t)] = m$$

and the autocovariance is a function of the time lag,

$$c(k) = \mathrm{E}[Y_{t+k} - m]\overline{[Y_t - m]} = \mathrm{E}[Y_{t+k}\overline{Y_t}] - |m|^2.$$

Observe that the variance $c(0) = \mathrm{E}[|Y_t - m|^2]$ is always real-valued. As in the real case, the autocorrelation function is obtained as the quotient (assuming, of course, $c(0) \neq 0$),

$$\rho(k) = \frac{c(k)}{c(0)}.$$

We have

$$c(-k) = \overline{c(k)}, \qquad |c(k)| \leq c(0) \tag{A.4}$$

and

$$\rho(-k) = \overline{\rho(k)}, \qquad |\rho(k)| \leq 1.$$

Observe that for all complex numbers $a_1, a_2, \ldots, a_N$, and integers $t_1, t_2, \ldots, t_N$, with $N \geq 1$,

$$0 \leq \text{Var}\left[\sum_{j=1}^{N} a_j Y_{t_j}\right] = \text{Cov}\left[\sum_{j=1}^{N} a_j Y_{t_j}, \sum_{l=1}^{N} a_l Y_{t_l}\right]$$

$$= \sum_{j=1}^{N}\sum_{l=1}^{N} a_j \overline{a_l} \text{Cov}[Y_{t_j}, Y_{t_l}] = \sum_{j=1}^{N}\sum_{l=1}^{N} a_j \overline{a_l} c(t_j - t_l). \qquad \text{(A.5)}$$

That is, $c(k)$ is *nonnegative definite*. The important implication of this ostensibly simple fact is that $c(k)$ admits a Fourier or spectral representation in terms of a uniquely defined bounded monotone nondecreasing real function F such that $F(-\pi) = 0$,

$$c(k) = \int_{-\pi}^{\pi} e^{ik\lambda} dF(\lambda), \qquad k = 0, \pm 1, \pm 2, \ldots \qquad \text{(A.6)}$$

This fact is attributed to the German mathematician G. Herglotz (1911) [184, p. 34], and $F(\lambda)$ is called *spectral distribution function*. Conversely, it is easy to see that a complex-valued sequence $c(k)$ which admits the Fourier representation (A.6) is nonnegative definite.

By extending the definition of $F(\lambda)$ such that $F(\lambda) = 0$ for $\lambda \leq -\pi$, and $F(\lambda) = F(\pi)$ for $\lambda \geq \pi$, $F(\lambda)$ becomes, except for a normalization by a constant, a probability distribution function. *Thus, in summary, the autocovariance of every weakly stationary process in discrete time admits a Fourier representation in terms of a uniquely defined distribution function which, up to a constant, is a probability distribution function supported on* $(-\pi, \pi]$.

As is the case with probability distributions, a spectral distribution function may have a *spectral density* $f(\lambda)$ such that

$$F(\lambda) = \int_{-\pi}^{\lambda} f(\omega) d\omega, \qquad f(\lambda) = F'(\lambda), \qquad -\pi \leq \lambda \leq \pi.$$

A sufficient condition for this is the absolute summability of the autocovariance, $\sum_{k=-\infty}^{\infty} |c(k)| \leq \infty$, in which case we have the inverse transform,

$$f(\lambda) = \frac{1}{2\pi} \sum_{k=-\infty}^{\infty} e^{-i\lambda k} c(k). \qquad \text{(A.7)}$$

Everything we said about discrete time extends naturally to continuous time as well. The continuous time version of (A.6) is referred to as the *Wiener-Khintchine relationship* or *Wiener-Khintchine theorem* [441, p. 93],

An I.I.D. Sequence: Suppose $\{Y_t\}$ constitutes a sequence of independently and identically distributed complex-valued random variables. Then $\{Y_t\}$ is strictly stationary. If in addition $\text{E}[Y_t] = m$, and $\text{Var}[Y_t] = \sigma^2 < \infty$, then,

$$c(k) = \begin{cases} \sigma^2, & \text{for } k = 0 \\ 0, & \text{for } k = \pm 1, \pm 2, \pm 3, \ldots \end{cases}$$

and the process is also wide sense stationary. In this case we can write

$$c(k) = \frac{\sigma^2}{2\pi} \int_{-\pi}^{\pi} e^{ik\lambda} d\lambda$$

and this defines uniquely the spectral distribution function as

$$F(\lambda) = \frac{\sigma^2}{2\pi}(\lambda + \pi), \qquad -\pi \leq \lambda \leq \pi,$$

which can also be expressed in terms of the spectral density $f(\lambda) = \sigma^2/2\pi$,

$$F(\lambda) = \int_{-\pi}^{\lambda} f(\omega) d\omega = \frac{\sigma^2}{2\pi} \int_{-\pi}^{\lambda} d\omega, \qquad -\pi \leq \lambda \leq \pi.$$

We observe, in particular, that for any sequence of uncorrelated real- or complex-valued random variables $\{u_t\}$ (this is a much weaker assumption than i.i.d.) with mean 0 and variance σ^2,

$$c_u(k) \equiv \mathrm{E}[u_t \overline{u_{t-k}}] = \begin{cases} \sigma^2, & \text{for } k = 0 \\ 0, & \text{for } k = \pm 1, \pm 2, \pm 3, \ldots \end{cases}$$

has the same autocovariance as does the preceding i.i.d. sequence. Therefore, such a sequence is weakly stationary with a flat spectral density $f(\lambda) = \sigma^2/2\pi$, $-\pi \leq \lambda \leq \pi$. For this reason a sequence of uncorrelated random variables with constant mean and variance is termed *white noise* by analogy with white light which consists of waves with the same power regardless of frequency. As a rule, we always take the mean of white noise to be 0.

The Spectrum and the Zero–Crossing Rate: The spectrum, or equivalently the autocorrelation, gives useful information about the oscillation of a stationary time series. But not necessarily complete information, since the spectrum does not determine the finite dimensional distributions. To see this, consider the case of zero-crossings by a zero-mean and unit-variance stationary Gaussian process in continuous time. The expected number of zero-crossings per unit time under mild conditions is given by *Rice's formula* as (see [252, Ch. 4]),

$$\frac{1}{\pi}\sqrt{-\rho''(0)}.$$

However, it is possible to construct examples of non–Gaussian processes where the true formula for the expected zero–crossing rate takes on the form

$$\frac{\kappa}{\pi}\sqrt{-\rho''(0)}$$

with κ smaller or greater than 1. For example, for odd powers of a zero-mean unit-variance Gaussian process $\kappa \leq 1$, and for products of independent Gaussian processes each with zero-mean and unit-variance $\kappa \geq 1$. This implies that there are two different processes of which one is Gaussian and both having the same spectrum, but the non-Gaussian process has a higher zero-crossing rate on average [28]. In numerous applications Rice's formula is used routinely ignoring the Gaussian requirement and thus running the risk of underestimating the oscillation rate.

Regarding oscillation, a related question is whether or not the *observed* rate of sign changes in a stationary time series converges to a *constant* as the series size increases. This surprisingly elusive problem has an answer in the Gaussian case in terms of the spectrum as follows. Let $Y_1, ..., Y_N$ be a real-valued stationary Gaussian time series with mean 0 and spectral distribution $F(\lambda)$. Also, let $X_t = 1$ when $Y_t \geq 0$ and $X_t = 0$ otherwise. The observed sign change rate, or zero-crossing rate in discrete time, $\hat{\gamma}$, is defined in terms of the indicators $d_t \equiv (X_t - X_{t-1})^2$,

$$\hat{\gamma} = \frac{1}{N-1} \sum_{t=2}^{N} d_t.$$

When the spectrum is continuous, $\hat{\gamma}$ converges to a constant almost surely (a.s.). However, two or more jumps (atoms) in $[0, \pi]$ in the spectrum of $\{Y_t\}$ prevent $\hat{\gamma}$ from converging to a *constant* almost surely, though it always converges almost surely to a random variable because of the Gaussian assumption and hence strict stationarity. For a single jump in $[0, \pi]$ in the spectrum there are cases when $\hat{\gamma}$ converges to a constant. This occurs, for example, when $\{Y_t\}$ is a pure random sinusoid [252, Ch. 6], [257].

The Stationary AR(1) Process: Let $\{\epsilon_t\}, t = 0, \pm 1, \pm 2, \ldots$, be a sequence of uncorrelated real-valued random variables with mean zero and variance σ_ϵ^2 and define a real-valued weakly stationary process $\{Y_t\}$ by the stochastic difference equation

$$Y_t = \phi_1 Y_{t-1} + \epsilon_t, \qquad t = 0, \pm 1, \pm 2, \ldots, \tag{A.8}$$

where $|\phi_1| < 1$. The process (A.8) is called a *first-order autoregressive process* and is commonly denoted by $AR(1)$. By repeated substitution we obtain

$$Y_t = \epsilon_t + \phi_1 \epsilon_{t-1} + \phi_1^2 \epsilon_{t-2} + \ldots + \phi_1^{p-1} \epsilon_{t-(p-1)} + \phi_1^p Y_{t-p}.$$

Observe now that since the process is weakly stationary, $\mathrm{E}[Y_t^2]$ is equal to a finite constant uniformly in t. This, coupled with the fact that $|\phi_1| < 1$, gives as $p \to \infty$,

$$\mathrm{E}[Y_t - (\epsilon_t + \phi_1 \epsilon_{t-1} + \phi_1^2 \epsilon_{t-2} + \ldots + \phi_1^{p-1} \epsilon_{t-(p-1)})]^2 = \phi_1^{2p} \mathrm{E}[Y_{t-p}^2] \to 0.$$

That is,

$$Y_t = \sum_{j=0}^{\infty} \phi_1^j \epsilon_{t-j}, \tag{A.9}$$

where the sum converges in mean square. In fact, with $|\phi_1| < 1$, the representation (A.9) is *the causal stationary solution* of the stochastic difference equation (A.8).

In other words, with $|\phi_1| < 1$, there exists a stationary solution in terms of present and past ϵ_t, and hence our stationarity assumption is not without foundation. Since limits in mean square and "E" commute, we find

$$\mathrm{E}[Y_t] = \mathrm{E}\left\{\lim_{n\to\infty}\sum_{j=0}^{n}\phi_1^j\epsilon_{t-j}\right\} = \lim_{n\to\infty}\mathrm{E}\left\{\sum_{j=0}^{n}\phi_1^j\epsilon_{t-j}\right\} = 0.$$

Thus $\mathrm{E}[Y_t] = 0$ for all t. Similarly, by the orthogonality of the ϵ_t ($\mathrm{E}[\epsilon_t\epsilon_s] = 0,\ s \neq t$)

$$\begin{aligned} c(0) = \mathrm{E}[Y_t^2] &= \mathrm{E}\left\{\lim_{m,n\to\infty}\sum_{j=0}^{m}\phi_1^j\epsilon_{t-j}\sum_{k=0}^{n}\phi_1^k\epsilon_{t-k}\right\} \\ &= \lim_{m,n\to\infty}\mathrm{E}\left\{\sum_{j=0}^{m}\phi_1^j\epsilon_{t-j}\sum_{k=0}^{n}\phi_1^k\epsilon_{t-k}\right\} \\ &= \lim_{m,n\to\infty}\sum_{j=0}^{m}\sum_{k=0}^{n}\phi_1^{j+k}\mathrm{E}[\epsilon_{t-j}\epsilon_{t-k}] \\ &= \sum_{j=0}^{\infty}\phi_1^{2j}\sigma_\epsilon^2 = \frac{\sigma_\epsilon^2}{1-\phi_1^2}. \end{aligned} \tag{A.10}$$

In the same way, using the representation (A.9) and the orthogonality of the ϵ_t, we obtain $\mathrm{E}[\epsilon_t Y_{t-k}] = 0,\ k = 1, 2, \ldots$, so that by multiplying both sides of (A.8) by Y_{t-k} and taking expectations,

$$c(k) = \phi_1 c(k-1).$$

This gives $\rho(1) = \phi_1$, and more generally, since $c(k) = c(-k)$,

$$\rho(k) = \phi_1^{|k|}, \qquad k = 0, \pm1, \pm2, \ldots \tag{A.11}$$

From this and (A.10) we obtain the autocovariance as

$$c(k) = \frac{\sigma_\epsilon^2\phi_1^{|k|}}{1-\phi_1^2}, \qquad k = 0, \pm1, \pm2, \ldots \tag{A.12}$$

Since the autocovariance is absolutely summable, a spectral density exists and is given by

$$\begin{aligned} f(\lambda) &= \frac{c(0)}{2\pi}\sum_{k=-\infty}^{\infty}\phi_1^{|k|}\cos(k\lambda) \\ &= \frac{c(0)}{2\pi}\left\{1 + \sum_{k=1}^{\infty}\phi_1^k[\exp(ik\lambda) + \exp(-ik\lambda)]\right\} \end{aligned}$$

$$
\begin{aligned}
&= \frac{c(0)}{2\pi}\left\{1+\frac{\phi_1 e^{i\lambda}}{1-\phi_1 e^{i\lambda}}+\frac{\phi_1 e^{-i\lambda}}{1-\phi_1 e^{-i\lambda}}\right\}\\
&= \frac{c(0)}{2\pi}\cdot\frac{1-\phi_1^2}{1-2\phi_1\cos(\lambda)+\phi_1^2}\\
&= \frac{\sigma_\epsilon^2}{2\pi}\cdot\frac{1}{1-2\phi_1\cos(\lambda)+\phi_1^2}, \qquad -\pi\le\lambda\le\pi. \qquad \text{(A.13)}
\end{aligned}
$$

Thus, the (weakly) stationary $AR(1)$ process has a continuous spectrum.

The $AR(1)$ process is special in that its parameter $\phi_1 (= \rho(1))$ single-handedly controls both the autocorrelation and the normalized spectral density (i.e. the density divided by $c(0)$ so that the total area under the curve is 1).

The Stationary AR(p) Process: The stationary $AR(1)$ process (A.8) can be generalized by extending the order of the stochastic difference equation,

$$Y_t = \phi_1 Y_{t-1} + \phi_2 Y_{t-2} + \ldots + \phi_p Y_{t-p} + \epsilon_t, \qquad t = 0, \pm1, \pm2, \ldots \qquad \text{(A.14)}$$

and $\{\epsilon_t\}$ is real-valued white noise as in the $AR(1)$ process. As such this defines a real-valued process referred to as an *autoregressive process of order* p, or simply $AR(p)$. In order to guarantee stationarity, the ϕ_j must satisfy certain restrictions expressed in terms of the roots (zeros) of the *characteristic equation*

$$\phi(y) \equiv 1 - \phi_1 z - \phi_2 z^2 - \ldots - \phi_p z^p = 0.$$

It turns out if the roots of the characteristic polynomial are all outside the unit circle, then (A.14) has a unique (weakly) stationary solution given by the infinite mean square convergent sum,

$$Y_t = \sum_{j=0}^{\infty} h_j \epsilon_{t-j}. \qquad \text{(A.15)}$$

This implies that $\mathrm{E}[Y_t] = 0$ for all t. The thing to notice is that only past and present ϵ_t are involved, and that (A.9) is only a special case with $h_j = \phi_1^j$. A representation of the form (A.15) is called an *infinite moving average*. To emphasize the fact that only past and present ϵ_t are involved, we say that the infinite moving average is *one-sided*, in which case the process is said to be *linear*.

With $k > 0$, multiply both sides of (A.14) by Y_{t-k} and take expectations. This gives a p'th order difference equation,

$$c(k) = \phi_1 c(k-1) + \phi_2 c(k-2) + \ldots + \phi_p c(k-p)$$

or, dividing by $c(0)$,

$$\rho(k) = \phi_1 \rho(k-1) + \phi_2 \rho(k-2) + \ldots + \phi_p \rho(k-p), \qquad k > 0. \qquad \text{(A.16)}$$

For $k = 1, 2, \ldots, p$, we obtain from (A.16) p linear equations called the *Yule-Walker* equations which allow us to determine the AR parameters $\phi_1, \phi_2, \ldots, \phi_p$ from $\rho(1), \rho(2), \ldots, \rho(p)$, where the latter are estimated from data.

The general solution of (A.16) is [61, p. 55],

$$\rho(k) = A_1 G_1^k + A_2 G_2^k + \ldots + A_p G_p^k,$$

where the G_j's are the reciprocal of the roots of $\phi(y) = 0$. Thus $|G_j| < 1$, $j = 1, \ldots, p$, which means that $\rho(k)$ falls off *exponentially fast* as $|k|$ increases. The fact that $\rho(k)$ is absolutely summable implies the existence of a spectral density.

ARMA Processes: An *autoregressive moving average* (ARMA) process is defined by as

$$Y_t = \phi_1 Y_{t-1} + \ldots + \phi_p Y_{t-p} + \epsilon_t - \theta_1 \epsilon_{t-1} - \ldots - \theta_q \epsilon_{t-q}, \qquad \text{(A.17)}$$

where $\{\epsilon_t\}$ is (real-valued) white noise with mean 0 and variance σ_ϵ^2, The process $\{Y_t\}$, $t = 0, \pm 1, \pm 2, \ldots$, is called *autoregressive moving average of order (p,q)* abbreviated to $ARMA(p,q)$ or *mixed ARMA(p,q)* [61]. With

$$\mathcal{P}(y) = 1 - \phi_1 y - \phi_2 y^2 - \ldots - \phi_p y^p$$

$$\mathcal{Q}(y) = 1 - \theta_1 y - \theta_2 y^2 - \ldots - \theta_q y^q$$

(A.17) becomes

$$\mathcal{P}(\mathcal{B}) Y_t = \mathcal{Q}(\mathcal{B}) \epsilon_t, \qquad t = 0, \pm 1, \pm 2, \ldots \qquad \text{(A.18)}$$

A sufficient condition for (wide sense) stationarity is that the roots of $\mathcal{P}(y)$ lie outside the unit circle. In this case, the spectral density of $\{Y_t\}$ is

$$f_y(\lambda) = \frac{|\mathcal{Q}(e^{-i\lambda})|^2}{|\mathcal{P}(e^{-i\lambda})|^2} f_\epsilon(\lambda) = \frac{\sigma_\epsilon^2}{2\pi} \cdot \frac{|\mathcal{Q}(e^{-i\lambda})|^2}{|\mathcal{P}(e^{-i\lambda})|^2}, \qquad -\pi \le \lambda \le \pi. \qquad \text{(A.19)}$$

This clearly is a *rational* function of $e^{-i\lambda}$. When also the roots of $\mathcal{Q}(y)$ lie outside the unit circle, the process is *invertible*, meaning that $\epsilon_t = [\mathcal{P}(\mathcal{B})/\mathcal{Q}(\mathcal{B})] Y_t$, which is seen to be an infinite autoregression.

Estimation of the Autocorrelation: Given a real-valued stationary time series, $Y_1, Y_2, \ldots, Y_N$, the autocovariance function $c(k)$ is usually estimated by the sample autocovariance [357, p. 330],

$$\hat{c}(k) = \frac{1}{N} \sum_{t=1}^{N-k} (Y_t - \hat{m})(Y_{t+k} - \hat{m}), \qquad k = 1, 2, 3, \ldots, \qquad \text{(A.20)}$$

where

$$\hat{m} = \frac{1}{N} \sum_{t=1}^{N} Y_t$$

is the *sample mean*. The autocorrelation $\rho(k)$ is often estimated by the *sample autocorrelation* (sample ACF)

$$\hat{\rho}(k) = \frac{\hat{c}(k)}{\hat{c}(0)}, \qquad k = 1, 2, 3, \ldots \qquad \text{(A.21)}$$

By substituting the $\hat{\rho}(k)$'s for the $\rho(k)$'s, we obtain estimates for the ϕ_k's by solving the Yule-Walker equations. The sample ACF may provide useful information even when the series is not technically stationary.

Curiously, the sample autocovariance $\hat{c}(k)$ is not necessarily consistent in the presence of a discrete spectral component (i.e. a sum of sinusoids), a fact often ignored. Consistency can be achieved when the discrete frequencies are well removed from 0 and $\pm\pi$, their differences are well removed from $\pm 2\pi$, and the corresponding amplitudes have the form $\sqrt{\mathrm{E}|\xi_j|^2} e^{i\phi_j}$, $j = 0, 1, 2, \ldots$, where the ϕ_j are random phases, compensating appropriately with additional components for real data [214]. *Thus, if the series is Gaussian and its spectrum is mixed due to the presence of sinusoidal components, its sample autocovariance is not consistent. The sample autocovariance of a stationary Gaussian time series is consistent when its spectrum is continuous.*

The sample ACF $\hat{\rho}(k)$ of Gaussian white noise has useful asymptotic properties. For large N and different lags $k_1, \ldots, k_m$, none equal to 0, $\hat{\rho}(k_1), \ldots, \hat{\rho}(k_m)$ tend to be independent random variables, each normally distributed with mean 0 and variance $1/N$ [63].

A useful alternative to the sample ACF $\hat{\rho}(k)$ is obtained by assuming a parametric form $\rho(k; \boldsymbol{\theta})$ and then estimating $\boldsymbol{\theta}$ by one of several statistical methods including maximum likelihood, least squares, and Bayesian procedures [98], [118]. An example is furnished by the Matérn correlation,

$$\rho(k; \boldsymbol{\theta}) = \begin{cases} \frac{1}{2^{\theta_2 - 1}\Gamma(\theta_2)} \left(\frac{k}{\theta_1}\right)^{\theta_2} \kappa_{\theta_2}\left(\frac{k}{\theta_1}\right) & \text{if } k \neq 0 \\ 1 & \text{if } k = 0, \end{cases}$$

where $\boldsymbol{\theta} = (\theta_1, \theta_2)$, $\theta_1 > 0, \theta_2 > 0$, and κ_{θ_2} is a modified Bessel function of the third kind of order θ_2. This type of correlation function is often used in modeling spatial data [193], [276], [399].

An alternative estimator for the variance $c(0)$ based on high peaks exceeding a fixed threshold is proposed in [27]. The *peak estimator* is defined as

$$\tilde{c}(0) = \frac{\sum_{j=1}^{n} X_j^2 - nL^2}{2n},$$

where $X_1, \ldots, X_n$ is a sample of n peaks greater then a fixed threshold L.

References

1. S. R. Adke and S. R. Deshmukh. Limit distributions of a high order Markov chain. *Journal of Royal Statistical Society, Series B*, 50:105–108, 1988.

2. A. Agresti. *Categorical Data Analysis.* Wiley, New York, 1990.

3. A. Agresti. Modeling ordered categorical data: Recent advances and future challenges. *Statistics in Medicine*, 18:2191–2207, 1999.

4. A. Agresti and R. Natarajan. Modeling clustered ordered categorical data: A survey. *International Statistical Review*, 69:345–371, 2001.

5. J. Aitchison and I. R. Dunsmore. *Statistical Prediction Analysis.* Cambridge University Press, Cambridge, 1975.

6. H. Akaike. Information theory and an extension of the maximum likelihood principle. In B. N. Petroc and F. Kaski, editors, *Second International Symposium in Information Theory*, pp. 267–281. Akademiai Kiado, Budapest, 1973.

7. H. Akaike. A new look at the statistical model identification. *IEEE Transactions on Automatic Control*, AC–19:716–723, 1974.

8. M. Al-Osh and A. A. Alzaid. Integer-valued moving average (INMA) process. *Statistische Hefte*, 29:281–300, 1988.

9. M. A. Al-Osh and A. A. Alzaid. First-order integer-valued autoregressive (INAR(1)) process. *Journal of Time Series Analysis*, 8:261–275, 1987.

10. M. A. Al-Osh and A. A. Alzaid. Binomial autoregressive moving average models. *Communications in Statistics. Stochastic Models*, 7:261–282, 1991.

11. A. Albert and J. A. Anderson. On the existence of maximum likelihood estimates in logistic regression models. *Biometrika*, 71:1–10, 1984.

12. J. H. Albert and S. Chib. Bayes inference via Gibbs sampling of autoregressive time series subject to Markov mean and variance shifts. *Journal of Business & Economic Statistics*, 11:1–15, 1993.

13. P. S. Albert, H. F. McFarland, M. E. Smith, and J .A. Frank. Time series for modelling counts from a relapsing-remitting disease: Application to modelling disease activity in multiple sclerosis. *Statistics in Medicine*, 13:453–466, 1994.

14. D. L. Alspach and H. W. Sorensen. Nonlinear Bayesian estimation using Gaussian sum approximations. *IEEE Transactions on Automatic Control*, 17:439–448, 1972.

15. A. A. Alzaid and M. Al-Osh. An integer-valued pth-order autoregressive structure (INAR(p)) process. *Journal of Applied Probability*, 27:314–324, 1990.

16. A. A. Alzaid and M. A. Al-Osh. Some autoregressive moving average processes with generalized Poisson marginal distributions. *Annals of the Institute of Statistical Mathematics*, 45:223–232, 1993.

17. J. R. M. Ameen and P. J. Harrison. Normal discount Bayesian models. In *Bayesian statistics, 2*, pp. 271–298. North-Holland, Amsterdam, 1985.

18. B. D. O. Anderson and J. B. Moore. *Optimal Filtering*. Prentice-Hall, Englewood Cliffs, NJ, 1979.

19. R. L. Anderson. Distribution of the serial correlation coefficient. *Annals of Mathematical Statistics*, 13:1–13, 1942.

20. T. W. Anderson. *The Statistical Analysis of Time Series*. Wiley, New York, 1971.

21. J. D. Aplevich. *The Essentials of Linear State–Space Systems*. Wiley, New York, 2000.

22. F. J. Aranda-Ordaz. On two families of transformations to additivity for binary response data. *Biometrika*, 68, 1981.

23. M. Ashby, J. Neuhaus, W. Hauck, P. Bacchetti, D. Heilbron, N. Jewell, M. Segal, and R. Fusaro. An annotated bibliography of methods for analyzing correlated categorical data. *Statistics in Medicine*, 11:67–99, 1992.

24. A. Azallini and A. W. Bowman. A look at some on the Old Faithful Geyser. *Applied Statistics*, 39:357–365, 1990.

25. A. Azzalini. Maximum likelihood estimation of order m for stationary stochastic process. *Biometrika*, 70:381–387, 1983.

26. I. Y. Bar-Itzhack. In-flight alignment of inertial navigation systems. In C. T. Leondes, editor, *Control and Dynamic Systems, Vol. 38, Advances in Aeronautical Systems*. Academic Press, San Diego, 1990.

27. J. T. Barnett, R. B. Clough, and B. Kedem. Power considerations in acoustic emission. *Journal of the Acoustical Society of America*, 98:2070–2081, 1995.

28. J. T. Barnett and B. Kedem. Zero-crossing rates of mixtures and products of Gaussian processes. *IEEE Transactions on Information Theory*, 44:1672–1677, 1998.

29. I. V. Basawa and R. L. S. Prakasa Rao. *Statistical Inference for Stochastic Processes*. Academic Press, London, 1980.

30. J. V. Basawa and D. J. Scott. *Asymptotic Optimal Inference for Non–ergodic Models*. Lecture Notes in Statistics. Springer, New York, 1983.

31. L. E. Baum. An inequality and associated maximization technique in statistical estimation for probabilistic functions of Markov processes. In *Inequalities, III (Proc. Third Symp. / UCLA, 1969; dedicated to the memory of Theodore S. Motzkin)*, pp. 1–8. Academic Press, New York, 1972.

32. L. E. Baum and J. A. Eagon. An inequality with applications to statistical estimation for probabilistic functions of Markov processes and to a model for ecology. *Bull. Amer. Math. Soc.*, 73:360–363, 1967.

33. L. E. Baum and T. Petrie. Statistical inference for probabilistic functions of finite state Markov chains. *Ann. Math. Statist.*, 37:1554–1563, 1966.

34. L. E. Baum, T. Petrie, G. Soules, and N. Weiss. A maximization technique occurring in the statistical analysis of probabilistic functions of Markov chains. *Ann. Math. Statist.*, 41:164–171, 1970.

35. R. A. Benjamin, M. A. Rigby, and M. D. Stasinopoulos. Modelling exponential family time series data. In B. Marx and H. Friedl, editors, *Proceedings of the 13th International Workshop on Statistical Modelling*, pp. 263–268. 1998.

36. R. A. Benjamin, M. A. Rigby, and M. D. Stasinopoulos. Generalized autoregressive moving average models. Internal Report 10/01, STORM, University of North London, UK, 2001.

37. A. Berchtold. The double chain Markov model. *Communications in Statistics. Theory and Methods*, 28:2569–2589, 1999.

38. A. Berchtold. Estimation of the mixture transition distribution model. *Journal of Time Series Analysis*, 22:379–397, 2001.

39. A. Berchtold and A. E. Raftery. The mixture transition (mtd) model for high–order Markov chains and non–Gaussian time series. Technical Report 360, Department of Statistics, University of Washington, Seattle, USA, 1999.

40. J. O. Berger, V. De Oliveira, and B. Sansó. Objective Bayesian analysis of spatially correlated data. *Journal of the American Statistical Association*, 96:1361–1374, 2001.

41. E. Berglund and K. Brännäs. Entry and exit of plants: A study based on swedish panel count data for municipalities. In *1995 Yearbook of the Finnish Statistical Society*, pp. 95–111, 1996.

42. E. Berglund and K. Brännäs. Plant's entry and exit in Swedish municipalities. *Annals of Regional Science*, 35:431–448, 2001.

43. J. Berkson. Application of the logistic function to bio-assay. *Journal of the American Statistical Association*, 39:357–365, 1944.

44. C. Berzuini, N. Best, W. Gilks, and C. Larizza. Dynamic conditional independence models and Markov chain Monte Carlo methods. *Journal of the American Statistical Association*, 92:1403–1412, 1997.

45. J. Besag. Spatial interaction and the statistical analysis of lattice systems. *Journal of the Royal Statistical Society, Series B*, 36:192–236, 1974, with discussion.

46. J. Besag. Efficiency of pseudo–likelihood estimation for simple Gaussian fields. *Biometrika*, 64:616–618, 1977.

47. P. J. Bickel and K. A. Doksum. An analysis of transformations revisited. *Journal of the American Statistical Association*, 76:296–311, 1981.

48. P. J. Bickel and Y. Ritov. Inference in hidden Markov models I: Local asymptotic normality in the stationary case. *Bernoulli*, 2:199–228, 1996.

49. P. J. Bickel, Y. Ritov, and T. Rydén. Asymptotic normality of the maximum-likelihood estimator for general hidden Markov models. *Annals of Statistics*, 26:1614–1635, 1998.

50. P. Billingsley. *Statistical Inference for Markov Processes*. University of Chicago Press, Chicago, 1961.

51. P. Billingsley. *Probability and Measure*. Wiley, New York, 2nd edition, 1986.

52. S. D. Bindel, V. De Oliveira, and B. Kedem. An implementation of the Bayesian transformed Gaussian spatial prediction model. Technical report, Department of Mathematics, University of Maryland at College Park, 1997. Available at `http://www.math.umd.edu/ bnk/btg_page.html`

53. Y. M. M. Bishop, S. E. Fienberg, and P. W. Holland. *Discrete Multivariate Analysis. Theory and Practice*. MIT Press, Cambridge, MA, 1975.

54. H. W. Block, N. A. Langberg, and D. S. Stoffer. Bivariate exponential and geometric autoregressive and autoregressive moving average models. *Advances in Applied Probability*, 20:798–821, 1988.

55. U. Böckenholt. Analyzing multiple emotions over time by autoregressive negative multinomial regression models. *Journal of the American Statistical Association*, 94:757–765, 1999.

56. T. Bollerslev. Generalized autoregressive conditional heteroskedasticity. *Journal of Econometrics*, 31:307–327, 1986.

57. T. Bollerslev, R. F. Engle, and D. B. Nelson. ARCH models. In R. F. Engle and D. L. McFadden, editors, *Handbook of Econometrics, Vol. IV*, pp. 2959–3038. North-Holland, Amsterdam, 1994.

58. E. G. Bonney. Logistic regression for dependent binary observations. *Biometrics*, 43:951–973, 1987.

59. S. Bose and B. Kedem. Non-dependence of the predictive distribution on the population size. *Statistics & Probability Letters*, 27:43–47, 1996.

60. G. E. P. Box and D. R. Cox. An analysis of transformations. *Journal of the Royal Statistical Society, Series B*, 26:211–252, 1964, with discussion.

61. G. E. P. Box and G. M. Jenkins. *Time Series Analysis: Forecasting and Control*. Holden-Day, San Francisco, 2nd edition, 1976.

62. G. E. P. Box and G. C. Tiao. Intervention analysis with applications to economics and environmental problems. *Journal of the American Statistical Association*, 70:70–79, 1975.

63. G. P. Box, G. M. Jenkins, and G. C. Reinsel. *Time Series Analysis, Forecasting and Control*. Prentice-Hall, Englewood Cliffs, NJ, 3rd edition, 1994.

64. E. Brännäs and K. Brännäs. A model of patch visit behaviour in fish. *Biometrical Journal*, 40:717–724, 1998.

65. K. Brännäs. Explanatory variables in the AR(1) count data model. Technical Report 381, Umeå Economic Studies, Umeå University, 1995.

66. K. Brännäs and A. Hall. Estimation in integer–valued moving average models. *Applied Stochastic Models in Business and Industry*, 17:277–291, 2001.

67. K. Brännäs and J. Hellström. Generalized integer–valued autoregression. *Econometric Reviews*, 20:425–443, 2001.

68. K. Brännäs and P. Johansson. Time series count regression. *Communications in Statistics–Theory and Methods*, 23(2907–2925), 1994.

69. D. R. Brillinger. *Time Series:Data Analysis and Theory*. Holt, Rinehart and Winston, New York, 1975.

70. D. R. Brillinger. An analysis of ordinal-valued time series. In *Athens Conference on Applied Probability and Time Series Analysis*, volume II:Time Series Analysis of *Lecture Notes in Statistics*, pp. 73–87, Springer, New York, 1996. In Memory of E.J. Hannan.

71. D. R. Brillinger, P. A. Morettin, R. A. Irizarry, and C. Chiann. Some wavelet-based analyses of Markov chain data. *Signal Processing*, 80:1607–1627, 2000.

72. P. J. Brockwell and R. A. Davis. *Time Series: Data Analysis and Theory*. Springer, New York, 2nd edition, 1991.

73. P. E. Brown, P. J. Diggle, M. E. Lord, and P. C. Young. Space–time callibration of radar rainfall data. *Applied Statistics*, 50:221–241, 2001.

74. P. J. Brown, N. D. Le, and J. V. Zidek. Multivariate spatial interpolation and exposure to air pollutants. *Canadian Journal of Statistics*, 22:489–509, 1994.

75. L. M. Brumback, B. A.and Ryan, J. D. Schwartz, L. M. Neas, P. C. Stark, and H. A. Burge. Transitional regression models with application to environmental time series. *Journal of the American Statistical Association*, 85:16–27, 2000.

76. P. Bühlmann and A. J. Wyner. Variable length Markov chains. *Annals of Statistics*, 27:480–513, 1999.

77. A. C. Cameron and L. Leon. Markov regression models for time series data, 1993. Presented at Western Economics Association Meeting, Lake Tahoe, NV.

78. A. C. Cameron and P. K. Trivedi. *Regression Analysis of Count Data*. Cambridge University Press, Cambridge, 1998.

79. M. J. Campbell. Time series regression for counts: An investigation into the relationship between sudden infant death syndrome and environmental temperature. *Journal of the Royal Statistical Society, Series A*, 157:191–208, 1994.

80. V. Carey, S. L. Zeger, and P. Diggle. Modeling multivariate binary data with alternating logistic regression. *Biometrika,*, 80:517–526, 1993.

81. C. Cargnoni, P. Müller, and M. West. Bayesian forecasting of multinomial time series through conditionally Gaussian dynamic models. *Journal of the American Statistical Association*, 92:640–647, 1997.

82. B. P. Carlin and N. G. Polson. Monte Carlo Bayesian methods for discrete regression models and categorical time series. In J. M. Bernardo et al., editors, *Bayesian Statistics 4*, pp. 577–586. Oxford University Press, Oxford, 1992.

83. B. P. Carlin, N. G. Polson, and D. S. Stoffer. A Monte Carlo approach to nonnormal and nonlinear state space modelling. *Journal of the American Statistical Association*, 75:493–500, 1992.

84. C. K. Carter and R. Kohn. On Gibbs sampling for state space models. *Biometrika*, 81:541–553, 1994.

85. C. K. Carter and R. Kohn. Markov chain Monte Carlo in conditionally Gaussian state space models. *Biometrika*, 83:589–601, 1996.

86. G. Casella and E. I. George. Explaining the Gibbs sampler. *American Statistician*, 3:167–174, 1992.

87. G. Chamberlain. Analysis of covariance with qualitative data. *Review of Economic Studies*, 47:225–238, 1980.

88. K. S. Chan and J. Ledolter. Monte Carlo EM estimation for time series models involving counts. *Journal of the American Statistical Association*, 90:242–252, 1995.

89. S. Chandrasekhar. Stochastic problems in physics and astronomy. *Reviews of Modern Physics*, 15:1–89, 1943.

90. T. J. Chang, J. W. Delleur, and M. L. Kavvas. Application of discrete autoregressive moving average models for estimation of daily runoff. *Journal of Hydrology*, 91:119–135, 1987.

91. T. J. Chang, J. W. Kavvas, and M. L. Delleur. Daily precipitation modeling by discrete autoregressive moving average process. *Water Resources Research*, 20:565–580, 1984.

92. T. J. Chang J. W. Kavvas, and M. L. Delleur. Modelling of sequences of wet and dry days by binary discrete autoregressive moving average processes. *Journal of Climate and Applied Meteorology*, 23:1367–1378, 1984.

93. M-H. Chen, D. K. Dey, and Q-M. Shao. A new skewed link model for dichotomous quantal response data. *Journal of the American Statistical Association*, 94:1172–1186, 1999.

94. R. Chen and J. S. Liu. Predictive updating methods with application to Bayesian classification. *Journal of the Royal Statistical Society, Series B*, 58:397–415, 1996.

95. S. Chib. Calculating posterior distributions and modal estimates in Markov mixture models. *Journal of Econometrics*, 75:79–97, 1996.

96. S. Chib and E. Greenberg. Hierarchical analysis of SUR models with extensions to correlated serial errors and time varying parameter models. *Journal of Econometrics*, 68:339–360, 1995.

97. S. Chib and E. Greenberg. Understanding the Metropolis–Hastings algorithm. *American Statistician*, 49:327–335, 1995.

98. J. P. Chilès and P. Delfiner. *Geostatistics: Modeling Spatial Uncertainty*. Wiley, New York, 1999.

99. B. Choi. *ARMA Model Identification*. Springer, New York, 1992.

100. J. Chollet. Some inequalities for principal submatrices. *American Mathematical Monthly*, 104:609–617, 1997.

101. D. G. Clayton. Repeated ordinal measurements: A generalized estimating equation approach. Technical report, Medical Research Council Biostatistics Unit, Cambridge, England, 1992.

102. R. Coe and R. D. Stern. A model fitting analysis of daily rainfall data. *Journal of Royal Statistical Society*, A47:1–34, 1984, with discussion.

103. J. W. Cooley and J. W. Tukey. An algorithm for the machine calculation of complex Fourier series. *Mathematics of Computation*, 19:297–301, 1965.

104. D. R. Cox. Regression models and life tables. *Journal of the Royal Statistical Society, Series B*, 74:187–220, 1972, with discussion.

105. D. R. Cox. Partial likelihood. *Biometrika*, 62:69–76, 1975.

106. D. R. Cox. Statistical analysis of time series: Some recent developments. *Scandinavian Journal of Statistics*, 8:93–115, 1981.

107. D. R. Cox and E. J. Snell. *The Analysis of Binary Data*. Chapman & Hall, London, 2nd edition, 1989.

108. D. R. Cox and E.J. Snell. A general definition of residuals. *Journal of the Royal Statistical Society, Series B*, 30:248–275, 1968, with discussion.

109. N. A. C. Cressie. *Statistics for Spatial Data*. Wiley, New York, 1993.

110. N. A. C. Cressie and T. R. C. Read. Multinomial goodness of fit tests. *Journal of the Royal Statistical Society, Series B*, 46:440–464, 1984.

111. E. L. Crow and K. Shimizu. *Lognormal Distributions, Theory and Applications*. Marcel Dekker, New York, 1988.

112. C. Czado. On link selection in generalized linear models. In L. Fahrmeir, editor, *Advances in GLIM and Statistical Modelling*, volume 78, pp. 60–65. Springer Lecture Notes in Statistics, 1992.

113. D. Martin D. Le Nhu and A. E. Raftery. Modeling flat stretches, burst and outliers in time series using mixture transition distribution models. *Journal of the American Statistical Association*, 91:1504–1515, 1996.

114. R. A. Davis, W. T. M. Dunsmuir, and Y. Wang. Modelling time series of count data. In S. Ghosh, editor, *Asymptotics, Nonparametric & Time Series*, pp. 63–114. Marcel Dekker, New York, 1999.

115. R. A. Davis, W. T. M. Dunsmuir, and Y. Wang. On autocorrelation in a Poisson regression model. *Biometrika*, 87:491–505, 2000.

116. P. de Jong and N. Shephard. The simulation smoother for time series models. *Biometrika*, 82:339–350, 1995.

117. V. De Oliveira. *Prediction in Some Classes of Non-Gaussian Random Fields*. PhD thesis, University of Maryland, Department of Mathematics, College Park, 1997.

118. V. De Oliveira, B. Kedem, and D. A. Short. Bayesian prediction of transformed Gaussian random fields. *Journal of the American Statistical Association*, 92:1422–1433, 1997.

119. J. W. Delleur, T. J. Chang, and M. L. Kavvas. Simulation models of sequences of wet and dry days. *Journal of Irrigation and Drainage Engineering*, 115:344–357, 1989.

120. A. P. Dempster, N. M. Laird, and D. B. Rubin. Maximum likelihood from incomplete data via the EM algorithm. *Journal of the Royal Statistical Society, Series B*, 39:1–38, 1977, with discussion.

121. D. K. Dey, S. K. Ghosh, and B. K. Mallick, editors. *Generalized Linear Models: A Bayesian Perspective*. Marcel Dekker, New York, 2000.

122. F. X. Diebold, J. Lee, and G. C. Weinbach. Regime switching with time–varying transition probabilities. In C. Hargreaves, editor, *Nonstationary Time Series and Cointegration*. Oxford University Press, Oxford, 1994.

123. J. P. Diggle, K-Y. Liang, and L. S. Zeger. *Analysis of Longitudinal Data*. Oxford University Press, New York, 1994.

124. A. J. Dobson. *An Introduction to Generalized Linear Models*. Chapman & Hall, London, 1990.

125. J. L. Doob. *Stochastic Processes*. Wiley, New York, 1953.

126. A. Doucet, N. de Freitas, and N. Gordon, editors. *Sequential Monte Carlo Methods in Practice*. Springer, New York, 2001.

127. A. Doucet, S. J. Godsill, and C. Andrieu. On sequential Monte Carlo methods for Bayesian filtering. *Statistics and Computing*, 10:197–208, 2000.

128. J. G. Du and Y. Li. The integer-valued autoregressive INAR(p) model. *Journal of Time Series Analysis*, 12:129–142, 1991.

129. J. Durbin and S. J. Koopman. Monte Carlo maximum likelihood estimation for non-Gaussian state space models. *Biometrika*, 84:669–684, 1997.

130. J. Durbin and S. J. Koopman. Time series analysis of non-Gaussian observations based on state space models from both classical and Bayesian perspective.

Journal of the Royal Statistical Society, Series B, 62:3–56, 2000, with discussion.

131. B. Efron. Double exponential families and their use in generalized linear regression. *Journal of the American Statistical Association*, 81:709–721, 1986.

132. R. J. Elliot, L. Aggoun, and J. B. Moore. *Hidden Markov Models: Estimation and Control*. Springer, New York, 1995.

133. R. F. Engle. Autoregressive conditional heteroscedasticity with estimates of the variance of United Kingdom inflation. *Econometrica*, 50:987–1007, 1982.

134. B. S. Everitt and D. J. Hand. *Finite Mixture Distributions*. Chapman & Hall, London, 1981.

135. L. Fahrmeir. Asymptotic testing theory for generalized linear models. *Statistics*, 18:65–76, 1987.

136. L. Fahrmeir. Posterior mode estimation by extended kalman filtering for multivariate dynamic generalized linear models. *Journal of the American Statistical Association*, 87:501–509, 1992.

137. L. Fahrmeir. State space modeling and conditional mode estimation for categorical time series. In D. Brillinger et al., editors, *New Directions in Time Series Analysis, Part I*, pp. 87–109. Springer, New York, 1992.

138. L. Fahrmeir, W. Hennevogl, and K. Klemme. Smoothing in dynamic generalized linear models by Gibbs sampling. In L. Fahrmeir et al., editors, *Advances in GLIM and Statistical Modelling*, pp. 85–90. Springer, New York, 1992.

139. L. Fahrmeir and H. Kaufmann. Consistency and asymptotic normality of the maximum likelihood estimates in generalized linear models. *Annals of Statistics*, 13:342–368, 1985.

140. L. Fahrmeir and H. Kaufmann. Regression models for nonstationary categorical time series. *Journal of Time Series Analysis*, 8:147–160, 1987.

141. L. Fahrmeir and H. Kaufmann. On Kalman filtering, posterior mode estimation and Fisher scoring in dynamic exponential family regression. *Metrika*, 38:37–60, 1991.

142. L. Fahrmeir and S. Lang. Bayesian inference for generalized additive mixed models based on Markov random field priors. *Applied Statistics*, 50:201–220, 2001.

143. L. Fahrmeir and L. Pritscher. Regression analysis of forest damage by marginal models for correlated ordinal responses. *Journal of Environmental and Ecological Statistics*, 3:257–268, 1996.

144. L. Fahrmeir and G. Tutz. *Multivariate Statistical Modelling Based on Generalized Linear Models*. Springer, New York, 2nd edition, 2001.

145. L. Fahrmeir and S. Wagenpfeil. Penalized likelihood estimation and iterative Kalman smoothing for non–Gaussian dynamic regression models. *Computational Statistics and Data Analysis*, 24:295–320, 1997.

146. A. J. Filardo. Business cycle phases and their transitional dynamics. *Journal of Business and Economics Statistics*, 9:299–308, 1994.

147. D. Finney. *Probit Analysis*. Cambridge University Press, Cambridge, 3rd edition, 1971.

148. D. Firth. Generalized linear models. In D. Hinkley et al., editors, *Statistical Theory and Modelling. In Honor of Sir David Cox, FRS*, chapter 3. Chapman & Hall, London, 1991.

149. G. M. Fitzmaurice, N. M. Laird, and A. G. Rotnitsky. Regression models for discrete longitudinal responses. *Statistical Science*, 8:284–309, 1993, with discussion.

150. K. Fokianos. Power divergence family of tests for categorical time series models. *Annals of the Institute of Statistical Mathematics*, 2002. to appear.

151. K. Fokianos. Truncated Poisson regression for time series of counts. *Scandinavian Journal of Statistics*, 28:645–659, 2001.

152. K. Fokianos and B. Kedem. Prediction and classification of non-stationary categorical time series. *Journal of Multivariate Analysis*, 67:277–296, 1998.

153. K. Fokianos and B. Kedem. A stochastic approximation algorithm for the adaptive control of time series following generalized linear models. *Journal of Time Series Analysis*, 20:289–308, 1999.

154. K. Fokianos, B. Kedem, and D. Short. Predicting precipitation level. *Journal of Geophysical Research*, 101:26473–26477, 1996.

155. K. Fokianos, B. Kedem, J. Qin, J. L. Haferman, and D. Short. On combining instruments. *Journal of Applied Meteorology*, 37:220–226, 1998.

156. K. Fokianos, B. Kedem, J. Qin, and D. A. Short. A semiparametric approach to the one–way layout. *Technometrics*, 43:56–64, 2001.

157. J. Franke and T. Seligmann. Conditional maximum likelihood estimates for (INAR(1)) processes and their application to modeling epileptic seizure counts. In T. Subba Rao, editor, *Developments in Time Series Analysis*, pp. 310–330. Chapman & Hall, London, 1993.

158. S. Frühwirth-Schnatter. Integration-based Kalman-filtering for a dynamic generalized linear trend model. *Computational Statistics & Data Analysis*, 13:447–459, 1992.

159. S. Frühwirth-Schnatter. Data augmentation and dynamic linear models. *Journal of Time Series Analysis*, 15:183–202, 1994.

160. S. Frühwirth-Schnatter. Markov chain Monte Carlo estimation of classical and dynamic switching mixture models. *Journal of the American Statistical Association*, 96:194–209, 2001.

161. W. A. Fuller. *Introduction to Statistical Time Series*. Wiley, New York, 2nd edition, 1996.

162. D. Gamerman. *Markov Chain Monte Carlo*. Chapman & Hall, London, 1997.

163. D. Gamerman. Markov chain Monte Carlo for dynamic generalised linear models. *Biometrika*, 85:215–227, 1998.

164. M. Gaudard, M. Karson, E. Linder, and D. Sinha. Bayesian spatial prediction. *Environmental and Ecological Statistics*, 6:147–179, 1999, with discussion.

165. D. P. Gaver and P. A. W. Lewis. First–order autoregressive gamma sequences and point processes. *Advances in Applied Probability*, 12:727–745, 1980.

166. A. Gelb. *Applied Optimal Estimation*. MIT Press, Cambridge, MA, 1974.

167. A. E. Gelfand and A. F. M. Smith. Sampling-based approaches to calculating marginal densities. *Journal of American Statistical Association*, 85:398–409, 1990.

168. S. Geman and D. Geman. Stochastic relaxation, Gibbs distributions, and the Bayesian restoration of images. *IEEE Transactions on Pattern Analysis and Machine Intelligence*, 6:721–741, 1984.

169. L. Gerencsér. Stability of random iterative mappings. In M. Dror et al., editors, *Modelling Uncertainty: An Examination of Stochastic Theory, Methods, and Applications*, pp. 359–371. Kluwer, Boston, 2002.

170. R. Gerlach, C. K. Carter, and R. Kohn. Efficient Bayesian inference for dynamic mixture models. *Journal of the American Statistical Association*, 95:819–828, 2000.

171. J. Geweke. Bayesian inference in econometric models using Monte Carlo integration. *Econometrica*, 57:1317–1339, 1989.

172. W. Gilks and P. Wild. Adaptive rejection sampling for Gibbs sampling. *Applied Statistics*, 4:337–348, 1992.

173. W. R. Gilks, S. Richardson, and D. J. Spiegelhalter, editors. *Markov Chain Monte Carlo in Practice*. Chapman & Hall, London, 1996.

174. R. D. Gill. Marginal partial likelihood. *Scandinavian Journal of Statistics*, 19:133–137, 1992.

175. V. P. Godambe. An optimal property of regular maximum-likelihood estimation. *Annals of Mathematical Statistics*, 31:1208–1211, 1960.

176. V. P. Godambe. *Estimating Functions*. Oxford Science Publications, Oxford, 1991.

177. V. P. Godambe and C. C. Heyde. Quasi–likelihood and optimal estimation. *International Statistical Review*, 55:231–244, 1987.

178. N. J. Gordon, D. J. Salmond, and A. F. M. Smith. Novel approach to nonlinear/non–Gaussian Bayesian state estimation. *IEE Proceedings F*, 140:107–113, 1993.

179. D. Goshen-Meskin and I. Y. Bar-Itzhack. Observability analysis of piece-wise constant systems I: Theory. *IEEE Transactions on Aerospace and Electronic Systems*, AES–28:1056–1067, 1992.

180. D. Goshen-Meskin and I. Y. Bar-Itzhack. Observability analysis of piece-wise constant systems II: Application to inertial navigation in-flight alignment. *IEEE Transactions on Aerospace and Electronic Systems*, AES–28:1068–1075, 1992.

181. C. Gouriéroux. *ARCH Models and Financial Applications*. Springer, New York, 1997.

182. C. Gouriéroux, A. Monfort, and A. Trognon. Estimation and test in probit models with serial correlation. In *Alternative Approaches to Time Series Analysis (Rouen, 1982)*, pp. 169–209. Publ. Fac. Univ. Saint-Louis, Brussels, 1984.

183. C. Gourieroux, A. Monfort, and A. Trognon. Pseudo-maximum likelihood methods: Theory. *Econometrica*, 52:681–700, 1984.

184. U. Grenander and M. Rosenblatt. *Statistical Analysis of Stationary Time Series*. Wiley, New York, 1957.

185. G. K. Grunwald, R. J. Hyndman, L. Tedesco, and R. L. Tweedie. Non-Gaussian conditional linear AR(1) models. *Australian & New Zealand Journal of Statistics*, 42:479–495, 2000.

186. V. M. Guerrero and R. A. Johnson. Use of the Box–Cox transformation with binary response models. *Biometrika*, 69:309–314, 1982.

187. P. Guttorp. *Statistical Inference for Branching Processes*. Wiley, New York, 1991.

188. P. Guttorp. *Stochastic Modelling of Scientific Data*. Chapman & Hall, London, 1995.

189. L. Györfi, G. Lugosi, and G. Morvai. A simple randomized algorithm for sequential prediction of ergodic time series. *IEEE Transactions on Information Theory*, 45:2642–2650, 1999.

190. S. J. Haberman. *The Analysis of Frequency Data*. University of Chicago Press, Chicago, 1974.

191. P. Hall and C. C. Heyde. *Martingale Limit Theory and Its Applications*. Academic Press, New York, 1980.

192. M. S. Handcock and J. R. Wallis. An approach to statistical spatial-temporal modeling of meteorological fields. *Journal of the American Statistical Association*, 89:368–390, 1994, with comments.

193. M. S. Handcock and M. L. Stein. A Bayesian analysis of kriging. *Technometrics*, 35:403–410, 1993.

194. J. E. Handschin. Monte Carlo techniques for prediction and filtering of non-linear stochastic processes. *Automatica*, 6:555–563, 1970.

195. J. E. Handschin and D. Q. Mayne. Monte Carlo techniques to estimate the conditional expectation in multi-stage non-linear filtering. *International Journal of Control*, 9:547–559, 1969.

196. E. J. Hannan and B. G. Quinn. The determination of the order of an autoregression. *Journal of Royal Statistical Society, Series B*, 41:190–195, 1979.

197. L. P. Hansen. Large sample properties of generalized method of moments estimators. *Econometrica*, 50:1029–1054, 1982.

198. P. J. Harrison and C. F. Stevens. Bayesian forecasting. *Journal of the Royal Statistical Society, Series B*, 38:205–247, 1976, with discussion.

199. J. A. Hartigan. Linear Bayesian methods. *Journal of the Royal Statistical Society, Series B*, 31:446–454, 1969.

200. A. Harvey. *Forecasting, Structural Time Series Models and the Kalman Filter*. Cambridge University Press, 1989.

201. A. C. Harvey and C. Fernandes. Time series models for count or qualitative observations. *Journal of Business & Economic Statistics*, 7:407–422, 1989, with discussion.

202. T. J. Hastie and R. J. Tibshirani. Generalized additive models. *Statistical Science*, 1:297–318, 1986, with discussion.

203. T. J. Hastie and R. J. Tibshirani. *Generalized Additive Models*. Chapman & Hall, London, 1990.

204. W. K. Hastings. Monte Carlo sampling methods using Markov chains and their applications. *Biometrika*, 57:97–109, 1970.

205. S. He and B. Kedem. Higher order crossings spectral analysis of an almost periodic random sequence in noise. *IEEE Transanctions on Information Theory*, 35:360–370, 1989.

206. P. J. Heagerty and S. L. Zeger. Marginal regression models for clustered ordinal measurements. *Journal of the American Statistical Association*, 91:1024–1036, 1996.

207. P. J. Heagerty and S. L. Zeger. Lorelogram: A regression approach to exploring dependence in longitudinal categorical responses. *Journal of the American Statistical Association*, 93:150–162, 1998.

208. J. J. Heckman. Dynamic discrete probability models. In C. F. Manski and D. McFadden, editors, *Structural Analysis of Discrete Data*, pp. 114–195. MIT Press, Cambridge, MA, 1981.

209. C. C. Heyde. *Quasi-Likelihood and Its Applications: A General Approach to Optimal Parameter Estimation*. Springer, New York, 1997.

210. C. C. Heyde and E. Seneta. Estimation theory for growth and immigration rates in a multiplicative process. *Journal of Applied Probability*, 9:235–256, 1972.

211. T. Higuchi. Monte Carlo filtering using the genetic algorithm operators. *Journal of Statistical Computation and Simulation*, 59:1–23, 1997.

212. P. Hodges and D. Hale. A computational method for estimating densities of non–Gaussian non–stationary univariate time series. *Journal of Time Series Analysis*, 14:163–178, 1993.

213. R. T. Holden. Time series analysis of contagious process. *Journal of the American Statistical Association*, 82:1019–1026, 1987.

214. C. Houdré and B. Kedem. A note on autocovariance estimation in the presence of discrete spectra. *Statistics & Probability Letters*, 24:1–8, 1995.

215. H.-C. Huang and N. Cressie. Spatio-temporal prediction of snow water equivalent using the Kalman filter. *Computational Statistics & Data Analysis*, 22:159–175, 1996.

216. M. Hürzeler and H. R. Künsch. Monte Carlo approximations for general state–space models. *Journal of Computational and Graphical Statistics*, 7:175–193, 1998.

217. C. E. Hutchinson. The Kalman filter applied to aerospace and electronic systems. *IEEE Transactions on Aerospace and Electronic Systems*, AES–20:500–504, 1984.

218. J. E. Hutton and P.I. Nelson. Quasi–likelihood estimation for semimartingales. *Stochastic Processes and Their Applications*, 22:245–257, 1986.

219. R. Hyndman and G. K. Grunwald. Generalized additive modeling of mixed distribution Markov models with application to Melbourne's rainfall. *Australian & New Zealand Journal of Statistics*, 42:145–158, 2000.

220. R. J. Hyndman. Non–parametric additive regression models for binary time series. In *Proceedings of the 1999 Australian Meeting of the Econometric Society*, University of Technology, Sydney, Australia, 1999.

221. P. A. Jacobs and P. A. W. Lewis. Discrete time series generated by mixtures. i. correlation and runs properties. *Journal of Royal Statistical Society, Series B*, 40:94–105, 1978.

222. P. A. Jacobs and P. A. W. Lewis. Discrete time series generated by mixtures. ii. asymptotic properties. *Journal of Royal Statistical Society, Series B*, 40:222–228, 1978.

223. P. A. Jacobs and P. A. W. Lewis. Stationary discrete autoregressive-moving average time series generated by mixtures. *Journal of Time Series Analysis*, 4:19–36, 1983.

224. J. Jacod. Partial likelihood processes and asymptotic normality. *Stochastic Processes and Their Applications*, 26:47–71, 1987.

225. J. Jacod. On partial likelihood. *Ann. Inst. H. Poincarè Probab. Statist.*, 26:299–329, 1990.

226. E. Jacquier, N. G. Polson, and P. E. Rossi. Bayesian analysis of stochastic volatility models. *Journal of Business & Economic Statistics*, 12:371–417, 1994, with discussion.

227. N. O. Jeffries. *Logistic Mixtures of Generalized Linear Model Time Series*. PhD thesis, University of Maryland, College Park, USA, 1998.

228. J. L. Jensen and N. V. Petersen. Asymptotic normality of the maximim likelihood estimator in state space models. *Annals of Statistics*, 27:514–535, 1999.

229. H. Joe. Time series models with univariate margins in the convolution–closed infinitely divisible class. *Journal of Applied Probability*, 33:664–677, 1996.

230. P. Johansson. Tests for serial correlation and overdispersion in a count data regression model. *Journal of Statistical Computation and Simulation*, 53:153–164, 1995.

231. N. L. Johnson, S. Kotz, and A. W. Kemp. *Univariate Discrete Distributions*. Wiley, New York, 2nd edition, 1992.

232. V. E. Johnson and J. H. Albert. *Ordinal Data Modelling*. Springer, New York, 1999.

233. R. H. Jones. *Longitudinal Data with Serial Correlation: A State-Space Approach*. Chapman & Hall, London, 1993.

234. B. Jørgensen. *The Theory of Dispersion Models*. Chapman & Hall, London, 1997.

235. B. Jørgensen, S. Lundbye-Christensen, P. X.-K. Song, and L. Sun. State-space models for multivariate longitudinal data of mixed types. *Canadian Journal of Statistics*, 24:385–402, 1996.

236. B. Jørgensen, S. Lundbye-Christensen, P. X.-K. Song, and L. Sun. A state space model for multivariate longitudinal count data. *Biometrika*, 86:169–181, 1999.

237. B. Jørgensen and P. X. Song. Stationary time series models with exponential dispersion model margins. *Journal of Applied Probability*, 35:78–92, 1998.

238. A. Kagan. Another look at the Cramér-Rao inequality. *American Statistician*, 55:211–212, 2001.

239. A. Kagan. A note on the logistic link function. *Biometrika*, 88:599–601, 2001.

240. A. Kagan and P.J. Smith. Multivariate normal distributions, Fisher information and matrix inequalities. *International Journal of Mathematical Education in Science and Technology*, 32:91–96, 2001.

241. T. Kailath. A view of three decades of linear filtering theory. *IEEE Transactions on Information Theory*, IT–20:146–181, 1974.

242. R. E. Kalman. A new approach to linear filtering and prediction problems. *Journal of Basic Engineering, Transactions ASME. Series D*, 82:35–45, 1960.

243. R. E. Kalman and R.S. Buccy. New results in linear filtering and prediction theory. *Journal of Basic Engineering, Transactions ASME. Series D*, 83:95–108, 1961.

244. M. Kanter. Autoregression for discrete processes mod 2. *Journal of Applied Probability*, 12:371–375, 1975.

245. S. Karlin and H. M. Taylor. *A First Course in Stochastic Processes*. Academic Press, 2nd edition, 1975.

246. H. Kaufmann. Regression models for nonstationary categorical time series: Asymptotic estimation theory. *Annals of Statistics*, 15:79–98, 1987.

247. H. Kaufmann. On existence and uniqueness of maximum likelihood estimates in quantal and ordinal response models. *Metrika*, 13:291–313, 1989.

248. R. Kay and S. Little. Transformation of the explanatory variables in the logistic regression model for binary data. *Biometrika*, 74:497–501, 1987.

249. U. Küchler and M. Sørensen. *Exponential Families of Stochastic Processes*. Springer, Berlin, 1997.

250. B. Kedem. Sufficient statistics associated with a two-state second-order Markov chain. *Biometrika*, 63:127–132, 1976.

251. B. Kedem. *Binary Time Series*. Marcel Dekker, New York, 1980.

252. B. Kedem. *Time Series Analysis by Higher Order Crossings.* IEEE Press, New York, 1994.

253. B. Kedem and L. S. Chiu. On the lognormality of rain rate. *Proceedings of the National Academy of Sciences, USA*, 84:901–905, 1987.

254. B. Kedem and K. Fokianos. Regression models for binary time series. In M. Dror et al., editors, *Modeling Uncertainty: An Examination of Stochastic Theory, Methods, and Applications*, pp. 185–199. Kluwer, Boston, 2002.

255. B. Kedem and B. Kozintsev. Graphical bootstrap. In *Proceedings of the Section on Statistics and the Environment*, pp. 30–32. American Statistical Association, Alexandria, VA, 2000.

256. B. Kedem and H. Pavlopoulos. On the threshold method for rainfall estimation: Choosing the optimal threshold level. *Journal of the American Statistical Association*, 86:626–633, 1991.

257. B. Kedem and E. V. Slud. On autocorrelation estimation in mixed-spectrum Gaussian processes. *Stochastic Processes and Their Applications*, 49:227–244, 1994.

258. D. M. Keenan. A time series analysis of binary data. *Journal of American Statistical Association*, 77:816–821, 1982.

259. T. D. Keenan and M. J. Manton. Darwin climate monitoring and research station: Observing precipitating systems in a monsoon environment. Technical Report 53, Bureau of Meteorology Research Centre, Victoria, Australia, 1996.

260. G. Kitagawa. Non-Gaussian state space modeling of nonstationary time series. *Journal of the American Statistical Association*, 82:1032–1063, 1987, with discussion.

261. G. Kitagawa. Non–Gaussian seasonal adjustment. *Computer and Mathematics with Applications*, 18:503–514, 1989.

262. G. Kitagawa. The two-filter formula for smoothing and an implementation of the Gaussian-sum smoother. *Annals of the Institute of Statistical Mathematics*, 46:605–623, 1994.

263. G. Kitagawa. Monte Carlo filter and smoother for non-Gaussian nonlinear state space models. *Journal of Computational and Graphical Statistics*, 5:1–25, 1996.

264. G. Kitagawa. A self–organizing state space model. *Journal of the American Statistical Association*, 93:1203–1215, 1998.

265. G. Kitagawa and W. Gersch. *Smoothness Priors Analysis of Time Series*, volume 116 of *Lecture Notes in Statistics.* Springer, New York, 1996.

266. G. Kitagawa and T. Higuchi, editors. *Nonlinear non-Gaussian Models and Related Filtering Methods*. Kluwer Academic Publishers, Tokyo, 2001. Selected papers from the International Symposium on Frontiers of Time Series Modeling, Tokyo, February 14–16, 2000, *Annals of the Institute for Statistical Mathematics*, 53, 2001, no. 1.

267. P. K. Kitanidis. *Introduction to Geostatistics: Applications in Hydrogeology*. Cambridge University Press, New York, 1997.

268. P. K. Kitanidis. Parameter uncertainty in estimation of spatial functions: Bayesian analysis. *Water Resources Research*, 22:449–507, 1986.

269. L. A. Klimko and P. I. Nelson. On conditional least squares estimation for stochastic processes. *Annals of Statistics*, 6:629–642, 1978.

270. L. Knorr-Held. Conditional prior proposals in dynamic models. *Scandinavian Journal of Statistics*, 26:129–144, 1999.

271. A. Kong, J. S. Liu, and W. H. Wong. Sequential imputations and Bayesian missing data problems. *Journal of the American Statistical Association*, 89:278–288, 1994.

272. L. H. Koopmans. *The Spectral Analysis of Time Series*. Academic Press, New York, 1974.

273. E. L. Korn and B. I. Graubard. *Analysis of Health Surveys*. Wiley, New York, 1999.

274. E. L. Korn and A. S. Whittemore. Methods for analyzing panel studies of acute health effects of air pollution. *Biometrics*, 35:795–802, 1979.

275. M. R. Kosorok and W.-H. Chao. The analysis of longitudinal ordinal response data in continuous time. *Journal of the American Statistical Association*, 91:807–817, 1996.

276. B. Kozintsev. *Computations with Gaussian Random Fields*. PhD thesis, University of Maryland, Department of Mathematics and Institute for Systems Research, College Park, 1999.

277. B. Kozintsev and B. Kedem. Generation of 'similar' images from a given discrete image. *Journal of Computational and Graphical Statistics*, 9:286–302, 2000.

278. A. Kozintseva. Comparison of three methods of spatial prediction. Master's thesis, University of Maryland, Department of Mathematics, College Park, 1999.

279. A.Y.C. Kuk and C. H. Chen. A mixture model combining logistic regression with proportional hazards regression. *Biometrika*, 79:531–541, 1992.

280. H. R. Künsch. State space and hidden Markov models. In E. Barndorff-Nielsen et al., editors, *Complex Stochastic Systems*, pp. 109–173, Chapman & Hall/CRC, Boca Raton, FL, 2001.

281. L. Kuo, J. Lee, P. Cheng, and J. Pai. Bayes inference for technological substitution data with data-based transformation. *Journal of Forecasting*, 16:65–82, 1997.

282. L. F. Léon and C. Tsai. Assessment of model adequacy for Markov regression time series models. *Biometrics*, 54:1165–1175, 1998.

283. T. Z. Lai and C. Z. Wei. Least squares estimation in stochastic regression models with applications to identification and control of dynamic systems. *Annals of Statistics*, 10:154–166, 1982.

284. N. M. Laird and J. H. Ware. Random–effects models for longitudinal data. *Biometrics*, 38:963–974, 1982.

285. D. Lambert. Zero–inflated Poisson regression with an application to defects is manufacturing. *Technometrics*, 34:1–14, 1992.

286. J. M. Landwehr, D. Pregibon, and A. C. Shoemaker. Graphical methods for assessing logistic regression models. *Journal of the American Statistical Association*, 79:61–71, 1984.

287. N. A. Langberg and D. S. Stoffer. Moving-average models with bivariate exponential and geometric distributions. *Journal of Applied Probability*, 24:48–61, 1987.

288. M. G. Larson and G. E. Dinse. A mixture model for the regression analysis of competing risks. *Journal of the Royal Statistical Society, Series C*, 34:201–211, 1985.

289. A. Latour. The multivariate GINAR(p) process. *Advances in Applied Probability*, 29:228–248, 1997.

290. A. Latour. Existence and stochastic structure of a non-negative integer-valued autoregressive process. *Journal of Time Series Analysis*, 19:439–455, 1998.

291. A. J. Lawrance. The innovation distribution of a gamma distributed autoregressive process. *Scandinavian Journal of Statistics*, 9:234–236, 1982.

292. A. J. Lawrance and P. A. W. Lewis. A moving average exponential point process (ema1). *Journal of Applied Probability*, 14:98–113, 1977.

293. N. D. Le, R. D. Martin, and A. E. Raftery. Modelling flat stretches, bursts, and outliers in time series using mixture transition distribution models. *Journal of the American Statistical Association*, 91:1504–1515, 1996.

294. B. G. Leroux. Maximum–likelihood estimation for hidden Markov models. *Stochastic Processes and Their Applications*, 40:127–143, 1992.

295. P. A. W. Lewis, E. McKenzie, and D. K. Hugus. Gamma processes. *Communications in Statistics–Stochastic Models*, 5:1–30, 1989.

296. T. H. Li and B. Kedem. Asymptotic analysis of a multiple frequency estimation method. *Journal of Multivariate Analysis*, 46:214–236, 1993.

297. T. H. Li and B. Kedem. Strong consistency of the contraction mapping method for frequency estimation. *IEEE Transactions on Information Theory*, 39:989–998, 1993.

298. W. K. Li. Testing model adequacy for some Markov regression models for time series. *Biometrika*, 78:83–89, 1991.

299. W. K. Li. Time series models based on generalized linear models: Some further results. *Biometrics*, 50:506–511, 1994.

300. K.-Y. Liang and S. L. Zeger. Longitudinal data analysis using generalized linear models. *Biometrika*, 73:13–22, 1986.

301. K.-Y. Liang and S. L. Zeger. A class of logistic regression models for multivariate binary time series. *Journal of American Statistical Association*, 84:447–451, 1989.

302. K.-Y. Liang, S. L. Zeger, and B. Qaqish. Multivariate regression analysis for categorical data. *Journal of the Royal Statistical Society, Series B*, 54:3–40, 1992. with discussion.

303. S. R. Lipsitz, N. M. Laird, and D. P. Harrington. Generalized estimating equations for correlated binary data: Using the odds ratio as a measure of association. *Biometrika*, 78:153–160, 1991.

304. J. S. Liu and R. Chen. Sequential Monte Carlo methods for dynamic systems. *Journal of the American Statistical Association*, 93:1032–1044, 1998.

305. R. D. Luce. *Individual Choice Behavior*. Wiley, New York, 1959.

306. I. L. MacDonald and W. Zucchini. *Hidden Markov and Other Models for Discrete-Valued Time Series*. Chapman & Hall, London, 1997.

307. S. Makridakis, S. C. Wheelwright, and R. J. Hyndman. *Forecasting: Methods and Applications*. Wiley New York, 3rd edition, 1998.

308. K. V. Mardia, C. Goodall, E. J. Redfern, and F. J. Alonso. The Kriged Kalman filter. *Test*, 7:217–285, 1998, with discussion.

309. D. E. K. Martin. On the distribution of the number of successes in fourth–or lower–order Markovian trials. *Computer & Operations Research*, 27:93–109, 2000.

310. W. P. McCormick and Y. S. Park. Asymptotic analysis of extremes from autoregressive negative binomial processes. *Journal of Applied Probability*, 29:904–920, 1992.

311. P. McCullagh. Regression models for ordinal data. *Journal of Royal Statistical Society, Series B*, 42:109–142, 1980, with discussion.

312. P. McCullagh. Quasi–likelihood functions. *Annals of Statistics*, 11:59–67, 1983.

313. P. McCullagh and J. A Nelder. *Generalized Linear Models*. Chapman & Hall, London, 2nd edition, 1989.

314. R. E. McCulloch and R. S. Tsay. Bayesian analysis of autoregressive time series via the Gibbs sampler. *Journal of Time Series Analysis*, 15:235–250, 1994.

315. D. McFadden. Conditional logit analysis of qualitative choice behavior. In P. Zarembka, editor, *Frontiers in Econometrics*, pp. 105–142. Academic Press, New York, 1973.

316. E. McKenzie. Some simple models for discrete variate time series. *Water Resources Bulletin*, 21:645–650, 1985.

317. E. McKenzie. Autoregressive moving-average processes with negative-binomial and geometric marginal distributions. *Advances in Applied Probability*, 18:679–705, 1986.

318. E. McKenzie. Some ARMA models for dependent sequences of Poisson counts. *Advances in Applied Probability*, 20:822–835, 1988.

319. G. McLachlan and K. E. Basford. *Mixture Models:Inference and Applications to Clustering*. Marcel Dekker, Inc, New York, 1988.

320. G. McLachlan and D. Peel. *Finite Mixture Models*. Wiley, New York, 2000.

321. G. J. McLachlan and T. Krishnan. *The EM Algorithm and Extensions*. Wiley, New York 1997.

322. N. Metropolis, A. Rosenbluth, W. Rosenbluth, M. Teller, and E. Teller. Equations of state calculations by fast computing machines. *Journal of Chemical Physics*, 21:1087–1091, 1953.

323. N. Metropolis and S. Ulam. The Monte Carlo method. *Journal of the American Statistical Association*, 44:335–341, 1949.

324. S. P. Meyn and R. L. Tweedie. *Markov Chains and Stochastic Stability*. Springer, London, 1993.

325. M. E. Miller, C. S. Davis, and R. J. Landis. The analysis of longitudinal polytomous data: Generalized estimated equations and connection with weighted least squares. *Biometrics*, 49:1033–1044, 1993.

326. R. G. Miller, Jr. *Survival Analysis*. Wiley, New York, 1981.

327. G. Molenberghs and E. Lessafre. Marginal modelling of multivariate categorical data. *Statistics in Medicine*, 18:2237–2255, 1999.

328. B. J. T. Morgan. Observations on quantitative analysis. *Biometrics*, 39:879–886, 1983.

329. G. Morvai, S. Yakowitz, and P. Algoet. Weakly convergent nonparametric forecasting of stationary time series. *IEEE Transactions on Information Theory*, 43:483–498, 1997.

330. G. Morvai, S. Yakowitz, and L. Györfi. Nonparametric inference for ergodic stationary time series. *Annals of Statistics*, 24:370–379, 1996.

331. L. R. Muenz and L.V. Rubinstein. Markov models for covariate dependence of binary sequences. *Biometrics*, 41:91–101, 1985.

332. P. Müller, M. West, and S. MacEachern. Bayesian models for non-linear autoregressions. *Journal of Time Series Analysis*, 18:593–614, 1997.

333. S. Murphy and B. Li. Projected partial likelihood and its application to longitudinal data. *Biometrika*, 82:399–406, 1995.

334. R. H. Myers, D. C. Montgomery, and G. G. Vining. *Generalized Linear Models*. Wiley, New York, 2002.

335. G. P. Nason, T. Sapatinas, and A. Sawczenko. Wavelet packet modeling of infant sleep state using heart rate data. *Sankya, Series B*, 63:199–217, 2001.

336. J. A. Nelder and R. W. M. Wedderburn. Generalized linear models. *Journal of the Royal Statistical Society, Series A*, 135:370–384, 1972.

337. J. Neyman and E. L. Scott. Constistent estimates based on partially consistent observations. *Econometrica*, 16:1–32, 1948.

338. Z. Ni and B. Kedem. Normal probabilities in the equicorrelated case. *Journal of Mathematical Analysis and Applications*, 246:280–295, 2000.

339. H. Omre. Bayesian kriging—merging observations and qualified guesses in kriging. *Mathematical Geology*, 19:25–39, 1987.

340. G. Osius and D. Rojek. Normal goodness of fit tests for multinomial models with large degrees of freedom. *Journal of the American Statistical Association*, 87:1145–1152, 1992.

341. A. B. Owen. Empirical likelihood ratio confidence intervals for a single functional. *Biometrika*, 75:237–249, 1988.

342. L. Pace and A. Salvan. *Principles of Statistical Inference*. World Scientific, Singapore, 1997.

343. E. Parzen. *Stochastic Processes*. Holden–Day, Oakland, CA, 1962.

344. V. L. Patterson, M. D. Hudlow, F. P. Pytlowany, P. J.and Richards, and J. D. Hoff. Gate radar rainfall processing system. EDIS 26, NOAA, Washington, DC, 1979.

345. G. G. S. Pegram. An autoregressive model for multilag Markov chains. *Journal of Applied Probability*, 17:350–362, 1980.

346. J. F. Pendergast, S. J. Gange, M. J. Lindstrom, M. A. Newton, M. Palta, and M.R. Fisher. A survey of methods for analyzing clustered binary response data. *International Statistical Review*, 64:1–30, 1996.

347. L. R. Pericchi. A Bayesian approach to transformations to normality. *Biometrika*, 68:35–43, 1981.

348. D. A. Pierce and D. W. Schafer. Residuals in generalized linear models. *Journal of the American Statistical Association*, 81:977–983, 1986.

349. M. K. Pitt and N. Shephard. Filtering via simulation: auxiliary particle filtering. *Journal of the American Statistical Association*, 94:590–599, 1999.

350. W. J. Pratt. Concavity of the log-likelihood. *Journal of the American Statistical Association*, 76:103–106, 1981.

351. D. Pregibon. Goodness of link tests for generalized linear models. *Applied Statistics*, 29:15–24, 1980.

352. D. Pregibon. Logistic regression diagnostics. *Annals of Statistics*, 9:705–724, 1981.

353. R. L. Prentice. A generalization of the probit and logit methods for dose response curves. *Biometrics*, 32:761–768, 1976.

354. R. L. Prentice. Correlated binary regression with covariates specific to each binary observation. *Biometrics*, 44:1033–1048, 1988.

355. R. L. Prentice and R. Pyke. Logistic disease incidence models and case–control studies. *Biometrika*, 66:403–411, 1979.

356. R. L. Prentice and L. P. Zhao. Estimating functions for parameters in means and covariates of multivariate discrete and continuous responses. *Biometrics*, 47(825–839), 1991.

357. M. B. Priestley. *Spectral Analysis and Time Series*. Academic Press, London, 1981.

358. M. B. Priestley. *Nonlinear and Nonstationary Time Series*. Academic Press, New York, 1988.

359. H. Pruscha. Categorical time series with a recursive scheme and with covariates. *Statistics*, 24:43–57, 1993.

360. M. P. Quine and E. Seneta. A limit theorem for the Galton-Watson process with immigration. *Australian Journal of Statistics*, 11:166–173, 1969.

361. B. G. Quinn and E. J. Hannan. *The Estimation and Tracking of Frequency*. Cambridge University Press, Cambridge, 2001.

362. A. E. Raftery. A model for high-order Markov chains. *Journal of the Royal Statistical Society. Series B*, 47:528–539, 1985.

363. A. E. Raftery. A new model for discrete-valued time series: Autocorrelations and extensions. In *Survey of Statistical Methods and Applications, 3–4 (Cagliari, 1983/84)*, pp. 149–162. Pitagora, Bologna, 1985.

364. A. E. Raftery and J. D. Banfield. Stopping the Gibbs sampler, the use of morphology and other issues in spatial statistics. *Annals of the Institute of Statistical Mathematics*, 43:32–43, 1991.

365. A. E. Raftery and S. Tavaré. Estimation and modelling repeated patterns in high order Markov chains with the mixture transition distribution model. *Applied Statistics*, 43:179–199, 1994.

366. C. R. Rao. *Linear Statistical Inference and Its Applications*. Wiley, New York, 2nd edition, 1973.

367. H. E. Rauch, F. Tung, and C. T. Striebel. Maximum likelihood estimates of linear dynamic systems. *American Institute of Aeronautics and Astronomics Journal*, 3:1445–1450, 1965.

368. T. R. C. Read and N. A. C. Cressie. *Goodness of Fit Statistics for Discrete Multivariate Data*. Springer, New York, 1988.

369. J. A. Rice and M. Rosenblatt. On frequency estimation. *Biometrika*, 75:477–484, 1988.

370. B. D. Ripley. *Stochastic Simulation*. Wiley, New York, 1987.

371. J. Rissanen. Modelling by shortest data description. *Automatica*, 14:465–471, 1978.

372. M. Rosenblatt. *Gaussian and Non-Gaussian Linear Time Series and Random Fields*. Springer, New York, 2000.

373. D. B. Rubin. Using the SIR algorithm to simulate posterior distributions. In J. M. Bernardo et al., editors, *Bayesian Statistics 3*, pp. 395–402. Oxford University Press, Oxford, 1988.

374. T. Rydén. Consistent and asymptotically normal parameter estimates for hidden Markov models. *Annals of Statistics*, 22:1884–1895, 1994.

375. B. Sansó and L. Guenni. A nonstationary multisite model for rainfall. *Journal of the American Statistical Association*, 95:1089–1100, 2000.

376. T. J. Santner and Duffy E. D. *Statistical Analysis of Discrete Data*. Springer, New York, 1989.

377. S. F. Schmidt. The Kalman filter: Its recognition and development for aerospace applications. *Journal of Guidance and Control*, 4:4–7, 1981.

378. D. Schoenfeld. Chi-square goodness-of-fit test for the proportional hazards regression model. *Biometrika*, 67:145–153, 1980.

379. A. Schuster. On the investigation of hidden periodicities with application to a supposed 26 day period of meteorological phenomena. *Terrestrial Magnetism*, 3:13–41, 1898.

380. G Schwarz. Estimating the dimension of a model. *Annals of Statistics*, 6:461–464, 1978.

381. R. Shepard. Statistical aspects of ARCH and stochastic volatility. In D. R. Cox et al., editors, *Time Series Models in Econometrics*, pp. 1–100. Chapman & Hall, London, 1996.

382. N. Shephard. Partial non-Gaussian state space. *Biometrika*, 81:115–131, 1994.

383. N. Shephard. Generalized linear autoregression. Technical report, Nuffield College, Oxford University, 1995. Available at `www.nuff.ox.ac.uk/economics/papers/1996/w8/glar.ps`

384. N. Shephard and M. K. Pitt. Likelihood analysis of non-Gaussian measurement time series. *Biometrika*, 84:653–657, 1997.

385. A. N. Shiryayev. *Probability*. Springer, New York, 1984.

386. D. A. Short, K. Shimizu, and B. Kedem. Optimal thresholds for the estimation of area rain-rate moments by the threshold method. *Journal of Applied Meteorology*, 32:182–192, 1993.

387. R. H. Shumway, A. S. Azari, and Y. Pawitan. Modeling mortality fluctuations in Los Angels as functions of pollution and weather effects. *Environmental Research*, 45:224–241, 1988.

388. R. H. Shumway and D. S. Stoffer. *Time Series Analysis and Its Applications*. Springer, New York, 2000.

389. M. J Silvapulle. On the existence of maximum likelihood estimates for the binomial response models. *Journal of Royal Statistical Society, Series B*, 43:310–313, 1981.

390. S. D. Silvey. A note on maximum likelihood in the case of dependent random variables. *Journal of the Royal Statistical Society, Series B*, 23:444–452, 1961.

391. A. C. Singh and G. R. Roberts. State space modeling of cross–classified time series of counts. *International Statistical Review*, 60:321–336, 1992.

392. E. V. Slud. Consistency and efficiency of inferences with the partial likelihood. *Biometrika*, 69:547–552, 1982.

393. E. V. Slud. Partial likelihood for continuous time stochastic processes. *Scandinavian Journal of Statistics*, 19:97–109, 1992.

394. E. V. Slud. A counting process model for the London mortality data. Technical Report 95–02, Department of Mathematics, University of Maryland at College Park, 1995.

395. E. V. Slud and B. Kedem. Partial likelihood analysis of logistic regression and autoregression. *Statistica Sinica*, 4:89–106, 1994.

396. A. F. M. Smith and M. West. Monitoring renal transplants: An application of the multi–process Kalman filter. *Biometrics*, 39:867–878, 1983.

397. E. J. Snell. A scaling procedure for ordered categorical data. *Biometrics*, 20:592–607, 1964.

398. K. S. Song and T. H. Li. A statistically and computationally efficient method for frequency estimation. *Stochastic Processes and Their Applications*, 86:29–47, 2000.

399. M. L. Stein. *Interpolation of Spatial Data: Some Theory for Kriging*. Springer, New York, 1999.

400. F. W. Steutel and K. van Harn. Discrete analogues of self-decomposability and stability. *Annals of Probability*, 7:893–899, 1979.

401. D. S. Stoffer, D. E. Tyler, and A. J. McDougall. Spectral analysis of categorical time series: Scaling and the spectral envelope. *Biometrika*, 80:611–622, 1993.

402. D. S. Stoffer, D. E. Tyler, and D. A. Wendt. The spectral envelope and its application. *Statistical Science*, 15:224–253, 2000.

403. O. D. Stram, J. L. Wei, and H. J. Ware. Analysis of repeated ordered categorical outcomes with possibly missing observations and time dependent covariates. *Journal of American Statistical Association*, 83:631–637, 1988.

404. R. L. Stratonovich. Application of the theory of Markov processes for optimum filtration of signals. *Radio Engineering and Electronic Physics*, 1:1–19, 1960. USSR.

405. R. L. Stratonovich. Detection and estimation of signals in noise when one or both are non-Gaussian. *Proceedings IEEE*, 58:670–679, 1970.

406. J. R. Stroud, P. Müller, and B. Sansó. Dynamic models for spatiotemporal data. *Journal of the Royal Statistical Society, Series B*, 63:673–689, 2001.

407. T. A. Stukel. Generalized logistic models. *Journal of the American Statistical Association*, 83:426–431, 1988.

408. B. C. Sutradhar and M. Kovacevic. Analysing ordinal longitudinal survey data: Generalised estimating equations approach. *Biometrika*, 87:837–848, 2000.

409. T. J. Sweeting. Consistent prior distributions for transformed models. In *Bayesian Statistics, 2 (Valencia, 1983)*, pp. 755–762. North-Holland, Amsterdam, 1985.

410. H. Tanizaki and R. S. Mariano. Nonlinear and non–Gaussian state–space modeling with Monte Carlo simulations. *Journal of Econometrics*, 83:263–290, 1998.

411. T. J. Thompson, P. J. Smith, and J. P. Boyle. Finite mixture models with concomitant information: Assessing diagnostic criteria for diabetes. *Applied Statistics*, 47:393–404, 1998.

412. L. Tierney. Markov chains for exploring posterior distributions. *Annals of Statistics*, 22:1701–1762, 1991, with discussion.

413. D. M. Titterington, A. F. M. Smith, and U. E. Makov. *Statistical Analysis of Finite Mixture Distributions*. Wiley, New York, 1985.

414. H. Tong. *Threshold Models in Nonlinear Time Series Analysis*. Lecture Notes in Statistics,21. Springer, New York, 1983.

415. H. Tong. *Nonlinear Time Series: A Dynamical System Approach*. Oxford University Press, New York, 1990.

416. R. S. Tsay. Regression models with time series errors. *Journal of the American Statistical Association*, 79:118–124, 1984.

417. K. van Harn. *Classifying Infinitely Divisible Distributions by Functional Equations*. Mathematisch Centrum, Amsterdam, 1978.

418. W. N. Venables and B. D. Ripley. *Modern Applied Statistics with S–PLUS*. Springer, New York, 3rd edition, 1999.

419. G. Verbeke and G. Molenberghs. *Linear Mixed Models for Longitudinal Data*. Springer, New York, 2000.

420. G. Verbeke, B. Spiessens, and E. Lessafre. Conditional linear mixed models. *American Statistician*, 55:25–34, 2001.

421. M. S. Waterman. *Introduction to Computational Biology: Maps, Sequences and Genomes*. Chapman & Hall, New York, 1995.

422. R. W. M. Wedderburn. Quasi–likelihood functions, generalized linear models and the Gaussian method. *Biometrika*, 61:439–447, 1974.

423. R. W. M. Wedderburn. On the existence and uniqueness of the maximum likelihood estimates. *Biometrika*, 63:27–32, 1976.

424. C. Z. Wei and J. Winnicki. Estimation of the means in the branching process with immigration. *Annals of Statistics*, 18:1757–1773, 1990.

425. A. S. Weigend and N. A. Gershenfeld. *Time Series Prediction: Forecasting the Future and Understanding the Past*. Addison-Wesley, Reading, MA, 1994.

426. M. West. Modeling with mixtures. In J. M. Bernardo et al., editors, *Bayesian Statistics 4*, pp. 577–586. Oxford University Press, Oxford, 1992, with discussion.

427. M. West and P.J. Harrison. *Bayesian Forecasting and Dynamic Models*. Springer, New York, 2nd edition, 1997.

428. M. West, P. J. Harrison, and H. S. Migon. Dynamic generalized linear models and Bayesian forecasting. *Journal of the American Statistical Association*, 80:73–97, 1985, with discussion.

429. H. White. Maximum likelihood estimation of misspecified models. *Econometrica*, 50:1–25, 1982.

430. A. S. Whittmore. Transformations to linearity in binary regression. *SIAM Journal on Applied Mathematics*, 43:703–710, 1983.

431. C. K. Wikle, M. Berliner, and N. Cressie. Hierarchical Bayesian space–time models. *Environmental and Ecological Statistics*, 5:117–124, 1998.

432. C. K. Wikle and N. Cressie. A dimension-reduced approach to space-time Kalman filtering. *Biometrika*, 86:815–829, 1999.

433. J. M. Williamson, K. M. Kim, and S. R. Lipsitz. Analyzing bivariate ordinal data using a global odds ratio. *Journal of the American Statistical Association*, 90:1432–1437, 1995.

434. R. Winkelmann. *Econometric Analysis of Count Data*. Springer, Berlin, 3rd edition, 2000.

435. J. Winnicki. *A useful estimation theory for the branching process with immigration*. PhD thesis, University of Maryland, College Park, USA, 1986.

436. C. S. Wong and W. K. Li. On a mixture autoregressive model. *Journal of the Royal Statistical Society*, B62:95–115, 2000.

437. C. S. Wong and W. K. Li. On a logistic mixture autoregressive model. *Biometrika*, 88:833–846, 2001.

438. C. S. Wong and W. K. Li. On a mixture autoregressive conditional heteroscedastic model. *Journal of the American Statistical Association*, 96:982–995, 2001.

439. W. H. Wong. Theory of partial likelihood. *Annals of Statistics*, 14:88–123, 1986.

440. C.-F. J. Wu. On the convergence properties of the EM algorithm. *Annals of Statistics*, 11:95–103, 1983.

441. A. M. Yaglom. *Correlation Theory of Stationary and Related Random Functions: Basic Results*. Springer, New York, 1987. Vol. I.

442. S. Yakowitz. Some contributions to a frequency location method due to He and Kedem. *IEEE Transactions on Information Theory*, 37:1177–1182, 1991.

443. S. Yakowitz, L. Györfi, J. Kieffer, and G. Morvai. Strongly consistent nonparametric forecasting and regression for stationary ergodic sequences. *Journal of Multivariate Analysis*, 71:24–41, 1999.

444. A. Zeevi, R. Meir, and R. J. Adler. Non-linear models for time series using mixtures of autoregressive models. Technical report, Faculty of Industrial Engineering and Management, Technion, Israel, 2000.

445. S. L. Zeger. A regression model for time series of counts. *Biometrika*, 75:621–629, 1988.

446. S. L. Zeger and K.-Y. Liang. Longitudinal data analysis for discrete and continuous outcomes. *Biometrics*, 42:121–130, 1986.

447. S. L. Zeger and B. Qaqish. Markov regression models for time series: A quasi-likelihood approach. *Biometrics*, 44:1019–1031, 1988.

448. A. Zellner. Bayesian and non-Bayesian analysis of the log-normal distribution and log-normal regression. *Journal of the American Statistical Association*, 66:327–330, 1971.

449. A. Zellner. *An Introduction to Bayesian Inference in Econometrics*. Wiley, New York, 1971.

450. L. P. Zhao and R. L. Prentice. Correlated binary regression using a quadratic exponential model. *Biometrika*, 77:642–648, 1990.

451. W. Zucchini and P. Guttorp. A hidden Markov model for space–time precipitation. *Water Resources Research*, 27:1917–1923, 1991.

Index

WILEY SERIES IN PROBABILITY AND STATISTICS

ESTABLISHED BY WALTER A. SHEWHART AND SAMUEL S. WILKS

Editors: *David J. Balding, Peter Bloomfield, Noel A. C. Cressie, Nicholas I. Fisher, Iain M. Johnstone, J. B. Kadane, Louise M. Ryan, David W. Scott, Adrian F. M. Smith, Jozef L. Teugels*

Editors Emeriti: *Vic Barnett, J. Stuart Hunter, David G. Kendall*

The ***Wiley Series in Probability and Statistics*** is well established and authoritative. It covers many topics of current research interest in both pure and applied statistics and probability theory. Written by leading statisticians and institutions, the titles span both state-of-the-art developments in the field and classical methods.

Reflecting the wide range of current research in statistics, the series encompasses applied, methodological and theoretical statistics, ranging from applications and new techniques made possible by advances in computerized practice to rigorous treatment of theoretical approaches.

This series provides essential and invaluable reading for all statisticians, whether in academia, industry, government, or research.

ABRAHAM and LEDOLTER · Statistical Methods for Forecasting
AGRESTI · Analysis of Ordinal Categorical Data
AGRESTI · An Introduction to Categorical Data Analysis
AGRESTI · Categorical Data Analysis
ANDĚL · Mathematics of Chance
ANDERSON · An Introduction to Multivariate Statistical Analysis, *Second Edition*
*ANDERSON · The Statistical Analysis of Time Series
ANDERSON, AUQUIER, HAUCK, OAKES, VANDAELE, and WEISBERG · Statistical Methods for Comparative Studies
ANDERSON and LOYNES · The Teaching of Practical Statistics
ARMITAGE and DAVID (editors) · Advances in Biometry
ARNOLD, BALAKRISHNAN, and NAGARAJA · Records
*ARTHANARI and DODGE · Mathematical Programming in Statistics
*BAILEY · The Elements of Stochastic Processes with Applications to the Natural Sciences
BALAKRISHNAN and KOUTRAS · Runs and Scans with Applications
BARNETT · Comparative Statistical Inference, *Third Edition*
BARNETT and LEWIS · Outliers in Statistical Data, *Third Edition*
BARTOSZYNSKI and NIEWIADOMSKA-BUGAJ · Probability and Statistical Inference
BASILEVSKY · Statistical Factor Analysis and Related Methods: Theory and Applications
BASU and RIGDON · Statistical Methods for the Reliability of Repairable Systems
BATES and WATTS · Nonlinear Regression Analysis and Its Applications
BECHHOFER, SANTNER, and GOLDSMAN · Design and Analysis of Experiments for Statistical Selection, Screening, and Multiple Comparisons
BELSLEY · Conditioning Diagnostics: Collinearity and Weak Data in Regression
BELSLEY, KUH, and WELSCH · Regression Diagnostics: Identifying Influential Data and Sources of Collinearity
BENDAT and PIERSOL · Random Data: Analysis and Measurement Procedures, *Third Edition*

*Now available in a lower priced paperback edition in the Wiley Classics Library.

BERRY, CHALONER, and GEWEKE · Bayesian Analysis in Statistics and Econometrics: Essays in Honor of Arnold Zellner
BERNARDO and SMITH · Bayesian Theory
BHAT · Elements of Applied Stochastic Processes, *Second Edition*
BHATTACHARYA and JOHNSON · Statistical Concepts and Methods
BHATTACHARYA and WAYMIRE · Stochastic Processes with Applications
BILLINGSLEY · Convergence of Probability Measures, *Second Edition*
BILLINGSLEY · Probability and Measure, *Third Edition*
BIRKES and DODGE · Alternative Methods of Regression
BLISCHKE AND MURTHY · Reliability: Modeling, Prediction, and Optimization
BLOOMFIELD · Fourier Analysis of Time Series: An Introduction, *Second Edition*
BOLLEN · Structural Equations with Latent Variables
BOROVKOV · Ergodicity and Stability of Stochastic Processes
BOULEAU · Numerical Methods for Stochastic Processes
BOX · Bayesian Inference in Statistical Analysis
BOX · R. A. Fisher, the Life of a Scientist
BOX and DRAPER · Empirical Model-Building and Response Surfaces
*BOX and DRAPER · Evolutionary Operation: A Statistical Method for Process Improvement
BOX, HUNTER, and HUNTER · Statistics for Experimenters: An Introduction to Design, Data Analysis, and Model Building
BOX and LUCEÑO · Statistical Control by Monitoring and Feedback Adjustment
BRANDIMARTE · Numerical Methods in Finance: A MATLAB-Based Introduction
BROWN and HOLLANDER · Statistics: A Biomedical Introduction
BRUNNER, DOMHOF, and LANGER · Nonparametric Analysis of Longitudinal Data in Factorial Experiments
BUCKLEW · Large Deviation Techniques in Decision, Simulation, and Estimation
CAIROLI and DALANG · Sequential Stochastic Optimization
CHAN · Time Series: Applications to Finance
CHATTERJEE and HADI · Sensitivity Analysis in Linear Regression
CHATTERJEE and PRICE · Regression Analysis by Example, *Third Edition*
CHERNICK · Bootstrap Methods: A Practitioner's Guide
CHILÈS and DELFINER · Geostatistics: Modeling Spatial Uncertainty
CHOW and LIU · Design and Analysis of Clinical Trials: Concepts and Methodologies
CLARKE and DISNEY · Probability and Random Processes: A First Course with Applications, *Second Edition*
*COCHRAN and COX · Experimental Designs, *Second Edition*
CONGDON · Bayesian Statistical Modelling
CONOVER · Practical Nonparametric Statistics, *Second Edition*
COOK · Regression Graphics
COOK and WEISBERG · Applied Regression Including Computing and Graphics
COOK and WEISBERG · An Introduction to Regression Graphics
CORNELL · Experiments with Mixtures, Designs, Models, and the Analysis of Mixture Data, *Third Edition*
COVER and THOMAS · Elements of Information Theory
COX · A Handbook of Introductory Statistical Methods
*COX · Planning of Experiments
CRESSIE · Statistics for Spatial Data, *Revised Edition*
CSÖRGŐ and HORVÁTH · Limit Theorems in Change Point Analysis
DANIEL · Applications of Statistics to Industrial Experimentation
DANIEL · Biostatistics: A Foundation for Analysis in the Health Sciences, *Sixth Edition*
*DANIEL · Fitting Equations to Data: Computer Analysis of Multifactor Data, *Second Edition*

*Now available in a lower priced paperback edition in the Wiley Classics Library.

DAVID · Order Statistics, *Second Edition*
*DEGROOT, FIENBERG, and KADANE · Statistics and the Law
DEL CASTILLO · Statistical Process Adjustment for Quality Control
DETTE and STUDDEN · The Theory of Canonical Moments with Applications in Statistics, Probability, and Analysis
DEY and MUKERJEE · Fractional Factorial Plans
DILLON and GOLDSTEIN · Multivariate Analysis: Methods and Applications
DODGE · Alternative Methods of Regression
*DODGE and ROMIG · Sampling Inspection Tables, *Second Edition*
*DOOB · Stochastic Processes
DOWDY and WEARDEN · Statistics for Research, *Second Edition*
DRAPER and SMITH · Applied Regression Analysis, *Third Edition*
DRYDEN and MARDIA · Statistical Shape Analysis
DUDEWICZ and MISHRA · Modern Mathematical Statistics
DUNN and CLARK · Applied Statistics: Analysis of Variance and Regression, *Second Edition*
DUNN and CLARK · Basic Statistics: A Primer for the Biomedical Sciences, *Third Edition*
DUPUIS and ELLIS · A Weak Convergence Approach to the Theory of Large Deviations
*ELANDT-JOHNSON and JOHNSON · Survival Models and Data Analysis
ETHIER and KURTZ · Markov Processes: Characterization and Convergence
EVANS, HASTINGS, and PEACOCK · Statistical Distributions, *Third Edition*
FELLER · An Introduction to Probability Theory and Its Applications, Volume I, *Third Edition,* Revised; Volume II, *Second Edition*
FISHER and VAN BELLE · Biostatistics: A Methodology for the Health Sciences
*FLEISS · The Design and Analysis of Clinical Experiments
FLEISS · Statistical Methods for Rates and Proportions, *Second Edition*
FLEMING and HARRINGTON · Counting Processes and Survival Analysis
FULLER · Introduction to Statistical Time Series, *Second Edition*
FULLER · Measurement Error Models
GALLANT · Nonlinear Statistical Models
GHOSH, MUKHOPADHYAY, and SEN · Sequential Estimation
GIFI · Nonlinear Multivariate Analysis
GLASSERMAN and YAO · Monotone Structure in Discrete-Event Systems
GNANADESIKAN · Methods for Statistical Data Analysis of Multivariate Observations, *Second Edition*
GOLDSTEIN and LEWIS · Assessment: Problems, Development, and Statistical Issues
GREENWOOD and NIKULIN · A Guide to Chi-Squared Testing
GROSS and HARRIS · Fundamentals of Queueing Theory, *Third Edition*
*HAHN · Statistical Models in Engineering
HAHN and MEEKER · Statistical Intervals: A Guide for Practitioners
HALD · A History of Probability and Statistics and their Applications Before 1750
HALD · A History of Mathematical Statistics from 1750 to 1930
HAMPEL · Robust Statistics: The Approach Based on Influence Functions
HANNAN and DEISTLER · The Statistical Theory of Linear Systems
HEIBERGER · Computation for the Analysis of Designed Experiments
HEDAYAT and SINHA · Design and Inference in Finite Population Sampling
HELLER · MACSYMA for Statisticians
HINKELMAN and KEMPTHORNE: · Design and Analysis of Experiments, Volume 1: Introduction to Experimental Design
HOAGLIN, MOSTELLER, and TUKEY · Exploratory Approach to Analysis of Variance
HOAGLIN, MOSTELLER, and TUKEY · Exploring Data Tables, Trends and Shapes

*Now available in a lower priced paperback edition in the Wiley Classics Library.

*HOAGLIN, MOSTELLER, and TUKEY · Understanding Robust and Exploratory Data Analysis
HOCHBERG and TAMHANE · Multiple Comparison Procedures
HOCKING · Methods and Applications of Linear Models: Regression and the Analysis of Variables
HOEL · Introduction to Mathematical Statistics, *Fifth Edition*
HOGG and KLUGMAN · Loss Distributions
HOLLANDER and WOLFE · Nonparametric Statistical Methods, *Second Edition*
HOSMER and LEMESHOW · Applied Logistic Regression, *Second Edition*
HOSMER and LEMESHOW · Applied Survival Analysis: Regression Modeling of Time to Event Data
HØYLAND and RAUSAND · System Reliability Theory: Models and Statistical Methods
HUBER · Robust Statistics
HUBERTY · Applied Discriminant Analysis
HUNT and KENNEDY · Financial Derivatives in Theory and Practice
HUSKOVA, BERAN, and DUPAC · Collected Works of Jaroslav Hajek—with Commentary
IMAN and CONOVER · A Modern Approach to Statistics
JACKSON · A User's Guide to Principle Components
JOHN · Statistical Methods in Engineering and Quality Assurance
JOHNSON · Multivariate Statistical Simulation
JOHNSON and BALAKRISHNAN · Advances in the Theory and Practice of Statistics: A Volume in Honor of Samuel Kotz
JUDGE, GRIFFITHS, HILL, LÜTKEPOHL, and LEE · The Theory and Practice of Econometrics, *Second Edition*
JOHNSON and KOTZ · Distributions in Statistics
JOHNSON and KOTZ (editors) · Leading Personalities in Statistical Sciences: From the Seventeenth Century to the Present
JOHNSON, KOTZ, and BALAKRISHNAN · Continuous Univariate Distributions, Volume 1, *Second Edition*
JOHNSON, KOTZ, and BALAKRISHNAN · Continuous Univariate Distributions, Volume 2, *Second Edition*
JOHNSON, KOTZ, and BALAKRISHNAN · Discrete Multivariate Distributions
JOHNSON, KOTZ, and KEMP · Univariate Discrete Distributions, *Second Edition*
JUREČKOVÁ and SEN · Robust Statistical Procedures: Aymptotics and Interrelations
JUREK and MASON · Operator-Limit Distributions in Probability Theory
KADANE · Bayesian Methods and Ethics in a Clinical Trial Design
KADANE AND SCHUM · A Probabilistic Analysis of the Sacco and Vanzetti Evidence
KALBFLEISCH and PRENTICE · The Statistical Analysis of Failure Time Data
KASS and VOS · Geometrical Foundations of Asymptotic Inference
KAUFMAN and ROUSSEEUW · Finding Groups in Data: An Introduction to Cluster Analysis
KEDEM and FOKIANOS · Regression Models for Time Series Analysis
KENDALL, BARDEN, CARNE, and LE · Shape and Shape Theory
KHURI · Advanced Calculus with Applications in Statistics
KHURI, MATHEW, and SINHA · Statistical Tests for Mixed Linear Models
KLUGMAN, PANJER, and WILLMOT · Loss Models: From Data to Decisions
KLUGMAN, PANJER, and WILLMOT · Solutions Manual to Accompany Loss Models: From Data to Decisions
KOTZ, BALAKRISHNAN, and JOHNSON · Continuous Multivariate Distributions, Volume 1, *Second Edition*
KOTZ and JOHNSON (editors) · Encyclopedia of Statistical Sciences: Volumes 1 to 9 with Index
KOTZ and JOHNSON (editors) · Encyclopedia of Statistical Sciences: Supplement Volume

*Now available in a lower priced paperback edition in the Wiley Classics Library.

KOTZ, READ, and BANKS (editors) · Encyclopedia of Statistical Sciences: Update Volume 1
KOTZ, READ, and BANKS (editors) · Encyclopedia of Statistical Sciences: Update Volume 2
KOVALENKO, KUZNETZOV, and PEGG · Mathematical Theory of Reliability of Time-Dependent Systems with Practical Applications
LACHIN · Biostatistical Methods: The Assessment of Relative Risks
LAD · Operational Subjective Statistical Methods: A Mathematical, Philosophical, and Historical Introduction
LAMPERTI · Probability: A Survey of the Mathematical Theory, *Second Edition*
LANGE, RYAN, BILLARD, BRILLINGER, CONQUEST, and GREENHOUSE · Case Studies in Biometry
LARSON · Introduction to Probability Theory and Statistical Inference, *Third Edition*
LAWLESS · Statistical Models and Methods for Lifetime Data
LAWSON · Statistical Methods in Spatial Epidemiology
LE · Applied Categorical Data Analysis
LE · Applied Survival Analysis
LEE · Statistical Methods for Survival Data Analysis, *Second Edition*
LEPAGE and BILLARD · Exploring the Limits of Bootstrap
LEYLAND and GOLDSTEIN (editors) · Multilevel Modelling of Health Statistics
LIAO · Statistical Group Comparison
LINDVALL · Lectures on the Coupling Method
LINHART and ZUCCHINI · Model Selection
LITTLE and RUBIN · Statistical Analysis with Missing Data
LLOYD · The Statistical Analysis of Categorical Data
MAGNUS and NEUDECKER · Matrix Differential Calculus with Applications in Statistics and Econometrics, *Revised Edition*
MALLER and ZHOU · Survival Analysis with Long Term Survivors
MALLOWS · Design, Data, and Analysis by Some Friends of Cuthbert Daniel
MANN, SCHAFER, and SINGPURWALLA · Methods for Statistical Analysis of Reliability and Life Data
MANTON, WOODBURY, and TOLLEY · Statistical Applications Using Fuzzy Sets
MARDIA and JUPP · Directional Statistics
MASON, GUNST, and HESS · Statistical Design and Analysis of Experiments with Applications to Engineering and Science
McCULLOCH and SEARLE · Generalized, Linear, and Mixed Models
McFADDEN · Management of Data in Clinical Trials
McLACHLAN · Discriminant Analysis and Statistical Pattern Recognition
McLACHLAN and KRISHNAN · The EM Algorithm and Extensions
McLACHLAN and PEEL · Finite Mixture Models
McNEIL · Epidemiological Research Methods
MEEKER and ESCOBAR · Statistical Methods for Reliability Data
MEERSCHAERT and SCHEFFLER · Limit Distributions for Sums of Independent Random Vectors: Heavy Tails in Theory and Practice
*MILLER · Survival Analysis, *Second Edition*
MONTGOMERY, PECK, and VINING · Introduction to Linear Regression Analysis, *Third Edition*
MORGENTHALER and TUKEY · Configural Polysampling: A Route to Practical Robustness
MUIRHEAD · Aspects of Multivariate Statistical Theory
MURRAY · X-STAT 2.0 Statistical Experimentation, Design Data Analysis, and Nonlinear Optimization
MYERS and MONTGOMERY · Response Surface Methodology: Process and Product Optimization Using Designed Experiments, *Second Edition*

*Now available in a lower priced paperback edition in the Wiley Classics Library.

MYERS, MONTGOMERY, and VINING · Generalized Linear Models. With Applications in Engineering and the Sciences
NELSON · Accelerated Testing, Statistical Models, Test Plans, and Data Analyses
NELSON · Applied Life Data Analysis
NEWMAN · Biostatistical Methods in Epidemiology
OCHI · Applied Probability and Stochastic Processes in Engineering and Physical Sciences
OKABE, BOOTS, SUGIHARA, and CHIU · Spatial Tesselations: Concepts and Applications of Voronoi Diagrams, *Second Edition*
OLIVER and SMITH · Influence Diagrams, Belief Nets and Decision Analysis
PANKRATZ · Forecasting with Dynamic Regression Models
PANKRATZ · Forecasting with Univariate Box-Jenkins Models: Concepts and Cases
*PARZEN · Modern Probability Theory and Its Applications
PEÑA, TIAO, and TSAY · A Course in Time Series Analysis
PIANTADOSI · Clinical Trials: A Methodologic Perspective
PORT · Theoretical Probability for Applications
POURAHMADI · Foundations of Time Series Analysis and Prediction Theory
PRESS · Bayesian Statistics: Principles, Models, and Applications
PRESS and TANUR · The Subjectivity of Scientists and the Bayesian Approach
PUKELSHEIM · Optimal Experimental Design
PURI, VILAPLANA, and WERTZ · New Perspectives in Theoretical and Applied Statistics
PUTERMAN · Markov Decision Processes: Discrete Stochastic Dynamic Programming
*RAO · Linear Statistical Inference and Its Applications, *Second Edition*
RENCHER · Linear Models in Statistics
RENCHER · Methods of Multivariate Analysis, *Second Edition*
RENCHER · Multivariate Statistical Inference with Applications
RIPLEY · Spatial Statistics
RIPLEY · Stochastic Simulation
ROBINSON · Practical Strategies for Experimenting
ROHATGI and SALEH · An Introduction to Probability and Statistics, *Second Edition*
ROLSKI, SCHMIDLI, SCHMIDT, and TEUGELS · Stochastic Processes for Insurance and Finance
ROSENBERGER and LACHIN · Randomization in Clinical Trials: Theory and Practice
ROSS · Introduction to Probability and Statistics for Engineers and Scientists
ROUSSEEUW and LEROY · Robust Regression and Outlier Detection
RUBIN · Multiple Imputation for Nonresponse in Surveys
RUBINSTEIN · Simulation and the Monte Carlo Method
RUBINSTEIN and MELAMED · Modern Simulation and Modeling
RYAN · Modern Regression Methods
RYAN · Statistical Methods for Quality Improvement, *Second Edition*
SALTELLI, CHAN, and SCOTT (editors) · Sensitivity Analysis
*SCHEFFE · The Analysis of Variance
SCHIMEK · Smoothing and Regression: Approaches, Computation, and Application
SCHOTT · Matrix Analysis for Statistics
SCHUSS · Theory and Applications of Stochastic Differential Equations
SCOTT · Multivariate Density Estimation: Theory, Practice, and Visualization
*SEARLE · Linear Models
SEARLE · Linear Models for Unbalanced Data
SEARLE · Matrix Algebra Useful for Statistics
SEARLE, CASELLA, and McCULLOCH · Variance Components
SEARLE and WILLETT · Matrix Algebra for Applied Economics
SEBER · Linear Regression Analysis
SEBER · Multivariate Observations
SEBER and WILD · Nonlinear Regression

*Now available in a lower priced paperback edition in the Wiley Classics Library.

SENNOTT · Stochastic Dynamic Programming and the Control of Queueing Systems
*SERFLING · Approximation Theorems of Mathematical Statistics
SHAFER and VOVK · Probability and Finance: It's Only a Game!
SMALL and McLEISH · Hilbert Space Methods in Probability and Statistical Inference
SRIVASTAVA · Methods of Multivariate Statistics
STAPLETON · Linear Statistical Models
STAUDTE and SHEATHER · Robust Estimation and Testing
STOYAN, KENDALL, and MECKE · Stochastic Geometry and Its Applications, *Second Edition*
STOYAN and STOYAN · Fractals, Random Shapes and Point Fields: Methods of Geometrical Statistics
STYAN · The Collected Papers of T. W. Anderson: 1943–1985
SUTTON, ABRAMS, JONES, SHELDON, and SONG · Methods for Meta-Analysis in Medical Research
TANAKA · Time Series Analysis: Nonstationary and Noninvertible Distribution Theory
THOMPSON · Empirical Model Building
THOMPSON · Sampling, *Second Edition*
THOMPSON · Simulation: A Modeler's Approach
THOMPSON and SEBER · Adaptive Sampling
TIAO, BISGAARD, HILL, PEÑA, and STIGLER (editors) · Box on Quality and Discovery: with Design, Control, and Robustness
TIERNEY · LISP-STAT: An Object-Oriented Environment for Statistical Computing and Dynamic Graphics
TSAY · Analysis of Financial Time Series
UPTON and FINGLETON · Spatial Data Analysis by Example, Volume II: Categorical and Directional Data
VAN BELLE · Statistical Rules of Thumb
VIDAKOVIC · Statistical Modeling by Wavelets
WEISBERG · Applied Linear Regression, *Second Edition*
WELSH · Aspects of Statistical Inference
WESTFALL and YOUNG · Resampling-Based Multiple Testing: Examples and Methods for p-Value Adjustment
WHITTAKER · Graphical Models in Applied Multivariate Statistics
WINKER · Optimization Heuristics in Economics: Applications of Threshold Accepting
WONNACOTT and WONNACOTT · Econometrics, *Second Edition*
WOODING · Planning Pharmaceutical Clinical Trials: Basic Statistical Principles
WOOLSON and CLARKE · Statistical Methods for the Analysis of Biomedical Data, *Second Edition*
WU and HAMADA · Experiments: Planning, Analysis, and Parameter Design Optimization
YANG · The Construction Theory of Denumerable Markov Processes
*ZELLNER · An Introduction to Bayesian Inference in Econometrics
ZHOU, OBUCHOWSKI, and McCLISH · Statistical Methods in Diagnostic Medicine

*Now available in a lower priced paperback edition in the Wiley Classics Library.